Contributions to Statistics

The series **Contributions to Statistics** contains publications in theoretical and applied statistics, including for example applications in medical statistics, biometrics, econometrics and computational statistics. These publications are primarily monographs and multiple author works containing new research results, but conference and congress reports are also considered.

Apart from the contribution to scientific progress presented, it is a notable characteristic of the series that publishing time is very short, permitting authors and editors to present their results without delay.

More information about this series at http://www.springer.com/series/2912

Müjgan Tez · Dietrich von Rosen
Editors

Trends and Perspectives
in Linear Statistical Inference

Proceedings of the LINSTAT2016 Meeting
Held 22–25 August 2016 in Istanbul, Turkey

 Springer

Editors
Müjgan Tez
Department of Statistics
Arts and Sciences Faculty
Marmara University
Istanbul
Turkey

Dietrich von Rosen
Department of Energy and Technology
Swedish University of Agricultural Sciences
Uppsala
Sweden

and

Department of Mathematics
Linköping University
Linköping
Sweden

ISSN 1431-1968
Contributions to Statistics
ISBN 978-3-319-73240-4   ISBN 978-3-319-73241-1   (eBook)
https://doi.org/10.1007/978-3-319-73241-1

Library of Congress Control Number: 2017964488

Printed on acid-free paper

This Springer imprint is published by Springer Nature
The registered company is Springer International Publishing AG
The registered company address is: Gewerbestrasse 11, 6330 Cham, Switzerland

Preface

The present volume contains the LINSTAT2016 conference proceedings held at Istanbul, Turkey, 22–25 August 2016, at the Congress and Culture Center of Istanbul University. The conference was organized by Marmara University. Istanbul offered a beautiful, interesting, and inspiring environment. The conference venue was close to the Taksim square, the heart of modern Istanbul, situated in the European part of Istanbul.

LINSTAT2016 was organized and supported by Marmara University. The event was sponsored by BAPKO (Scientific Research Projects Unit-FEN-L -110316-0099) and cooperative partners were Istanbul University and Turkish airlines.

The Scientific Committee was chaired by Müjgan Tez, while the head of the Organizing Committee was Birsen Eygi Erdoğan. The aim of the conference was to bring together researchers sharing an interest in a variety of aspects of statistics and its applications and offer them a possibility to discuss current developments in these topics.

During the 4 days of conference, there were 6 invited lectures and 5 special sessions. In total there were 17 sessions. LINSTAT2016 hosted more than 60 participants from 12 countries. The Scientific Committee awarded 3 talks and 1 poster presentation delivered by Ph.D. students or young scientists. The winners were: first prize—Omer Altindag (Turkey); second prize—Huruy Debessay Asfha (Turkey); third prize—Tugba Sokut Acar (Turkey); and the prize for best poster—Ayca Pamukcu (Turkey). Winners of Young Scientists Awards will be Invited Speakers at the next edition of LinStat. The conference was a continuation of earlier international meetings, which provided a good occasion for a review of important research in statistics. The next meeting in the LinStat series will be held at Bedlewo, Poland, 2018.

This special volume contains selected articles from the LINSTAT2016 conference. They are mainly concerned with estimation, prediction, and testing in linear univariate and multivariate models, statistical inference in a mixed linear model—estimation and testing of variance components, design and analysis of experiments, including optimality and comparison of linear experiments. For those interested in linear statistical inference, we hope this book will be an interesting contribution. In total, 13 articles were finally accepted for publication. Each article was independently refereed by two anonymous referees.

We would like to offer our sincere thanks to all the contributors. The time spent by numerous referees is also highly appreciated. On behalf of the Scientific Committee, we express our special gratitude to the local organizers who created a wonderful and very interesting conference.

We hope that this Special Issue will stimulate future research within the field of linear statistical inference.

Istanbul, Turkey   Müjgan Tez
Uppsala/Linköping, Sweden   Dietrich von Rosen
November 2017

Contents

Contributors

Fatma Gül Akgül Department of Computer Engineering, Artvin Çoruh University, Artvin, Turkey

Süreyya Özöğür Akyüz Faculty of Engineering and Natural Sciences, Bahçeşehir University, Istanbul, Turkey

Yasin Asar Department of Mathematics-Computer Sciences, Necmettin Erbakan University, Konya, Turkey

Huruy Debessay Asfha Department of Statistics, Graduate School of Sciences, Anadolu University, Eskisehir, Turkey

N. Balakrishnan Department of Mathematics and Statistics, McMaster University, Hamilton, ON, Canada

İnci Batmaz Department of Mathematics, Imperial College London, London, UK

D. Dai Department of Economics and Statistics, Linnaeus University, Växjö, Sweden

Ali Zafer Dalar Department of Statistics, Forecast Research Laboratory, Giresun University, Giresun, Turkey

Erol Eğrioğlu Department of Statistics, Forecast Research Laboratory, Giresun University, Giresun, Turkey

Birsen Eygi Erdogan Department of Statistics, Marmara University, Istanbul, Turkey

Serkan Eryilmaz Department of Industrial Engineering, Atilim University, Ankara, Turkey

Xing Fang Geodetic Science Program, School of Earth Sciences, The Ohio State University, Columbus, OH, USA; School of Geodesy and Geomatics, Wuhan University, Wuhan, People's Republic of China

T. Holgersson Department of Economics and Statistics, Linnaeus University, Växjö, Sweden

Ozlem Ilk Department of Statistics, Middle East Technical University, Ankara, Turkey

Jarkko Isotalo Department of Forest Sciences, University of Helsinki, Helsinki, Finland

Betul Kan Kilinc Department of Statistics, Faculty of Science, Anadolu University, Eskisehir, Turkey

Cihangir Kan Department of Mathematical Sciences, Xi'an Jiaotong-Liverpool University, Suzhou, China

Tuğba Kapucu Department of Statistics, Middle East Technical University, Ankara, Turkey

Augustyn Markiewicz Department of Mathematical and Statistical Methods, Poznań University of Life Sciences, Poznań, Poland

Timothy E. O'Brien Department of Mathematics & Statistics and Institute of Environmental Sustainability, Loyola University Chicago, Chicago, IL, USA

Simo Puntanen Faculty of Natural Sciences, University of Tampere, Tampere, Finland

Burkhard Schaffrin Geodetic Science Program, School of Earth Sciences, The Ohio State University, Columbus, OH, USA

Birdal Şenoğlu Department of Statistics, Ankara University, Ankara, Turkey

Kyle Snow Geodetic Science Program, School of Earth Sciences, The Ohio State University, Columbus, OH, USA; Topcon Positioning Systems, Inc., Columbus, OH, USA

Feng Su Department of Basic Courses, Guangzhou Maritime College, Guangzhou, Guangdong, China

Müjgan Tez Statistics Department, Faculty of Arts and Sciences, Marmara University, Istanbul, Turkey

Özlem Türkşen Statistics Department, Faculty of Science, Ankara University, Ankara, Turkey

Xiaojun Zhu Department of Mathematical Sciences, Xi'an Jiaotong-Liverpool University, Suzhou, People's Republic of China

Comparison of Estimation Methods for Inverse Weibull Distribution

Fatma Gül Akgül and Birdal Şenoğlu

Abstract The aim of this chapter is to estimate unknown parameters of inverse Weibull (IW) distribution using eight different estimation methods: maximum likelihood (ML), least squares (LS), weighted least squares (WLS), percentile (PC), maximum product of spacing (MPS), probability weighted moments (PWM), Cramér–von Mises (CM), and Anderson-Darling (AD). The performances of these estimation methods are compared via an extensive Monte Carlo simulation study. Their robustness properties are also investigated. At the end of the study, two real data sets are analyzed for illustration and comparison purposes.

Keywords Inverse Weibull distribution · Estimation methods
Efficiency · Monte Carlo simulation · Robustness

1 Introduction

The inverse Weibull (IW) distribution was introduced by Keller and Kamath (1982) as a suitable model for describing the degeneration phenomena of mechanical components, such as the dynamic components of diesel engines. Since then it has common usage in many areas, such as survival analysis (Erto and Rapone 1984), mechanics (Murthy et al. 2004), earthquake (Pasari and Dikshit 2014), and wind speed (Akgül et al. 2016). Besides these studies, it has received considerable attention in recent years particularly for modeling extreme events, since it has large right tail probability, see Rinne (2009).

F. G. Akgül (✉)
Department of Computer Engineering, Artvin Çoruh University, 08000 Artvin, Turkey
e-mail: ftm.gul.fuz@artvin.edu.tr

B. Şenoğlu
Department of Statistics, Ankara University, 06100 Ankara, Turkey
e-mail: senoglu@science.ankara.edu.tr

© Springer International Publishing AG 2018   1
M. Tez and D. von Rosen (eds.), *Trends and Perspectives
in Linear Statistical Inference*, Contributions to Statistics,
https://doi.org/10.1007/978-3-319-73241-1_1

Several methods have been used for estimating unknown parameters of IW distribution. For example, Calabria and Pulcini (1990) considered estimation of the parameters of IW distribution using the maximum likelihood (ML) and the least squares (LS) methodologies. Singh et al. (2013) described the Bayes estimators of the parameters of IW distribution for complete, type I and type II censored samples under general entropy and squared loss functions. Akgül et al. (2016) obtained the modified maximum likelihood (MML) estimators of the parameters of IW distribution and compared them with the ML and the LS estimators.

In this study, we use eight different estimation methods to estimate unknown parameters of IW distribution. They are ML, LS, weighted least squares (WLS), percentile (PC), maximum product of spacing (MPS), probability weighted moments (PWM), Cramér–von Mises (CM), and Anderson-Darling (AD). It can be said that since IW distribution is a special case of generalized extreme value (GEV) distribution, the MPS, the PWM, and the PC estimators of the parameters of IW distribution are obtained across the entire GEV distribution, see Hosking et al. (1985), Soukissian and Tsalis (2015) and Ashoori et al. (2017). However, to the best of our knowledge, this is the first study using eight different estimation methods for estimating the parameters of IW distribution in the same study. The performance of all of these estimators is compared via an extensive Monte Carlo simulation study. Their robustness properties are also investigated. At the end of the study, two real data sets are analyzed for illustrative purposes.

The rest of the chapter is organized as follows: In Sect. 2, we give certain distributional properties of IW distribution. Estimation methods are considered in Sect. 3. A Monte Carlo simulation study is performed to compare the performance of the proposed estimators in Sect. 4. The robustness properties of the estimators are investigated in Sect. 5. In Sect. 6, two real data sets are used to show the implementation of the estimation methods. Comments and conclusions are given in the final section.

2   IW Distribution

Let X be a random variable from IW distribution with shape parameter α and scale parameter β. The cumulative density function (cdf) and the probability density function (pdf) of IW distribution are given below:

$$F(x,\alpha,\beta)=e^{-(x/\beta)^{-\alpha}}, \quad x>0, \quad \alpha,\beta>0 \tag{1}$$

and

$$f(x,\alpha,\beta)=\tfrac{\alpha}{\beta}\left(\tfrac{x}{\beta}\right)^{-(\alpha+1)}e^{-(x/\beta)^{-\alpha}}, \quad x>0, \tag{2}$$

respectively. If the random variable X has an IW distribution with parameters α and β, it is denoted by $X \sim IW(\alpha, \beta)$. It should be stated that IW distribution is known by different names such as Fréchet (Fréchet 1927), complementary Weibull (Drapella 1993), reciprocal Weibull (Mudholkar and Kollia 1994), or reverse Weibull (Murthy et al. 2004) in literature.

The survival and hazard function of IW distribution are given by

$$S(x) = 1 - e^{-(x/\beta)^{-\alpha}}, \quad x > 0 \tag{3}$$

and

$$h(x) = \frac{\frac{\alpha}{\beta}\left(\frac{x}{\beta}\right)^{-(\alpha+1)} e^{-(x/\beta)^{-\alpha}}}{1 - e^{-(x/\beta)^{-\alpha}}}, \quad x > 0, \tag{4}$$

respectively. The k-th moment around zero is calculated as shown below:

$$E(X^k) = \int_0^\infty x^k \frac{\alpha}{\beta}\left(\frac{x}{\beta}\right)^{-(\alpha+1)} e^{-(x/\beta)^{-\alpha}} dx = \beta^k \Gamma\left(\frac{\alpha-k}{k}\right), \tag{5}$$

where $\alpha > k$. Hence, the expected value of X is

$$E(X) = \beta \Gamma\left(\frac{\alpha-1}{\alpha}\right)$$

and the variance of X is

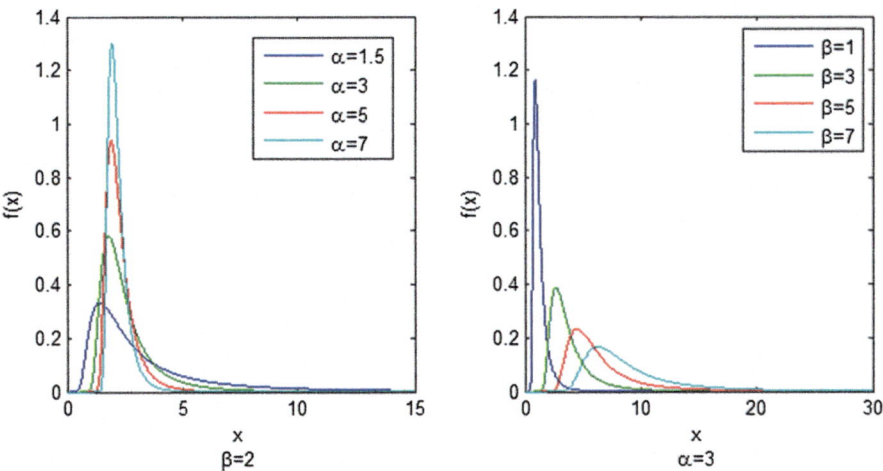

Fig. 1 Plots of the pdf of IW distribution for certain representative parameter values

$$Var(X) = \beta^2 \left[\Gamma\left(\frac{\alpha-2}{\alpha}\right) - \Gamma^2\left(\frac{\alpha-1}{\alpha}\right) \right].$$

The moment generating function and the cumulant generating function of X are defined as follows:

$$M_X(t) = \sum_{k=0}^{n} \frac{t^k}{k!} \beta^k \Gamma\left(\frac{\alpha-k}{k}\right) \tag{6}$$

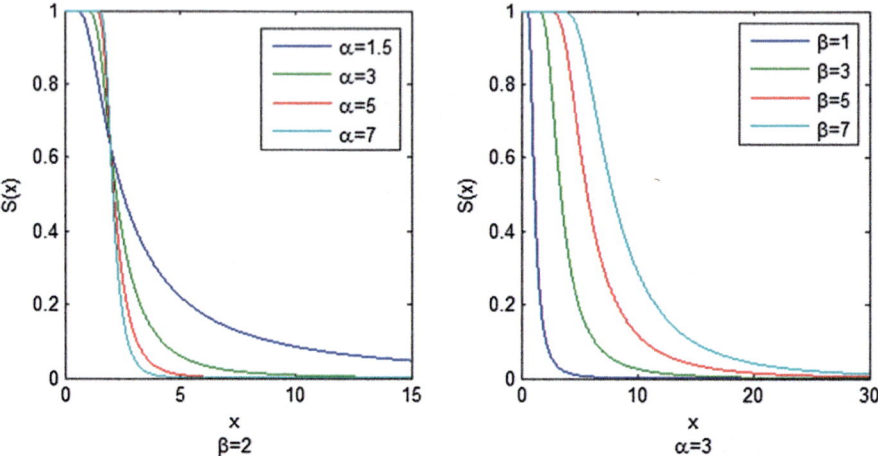

Fig. 2 Plots of the survival function of IW distribution for certain representative parameter values

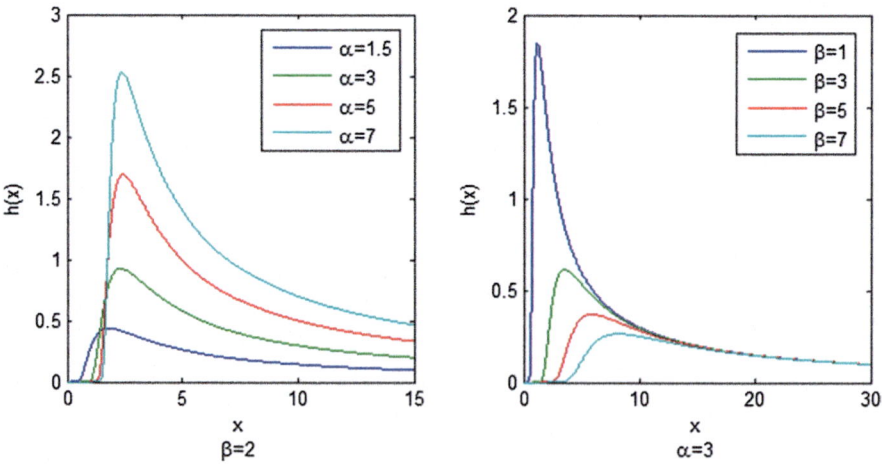

Fig. 3 Plots of the hazard function of IW distribution for certain representative parameter values

and

$$K_X(t) = \log \left[\sum_{k=0}^{n} \frac{t^k}{k!} \beta^k \Gamma \left(\frac{\alpha - k}{k} \right) \right],\tag{7}$$

respectively. Here, $|t| < 1$.

Plots of the pdf, survival, and hazard functions of IW distribution for various different parameter values are displayed in Figs. 1, 2, and 3.

3 Parameter Estimation Methods

Parameter estimation methods for estimating the shape parameter α and the scale parameter β of IW distribution are described in the following subsections.

3.1 The Maximum Likelihood Method

Let $x = (x_1, x_2, \ldots, x_n)$ be a random sample of size n from IW distribution with parameters (α, β). From Eq. (2), the likelihood and the log-likelihood functions are written as follows:

$$L(\alpha, \beta | x) = \frac{\alpha^n}{\beta^{n\alpha}} \prod_{i=1}^{n} x_i^{-(\alpha+1)} e^{-\sum_{i=1}^{n}(x_i/\beta)^{-\alpha}}\tag{8}$$

and

$$l(\alpha, \beta | x) = n \ln \alpha + n\alpha \ln \beta - (\alpha + 1) \sum_{i=1}^{n} \ln x_i - \beta^\alpha \sum_{i=1}^{n} x_i^{-\alpha},\tag{9}$$

respectively. By taking the derivatives of (9) with respect to the unknown parameters α and β and equating them to zero, we obtain the following likelihood equations:

$$\frac{\partial l}{\partial \alpha} = \frac{n}{\alpha} + n \ln \beta - \sum_{i=1}^{n} \ln x_i - \beta^\alpha \ln \beta \sum_{i=1}^{n} x_i^{-\alpha} + \beta^\alpha \sum_{i=1}^{n} x_i^{-\alpha} \ln x_i = 0,\tag{10}$$

$$\frac{\partial l}{\partial \beta} = \frac{n\alpha}{\beta} - \alpha \beta^{\alpha-1} \sum_{i=1}^{n} x_i^{-\alpha} = 0.\tag{11}$$

From (11), we obtain

$$\beta(\alpha) = \left(\frac{n}{\sum_{i=1}^{n} x_i^{-\alpha}}\right)^{1/\alpha}.$$ (12)

By putting (12) into (10), we obtain

$$\frac{n}{\alpha} - \sum_{i=1}^{n} \ln x_i + \frac{n}{\sum_{i=1}^{n} x_i^{-\alpha}} \sum_{i=1}^{n} x_i^{-\alpha} \ln x_i = 0.$$ (13)

Hence, $\hat{\alpha}$ can be obtained as a solution of the nonlinear equation of the form $h(\alpha) = \alpha$ where

$$h(\alpha) = \frac{n \sum_{i=1}^{n} x_i^{-\alpha}}{\sum_{i=1}^{n} x_i^{-\alpha} \sum_{i=1}^{n} \ln x_i - n \sum_{i=1}^{n} x_i^{-\alpha} \ln x_i}.$$ (14)

Note that (14) cannot be solved explicitly and therefore, we resort to iterative methods. Here, the Newton–Raphson algorithm is used to solve (14).

3.2 The Least Squares Method

The LS estimators of the unknown parameters are obtained by minimizing the following equation with respect to the parameters of interest:

$$\sum_{i=1}^{n} \left(F\left(x_{(i)}\right) - \frac{i}{n+1}\right)^2.$$ (15)

Here, $x_1, \ldots, x_n$ is a random sample from the distribution function $F(.)$, $x_{(1)} \leq \cdots \leq x_{(n)}$ denotes the corresponding order statistics and $\frac{i}{n+1}, (i = 1, \ldots, n)$ are the expected values of $F\left(x_{(i)}\right)$. Therefore, in the case of IW distribution, the LS estimators of the unknown parameters are obtained by minimizing (15) with respect to α and β. They are obtained by solving the following nonlinear equations:

$$\sum_{i=1}^{n} \left(F\left(x_{(i)}, \alpha, \beta\right) - \frac{i}{n+1}\right) F'_\alpha\left(x_{(i)}, \alpha, \beta\right) = 0,$$ (16)

$$\sum_{i=1}^{n} \left(F\left(x_{(i)}, \alpha, \beta\right) - \frac{i}{n+1}\right) F'_\beta\left(x_{(i)}, \alpha, \beta\right) = 0.$$ (17)

Here, $F(x, \alpha, \beta)$ is given in (1),

$$F'_\alpha(x, \alpha, \beta) = \left(\frac{x}{\beta}\right)^{-\alpha} \ln\left(\frac{x}{\beta}\right) e^{-(x/\beta)^{-\alpha}} \quad (18)$$

and

$$F'_\beta(x, \alpha, \beta) = -\left(\frac{\alpha}{\beta}\right)\left(\frac{x}{\beta}\right)^{-\alpha} e^{-(x/\beta)^{-\alpha}}. \quad (19)$$

3.3 The Weighted Least Squares Method

The WLS estimators of the unknown parameters are obtained by minimizing the following equation with respect to the parameters of interest:

$$\sum_{i=1}^{n} w_i \left(F(x_{(i)}) - \frac{i}{n+1} \right)^2. \quad (20)$$

Here, w_i, $(i = 1, \ldots, n)$ are equal to $1/V(F(x_{(i)})) = (n+1)^2(n+2)/i(n-i+1)$. Therefore, in the case of IW distribution, the WLS estimators of the unknown parameters are obtained by minimizing (20) with respect to α and β and they are the solutions of the following equations:

$$\sum_{i=1}^{n} w_i \left(F(x_{(i)}, \alpha, \beta) - \frac{i}{n+1} \right) F'_\alpha(x_{(i)}, \alpha, \beta) = 0, \quad (21)$$

$$\sum_{i=1}^{n} w_i \left(F(x_{(i)}, \alpha, \beta) - \frac{i}{n+1} \right) F'_\beta(x_{(i)}, \alpha, \beta) = 0, \quad (22)$$

where $F'_\alpha(x, \alpha, \beta)$ are $F'_\beta(x, \alpha, \beta)$ are defined as in (18) and (19), respectively.

3.4 The Percentile Method

The PC estimators of the unknown parameters of IW distribution are obtained by applying the method suggested by Kao (1958), since this distribution has an explicit cdf. Let p_i be an estimate of $F(x_{(i)})$ and $x(p_i)$ be the p_i-th quantile of $F(x_{(i)})$, the PC estimators of the unknown parameters are obtained by minimizing the following equation with respect to the parameters of interest:

$$\sum_{i=1}^{n} \left(x_{(i)} - x(p_i) \right)^2. \tag{23}$$

Here, $x_{(i)}$'s are ordered sample observations. There exist several estimators of p_i's in the literature. In this study, p_i are taken as $\frac{i}{n+1}$. Therefore, in case of IW distribution, the PC estimators of the unknown parameters are obtained by minimizing

$$\sum_{i=1}^{n} \left(x_{(i)} - \left(- \ln \left(\frac{i}{n+1} \right) \right)^{-1/\alpha} \beta \right)^2 \tag{24}$$

with respect to α and β.

3.5   The Maximum Product of Spacing Method

The MPS estimators of the unknown parameters of IW distribution are obtained using the methodology proposed by Cheng and Amin (1983). This method is based on the idea of differences between the values of the cdf at consecutive data points. The MPS estimators of the parameters are obtained by maximizing the following geometric mean (GM) of the differences with respect to parameters of interest

$$GM = \left(\prod_{i=1}^{n+1} D_i \right)^{1/(n+1)}. \tag{25}$$

Here, D_i is defined as

$$D_i = \int_{x_{(i-1)}}^{x_{(i)}} f(x, \alpha, \beta) dx, \, i = 1, \ldots, n+1, \tag{26}$$

in other words $D_i = F\left(x_{(i)}, \alpha, \beta \right) - F\left(x_{(i-1)}, \alpha, \beta \right)$ where $0 = F\left(x_{(0)}, \alpha, \beta \right)$ $< F\left(x_{(1)}, \alpha, \beta \right) < \cdots < F\left(x_{(n)}, \alpha, \beta \right) < F\left(x_{(n+1)}, \alpha, \beta \right) = 1$. By taking a logarithm of (25), we have

$$\ln GM = \frac{1}{n+1} \sum_{i=1}^{n+1} \ln \left[F\left(x_{(i)}, \alpha, \beta \right) - F\left(x_{(i-1)}, \alpha, \beta \right) \right]. \tag{27}$$

The MPS estimators of α and β can be obtained by solving the following equations:

$$\frac{\partial \ln GM}{\partial \alpha} = \frac{1}{n+1} \sum_{i=1}^{n+1} \left[\frac{F_\alpha'(x_{(i)}, \alpha, \beta) - F_\alpha'(x_{(i-1)}, \alpha, \beta)}{F(x_{(i)}, \alpha, \beta) - F(x_{(i-1)}, \alpha, \beta)} \right] = 0, \qquad (28)$$

$$\frac{\partial \ln GM}{\partial \beta} = \frac{1}{n+1} \sum_{i=1}^{n+1} \left[\frac{F_\beta'(x_{(i)}, \alpha, \beta) - F_\beta'(x_{(i-1)}, \alpha, \beta)}{F(x_{(i)}, \alpha, \beta) - F(x_{(i-1)}, \alpha, \beta)} \right] = 0 \qquad (29)$$

where $F_\alpha'(x, \alpha, \beta)$ and $F_\beta'(x, \alpha, \beta)$ are defined as in (18) and (19), respectively.

3.6 The Probability Weighted Moments Method

The PWM estimators of the unknown parameters of IW distribution are obtained using the methodology originated by Greenwood et al. (1979). If X is a random variable from the distribution of F, then the general expression for the weighted moments of i, r, and s is defined as

$$M(i, r, s) = E\left[X^i F(X)^r (1 - F(X))^s \right], \qquad (30)$$

where i, r, and s are the integer numbers. For the special case $i = 1$, $s = 0$, the PWM of IW distribution is calculated as given below:

$$\nu_r = M(1, r, 0) = E(XF(X)^r) = \int_0^\infty xF(x)^r f(x)dx$$
$$= \beta(r+1)^{1/\alpha - 1} \Gamma\left(1 - \frac{1}{\alpha}\right). \qquad (31)$$

For an ordered sample $x_{(1)} \leq x_{(2)} \leq \ldots \leq x_{(n)}$ of size n, the unbiased estimator of ν_r, say $\hat{\nu}_r$, is obtained from the following equation:

$$\hat{\nu}_r = \frac{1}{n} \sum_{j=1}^n \frac{(j-1)(j-2)\ldots(j-r)}{(n-1)(n-2)\ldots(n-r)} x_{(j)}, \qquad (32)$$

see Landwehr et al. (1979). It should be noted that Eq. (32) can be called sample PWM.

By equating the sample PWM to the population PWM, the PWM estimators of the shape parameter α and the scale parameter β of IW distribution are obtained as given below:

$$\hat{\alpha} = \frac{\log 2}{\log(2\hat{\nu}_1/\hat{\nu}_0)} \quad \text{and} \quad \beta = \frac{\hat{\nu}_0}{\Gamma(1 - 1/\hat{\alpha})}, \qquad (33)$$

respectively.

3.7 The Minimum Distance Methods

Now, we obtain the estimators of the shape and scale parameters of IW distribution based on minimizing empirical distribution function statistics. These methods can be considered as an informal procedure to minimize the distance between the empirical and estimated cdfs, see Luceño (2006, 2008). The estimators based on the general class of minimum distance estimators were first considered by Wolfowitz (1953, 1957). It should be noted that these estimators are also called as maximum goodness of fit estimators by Luceño (2006, 2008) to avoid confusion with other minimum distance methods which are not related to empirical distribution function. Luceño (2006, 2008) used minimum distance methods to obtain the estimators of unknown parameters of generalized Pareto and Weibull distributions, respectively. In addition to these studies, Santo and Mazucheli (2015) and Louzada et al. (2016) considered the estimation of the parameters of Marshall–Olkin extended Lindley and extended exponential geometric distributions by using goodness of fit estimators, respectively.

3.7.1 The Cramér–Von Mises Method

The CM estimators of the unknown parameters of IW distribution are obtained by minimizing the following equation with respect to the parameters α and β:

$$C(\alpha, \beta | x) = \frac{1}{12n} + \sum_{i=1}^{n} \left(F\left(x_{(i)}, \alpha, \beta\right) - \frac{2i-1}{2n} \right)^2. \tag{34}$$

It is clear that the CM estimators of the parameters are obtained by solving the following equations:

$$\sum_{i=1}^{n} \left(F\left(x_{(i)}, \alpha, \beta\right) - \frac{2i-1}{2n} \right) F'_{\alpha}\left(x_{(i)}, \alpha, \beta\right) = 0 \tag{35}$$

and

$$\sum_{i=1}^{n} \left(F\left(x_{(i)}, \alpha, \beta\right) - \frac{2i-1}{2n} \right) F'_{\beta}\left(x_{(i)}, \alpha, \beta\right) = 0, \tag{36}$$

where $F'_{\alpha}(x, \alpha, \beta)$ and $F'_{\beta}(x, \alpha, \beta)$ are defined as in (18) and (19), respectively.

3.7.2 The Anderson-Darling Method

The AD estimators of the unknown parameters of IW distribution are obtained by minimizing the following equation with respect to the parameters α and β:

$$A(\alpha,\beta|x) = -n - \frac{1}{n}\sum_{i=1}^{n}(2i-1)\log\left[F\left(x_{(i)},\alpha,\beta\right)\left(1 - F\left(x_{(n-i+1)},\alpha,\beta\right)\right)\right]. \quad (37)$$

The AD estimators of the parameters are obtained by solving the following equations:

$$\sum_{i=1}^{n}(2i-1)\left[\frac{F'_{\alpha}\left(x_{(i)},\alpha,\beta\right)}{F\left(x_{(i)},\alpha,\beta\right)} - \frac{F'_{\alpha}\left(x_{(n-i+1)},\alpha,\beta\right)}{1 - F\left(x_{(n-i+1)},\alpha,\beta\right)}\right] = 0 \quad (38)$$

and

$$\sum_{i=1}^{n}(2i-1)\left[\frac{F'_{\beta}\left(x_{(i)},\alpha,\beta\right)}{F\left(x_{(i)},\alpha,\beta\right)} - \frac{F'_{\beta}\left(x_{(n-i+1)},\alpha,\beta\right)}{1 - F\left(x_{(n-i+1)},\alpha,\beta\right)}\right] = 0, \quad (39)$$

where $F'_{\alpha}(x,\alpha,\beta)$ and $F'_{\beta}(x,\alpha,\beta)$ are defined as in (18) and (19), respectively.

4 Simulation Study

In this section, we perform a Monte Carlo simulation study to compare the performances of the ML, LS, WLS, PC, MPS, PWM, CM, and AD estimators of the unknown parameters α and β. The comparisons are made based on bias and mean square error (MSE) criteria. In addition, we use the deficiency (*Def*) criterion to compare the efficiencies of estimators α and β simultaneously, see Tiku and Akkaya (2004). *Def* is defined as given below:

$$Def(\hat{\alpha},\beta) = MSE(\hat{\alpha}) + MSE(\beta).$$

In the simulation setup, we use different sample sizes and different parameter settings shown below:

$$n = 25, 50, 100, 200, 500, \quad \alpha = 1.5, 3, 5 \quad \text{and} \quad \beta = 2.$$

The means, MSEs, and *Def* values are calculated based on $\lceil 100,000/n \rceil$ (integer value) Monte Carlo runs. All computations are conducted using MATLAB2013a. The results are reported in Tables 1, 2, and 3.

In the context of bias, the following conclusions can be made from the simulation results. The PC and the PWM estimators of α do not perform well for all sample sizes when $\alpha = 1.5$ and 5. On the other hand, the PC estimator of β has larger bias for all scenarios. The LS estimators of α and β outperform other estimators for small and moderate sample sizes. Although the ML and the MPS estimators produce larger bias for small sample sizes, their biases decrease with

Table 1 Simulated mean, MSE, and Def values for the estimators of α and β; $\alpha = 1.5$, $\beta = 2$

		$\hat{\alpha}$		β		
n	Method	Mean	MSE	Mean	MSE	Def
25	ML	1.5878	0.0764	2.0468	0.0904	0.1668[3]
	LS	1.4960	0.0977	2.0247	0.0945	0.1922[5]
	WLS	1.5112	0.0854	2.0300	0.0914	0.1768[4]
	PC	1.3661	0.3031	1.8008	0.7457	1.0488[8]
	MPS	1.4119	0.0623	2.0182	0.0853	0.1476[1]
	PWM	1.6973	0.1309	1.9606	0.1025	0.2334[7]
	CM	1.5919	0.1236	2.0501	0.1001	0.2237[6]
	AD	1.5254	0.0712	2.0368	0.0908	0.1620[2]
50	ML	1.5397	0.0327	2.0208	0.0431	0.0757[2]
	LS	1.4941	0.0431	2.0071	0.0473	0.0904[5]
	WLS	1.5055	0.0371	2.0116	0.0448	0.0819[4]
	PC	1.3324	0.2509	1.7230	0.7484	0.9993[8]
	MPS	1.4352	0.0313	2.0057	0.0418	0.0731[1]
	PWM	1.6418	0.0807	1.9896	0.0537	0.1344[7]
	CM	1.5404	0.0481	2.0194	0.0486	0.0966[6]
	AD	1.5081	0.0339	2.0138	0.0442	0.0780[3]
100	ML	1.5186	0.0153	2.0127	0.0204	0.0357[2]
	LS	1.4967	0.0226	2.0076	0.0217	0.0443[5]
	WLS	1.5050	0.0184	2.0100	0.0208	0.0392[4]
	PC	1.3381	0.2069	1.6963	0.7319	0.9388[8]
	MPS	1.4570	0.0156	2.0045	0.0200	0.0357[1]
	PWM	1.6126	0.0515	2.0100	0.0286	0.0801[7]
	CM	1.5194	0.0238	2.0137	0.0220	0.0458[6]
	AD	1.5040	0.0175	2.0103	0.0207	0.0382[3]
200	ML	1.5145	0.0074	1.9967	0.0110	0.0183[2]
	LS	1.5035	0.0110	1.9941	0.0117	0.0227[5]
	WLS	1.5087	0.0090	1.9951	0.0112	0.0202[4]
	PC	1.3172	0.2017	1.5914	0.8717	1.0734[8]
	MPS	1.4780	0.0072	1.9925	0.0110	0.0182[1]
	PWM	1.5816	0.0321	2.0003	0.0169	0.0491[7]
	CM	1.5148	0.0114	1.9971	0.0118	0.0231[6]
	AD	1.5074	0.0087	1.9953	0.0112	0.0199[3]
500	ML	1.4959	0.0030	2.0011	0.0045	0.0075[1]
	LS	1.4854	0.0045	1.9986	0.0047	0.0092[6]
	WLS	1.4911	0.0034	2.0005	0.0045	0.0080[3]
	PC	1.3269	0.1683	1.6028	0.8672	1.0355[8]
	MPS	1.4786	0.0034	1.9993	0.0045	0.0078[2]
	PWM	1.5516	0.0171	2.0127	0.0080	0.0251[7]
	CM	1.4899	0.0044	1.9998	0.0047	0.0091[5]
	AD	1.4905	0.0035	2.0005	0.0045	0.0080[4]

Table 2 Simulated mean, MSE, and Def values for the estimators of α and β; $\alpha = 3$, $\beta = 2$

		$\hat{\alpha}$		$\hat{\beta}$		
n	Method	Mean	MSE	Mean	MSE	Def
25	ML	3.1741	0.3102	2.0146	0.0216	0.3318[4]
	LS	2.9983	0.3939	2.0036	0.0230	0.4169[6]
	WLS	3.0257	0.3516	2.0062	0.0221	0.3736[5]
	PC	2.7252	0.7546	1.9088	0.1088	0.8634[8]
	MPS	2.8218	0.2534	2.0005	0.0209	0.2744[1]
	PWM	2.8744	0.2605	1.9085	0.0285	0.2890[2]
	CM	3.1911	0.5000	2.0161	0.0238	0.5238[7]
	AD	3.0526	0.2882	2.0096	0.0219	0.3101[3]
50	ML	3.0844	0.1280	2.0104	0.0103	0.1383[2]
	LS	2.9995	0.1812	2.0046	0.0112	0.1924[6]
	WLS	3.0212	0.1507	2.0065	0.0106	0.1613[4]
	PC	2.7269	0.5388	1.9005	0.0890	0.6278[8]
	MPS	2.8751	0.1210	2.0028	0.0101	0.1311[1]
	PWM	2.9516	0.1723	1.9568	0.0119	0.1842[5]
	CM	3.0922	0.2041	2.0108	0.0114	0.2155[7]
	AD	3.0272	0.1369	2.0077	0.0105	0.1475[3]
100	ML	3.0536	0.0614	2.0041	0.0048	0.0662[2]
	LS	3.0199	0.0877	2.0014	0.0054	0.0931[5]
	WLS	3.0316	0.0727	2.0026	0.0050	0.0777[4]
	PC	2.7732	0.3708	1.9102	0.0628	0.4335[8]
	MPS	2.9291	0.0593	2.0000	0.0048	0.0641[1]
	PWM	2.9979	0.1071	1.9774	0.0055	0.1125[7]
	CM	3.0658	0.0949	2.0045	0.0054	0.1004[6]
	AD	3.0308	0.0692	2.0030	0.0050	0.0742[3]
200	ML	3.0186	0.0262	2.0005	0.0027	0.0289[1]
	LS	2.9920	0.0412	1.9985	0.0029	0.0441[5]
	WLS	3.0048	0.0327	1.9995	0.0027	0.0354[4]
	PC	2.7557	0.2973	1.8965	0.0690	0.3662[8]
	MPS	2.9465	0.0274	1.9984	0.0027	0.0301[2]
	PWM	2.9856	0.0569	1.9867	0.0029	0.0598[7]
	CM	3.0145	0.0421	2.0000	0.0029	0.0450[6]
	AD	3.0031	0.0320	1.9995	0.0027	0.0347[3]
500	ML	3.0114	0.0131	2.0022	0.0011	0.0142[1]
	LS	3.0049	0.0183	2.0019	0.0012	0.0196[5]
	WLS	3.0086	0.0156	2.0021	0.0012	0.0167[4]
	PC	2.8140	0.1842	1.9170	0.0429	0.2271[8]
	MPS	2.9764	0.0133	2.0013	0.0011	0.0144[2]
	PWM	2.9973	0.0333	1.9965	0.0012	0.0346[7]
	CM	3.0139	0.0186	2.0025	0.0013	0.0199[6]
	AD	3.0069	0.0155	2.0020	0.0012	0.0166[3]

Table 3 Simulated mean, MSE, and Def values for the estimators of α and β; $\alpha = 5$, $\beta = 2$

		$\hat{\alpha}$		β		
n	Method	Mean	MSE	Mean	MSE	Def
25	ML	5.2883	0.8639	2.0072	0.0075	0.8714[3]
	LS	4.9859	1.0722	2.0006	0.0080	1.0802[6]
	WLS	5.0351	0.9454	2.0023	0.0077	0.9531[4]
	PC	4.6214	1.5396	1.9778	0.0145	1.5541[8]
	MPS	4.7026	0.7059	1.9987	0.0073	0.7132[1]
	PWM	4.2639	0.9512	1.9148	0.0141	0.9652[5]
	CM	5.3047	1.3531	2.0080	0.0082	1.3613[7]
	AD	5.0849	0.8105	2.0042	0.0076	0.8181[2]
50	ML	5.1316	0.3595	2.0030	0.0037	0.3632[2]
	LS	4.9745	0.5037	1.9998	0.0040	0.5077[6]
	WLS	5.0193	0.4215	2.0010	0.0038	0.4252[4]
	PC	4.6602	0.9839	1.9769	0.0091	0.9929[8]
	MPS	4.7847	0.3439	1.9984	0.0037	0.3476[1]
	PWM	4.6131	0.4773	1.9572	0.0054	0.4827[5]
	CM	5.1281	0.5594	2.0035	0.0040	0.5634[7]
	AD	5.0291	0.3874	2.0016	0.0037	0.3911[3]
100	ML	5.0731	0.1835	2.0012	0.0017	0.1852[2]
	LS	5.0011	0.2506	1.9995	0.0019	0.2525[5]
	WLS	5.0228	0.2075	2.0002	0.0018	0.2093[4]
	PC	4.7022	0.6314	1.9789	0.0042	0.6355[8]
	MPS	4.8677	0.1831	1.9988	0.0017	0.1848[1]
	PWM	4.8023	0.2556	1.9782	0.0022	0.2578[6]
	CM	5.0773	0.2662	2.0013	0.0019	0.2682[7]
	AD	5.0231	0.1983	2.0004	0.0018	0.2001[3]
200	ML	5.0181	0.0755	2.0017	0.0009	0.0764[1]
	LS	4.9680	0.1152	2.0000	0.0010	0.1162[5]
	WLS	4.9924	0.0932	2.0010	0.0009	0.0941[4]
	PC	4.7506	0.3563	1.9848	0.0023	0.3587[8]
	MPS	4.8974	0.0819	2.0004	0.0009	0.0828[2]
	PWM	4.8829	0.1285	1.9901	0.0010	0.1294[7]
	CM	5.0053	0.1163	2.0009	0.0010	0.1173[6]
	AD	4.9889	0.0910	2.0010	0.0009	0.0919[3]
500	ML	5.0406	0.0350	1.9994	0.0003	0.0353[2]
	LS	5.0257	0.0510	1.9989	0.0004	0.0514[6]
	WLS	5.0327	0.0410	1.9992	0.0004	0.0413[4]
	PC	4.8565	0.1879	1.9864	0.0013	0.1892[8]
	MPS	4.9825	0.0329	1.9989	0.0003	0.0333[1]
	PWM	4.9828	0.0491	1.9948	0.0004	0.0495[5]
	CM	5.0408	0.0524	1.9993	0.0004	0.0528[7]
	AD	5.0310	0.0402	1.9992	0.0004	0.0406[3]

increasing values of the sample sizes. It should also be stated that the WLS, CM, and AD estimators have negligible bias for moderate and large sample sizes.

In view of the MSE, it is clear to say that the MPS estimators show the best performance for all sample sizes and parameter settings. They are followed by the ML and AD estimators. It should be noted that the MPS and ML estimators are close to each other for moderate and large sample sizes. The WLS, CM, and LS estimators demonstrate better performance when the sample size increases. However, the PC and PWM estimators do not perform well and the PC estimators show the worst performance with the highest MSE values.

To compare the simultaneous efficiencies of the estimators, we use the *Def* criterion. It can be seen from Tables 1, 2, and 3 that the *Def* values are ranked from the smallest to the largest for each estimator. It is clear to say that the MPS and ML estimators demonstrate the strongest performance with the lowest *Def* values. This is followed by the AD estimators. Similar to the MSE criterion, the WLS, CM, and LS estimators show better performance than the PC and PWM estimators with smaller *Def* values. The PC estimators show the worst performance with the highest *Def* values among the others.

5 Robustness Property

In this section, we compare the efficiencies of the ML, LS, WLS, PC, MPS, PWM, CM, and AD estimators of α and β when the data contains outliers. To do this, we assume that the underlying distribution is $IW(\alpha = 3, \beta = 2)$ and add r outliers to the observations in the true model. Here, $r = [\![0.5 + 0.1n]\!]$ (integer value). Simulation results are reported in Table 4, see also Table 2 in the context of no outliers.

It is clear from Table 4 that the PC and PWM estimators of the parameters are the most sensitive estimators to the outliers. Despite being the most efficient estimator under the true model, the MPS estimators lose their efficiency drastically when the data set contains outliers. On the other hand, the ML and AD estimators are resistant to outliers for small and moderate sample sizes. However, the WLS estimator is more preferable than the others in terms of *Def* criterion for large sample sizes.

6 Real Data Examples

In this section, we analyze two different data sets taken from the literature to illustrate the modeling performance of IW distribution. We also compare the performance of the proposed estimators of the unknown parameters.

Table 4 Simulated mean, MSE, and Def values for the estimators of α and β with outliers; $\alpha = 3$, $\beta = 2$

		$\hat{\alpha}$		β		
n	Method	Mean	MSE	Mean	MSE	Def
25	ML	2.7376	0.1866	2.0594	0.0243	0.2109[1]
	LS	2.9493	0.4185	2.0035	0.0224	0.4409[4]
	WLS	2.9177	0.3838	2.0098	0.0218	0.4056[3]
	PC	1.7649	1.6002	1.8287	0.1044	1.7046[8]
	MPS	2.4148	0.4318	2.0452	0.0224	0.4541[5]
	PWM	2.1742	0.7201	1.9064	0.0284	0.7485[7]
	CM	3.1402	0.4985	2.0162	0.0233	0.5218[6]
	AD	2.6942	0.2446	2.0192	0.0219	0.2666[2]
50	ML	2.7652	0.1196	2.0425	0.0122	0.1317[1]
	LS	2.9729	0.1866	2.0033	0.0113	0.1979[4]
	WLS	2.9539	0.1649	2.0073	0.0108	0.1757[3]
	PC	1.9470	1.1907	1.8497	0.0870	1.2777[8]
	MPS	2.5709	0.2394	2.0350	0.0115	0.2509[6]
	PWM	2.2968	0.5281	1.9533	0.0124	0.5405[7]
	CM	3.0645	0.2034	2.0095	0.0116	0.2150[5]
	AD	2.7868	0.1343	2.0118	0.0109	0.1452[2]
100	ML	2.7355	0.0999	2.0365	0.0063	0.1062[5]
	LS	2.9635	0.0851	1.9993	0.0054	0.0904[2]
	WLS	2.9461	0.0750	2.0030	0.0052	0.0802[1]
	PC	2.1260	0.8291	1.9476	0.0468	0.8759[8]
	MPS	2.6236	0.1693	2.0323	0.0060	0.1753[6]
	PWM	2.3288	0.4699	1.9792	0.0055	0.4754[7]
	CM	3.0081	0.0867	2.0023	0.0054	0.0921[3]
	AD	2.7747	0.0926	2.0070	0.0052	0.0978[4]
200	ML	2.7254	0.0914	2.0338	0.0034	0.0947[5]
	LS	2.9805	0.0441	1.9990	0.0024	0.0465[2]
	WLS	2.9586	0.0387	2.0018	0.0023	0.0410[1]
	PC	2.2731	0.5810	2.0248	0.0270	0.6080[8]
	MPS	2.6612	0.1301	2.0315	0.0032	0.1333[6]
	PWM	2.3478	0.4372	1.9928	0.0023	0.4395[7]
	CM	3.0028	0.0445	2.0005	0.0024	0.0469[3]
	AD	2.7796	0.0711	2.0054	0.0023	0.0735[4]
500	ML	2.7212	0.0830	2.0335	0.0021	0.0851[5]
	LS	2.9845	0.0171	1.9996	0.0010	0.0181[3]
	WLS	2.9592	0.0150	2.0016	0.0010	0.0160[1]
	PC	2.4136	0.3842	2.0995	0.0294	0.4136[7]
	MPS	2.6913	0.1004	2.0325	0.0020	0.1025[6]
	PWM	2.3513	0.4272	2.0012	0.0010	0.4282[8]
	CM	2.9934	0.0170	2.0002	0.0010	0.0181[2]
	AD	2.7801	0.0564	2.0052	0.0010	0.0574[4]

6.1   The Rainfall Data

We first analyze rainfall data taken from Asgharzadeh et al. (2016). This data set concerns seasonal rainfall in inches recorded at Los Angeles Civic Center from 1962 to 2012. The data are as follows:

x_i: 08.38 07.93 13.68 20.44 22.00 16.58 27.47 07.74 12.32 07.17 21.26 14.92 14.35
07.21 12.30 33.44 19.67 26.98 08.96 10.71 31.28 10.43 12.82 17.86 07.66 12.48
08.08 07.35 11.99 21.00 27.36 08.11 24.35 12.44 12.40 31.01 09.09 11.57 17.94
04.42 16.42 09.25 37.96 13.19 03.21 13.53 09.08 16.36 20.20 08.69

First, we want to ascertain whether or not this data set can be modeled with IW distribution. To do this, we draw an IW Q-Q plot of 50 observations, see Fig. 4a. It should be noted that the ML estimates of the parameters are used to construct the IW Q-Q plot of the observations. We also draw the fitted distribution function and empirical distribution for the rainfall data, see Fig. 4b. It is clear from these figures that IW distribution provides a good fit for the rainfall data.

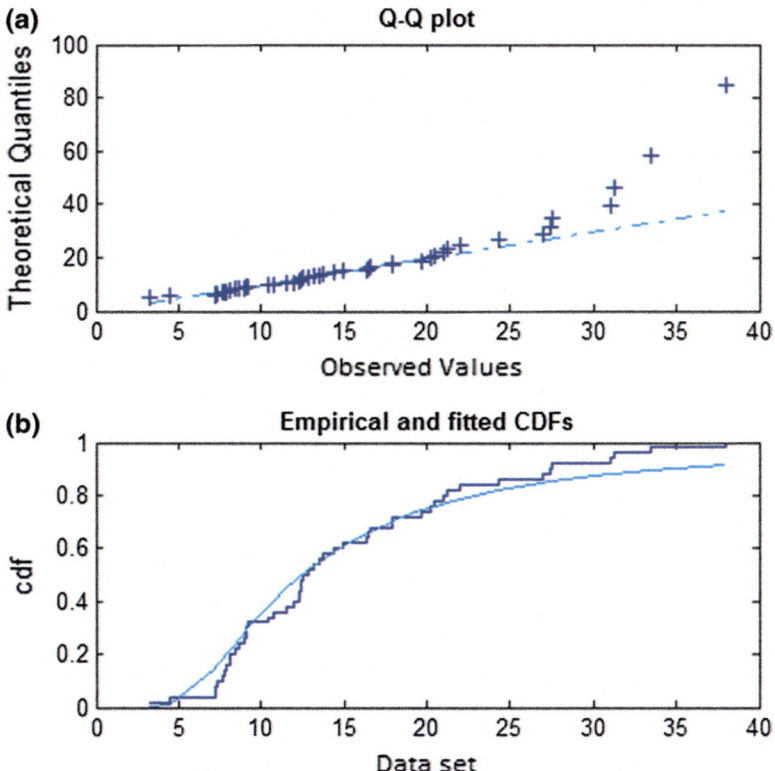

Fig. 4 Diagnostic plots for rainfall data

Then, we estimate the unknown parameters of IW distribution using the estimation methods defined in Sect. 3. Furthermore, to compare the performance of these methods, we use model selection criteria. These are given below

Akaike information criterion: $AIC = -2 \ln L(\theta) + 2k$,
Bayesian information criterion: $BIC = -2 \ln L(\theta) + k \ln n$,
Corrected AIC: $AICc = -2 \ln L(\theta) + \frac{2kn}{n-k-1}$,
Hannan–Quinn criterion: $HQC = -2 \ln L(\theta) + 2k \ln \ln n$.

Here, $\theta = (\alpha, \beta)$ and k is the number of the unknown parameters. It is obvious that smaller values of these criteria show better fit. For more detailed information regarding these criteria, see Burnham and Anderson (2004) and Bagheri et al. (2016). The results are reported in Table 5.

Besides the model selection criteria, we draw the histogram of the rainfall data and the fitted pdfs of IW distribution based on the different estimation methods, see Fig. 5.

It is clear from Table 5 and Fig. 5 that ML estimator is the best estimator with the smallest values for all model selection criteria. It is followed by the MPS estimator. However, the PWM and PC estimators show the worst performance.

6.2 The Failure Times Data

Our second data set concerns the times between failures of secondary reactor pumps taken from Sharma et al. (2014). The data are given as follows:

x_i: 2.160 0.150 4.082 0.746 0.358 0.199 0.402 0.101 0.605 0.954 1.359 0.273 0.491
3.465 0.070 6.560 1.060 0.062 4.992 0.614 5.320 0.347 1.921

Similar to the first example, Fig. 6a displays an IW Q-Q plot of 23 observations. We also draw the fitted distribution function and empirical distribution for data set

Table 5 Parameter estimates and model selection criteria for rainfall data

Estimation methods	$\hat{\alpha}$	$\hat{\beta}$	AIC	BIC	AICc	HQC
ML	1.8519	10.2336	350.3042	354.1283	350.5595	351.7604
LS	2.0841	10.7484	353.4826	357.3066	353.7379	354.9388
WLS	2.2254	10.7368	357.1555	360.9796	357.4109	358.6118
PC	2.9199	11.7085	430.5924	434.4165	430.8477	432.0486
MPS	1.7312	10.1138	350.8805	354.7045	351.1358	352.3367
PWM	2.6127	10.4906	373.4062	377.2303	373.6615	374.8624
CM	2.1403	10.7917	354.9688	358.7928	355.2241	356.4250
AD	2.0923	10.6512	353.3070	357.1310	353.5623	354.7632

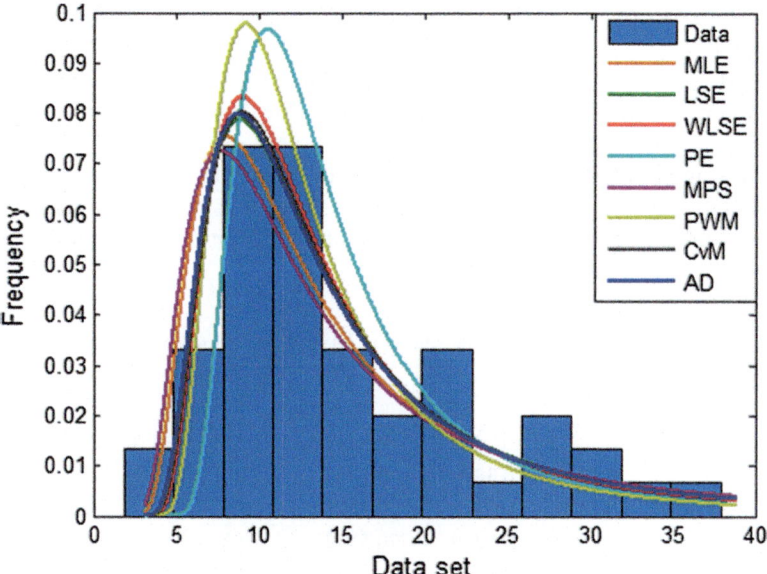

Fig. 5 Histogram of rainfall data with fitted pdfs

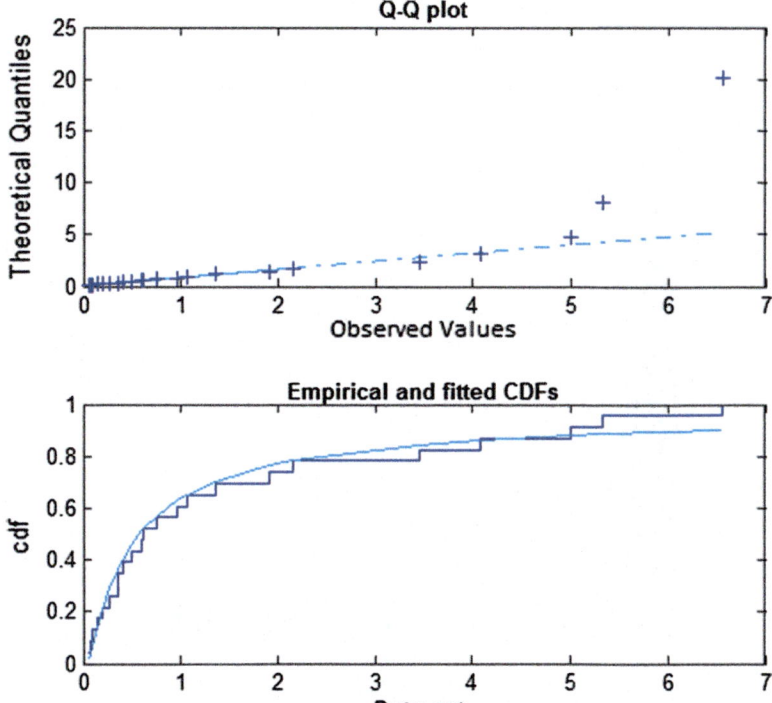

Fig. 6 Diagnostic plots for failure times data

in Fig. 6b. It is clear from these figures that this data set can be modeled by IW distribution.

Estimates of the unknown parameters of IW distribution based on eight different estimation methods and model selection criteria for these methods are reported in Table 6.

The histogram of the failure times data with fitted pdfs is drawn in Fig. 7. It is obvious from Fig. 7 and Table 6 that the ML estimator is the best estimator with the smallest model selection criteria value. Similarly, as in the first application, the

Table 6 Parameter estimates and model selection criteria for failure times data

Estimation methods	$\hat{\alpha}$	$\hat{\beta}$	AIC	BIC	AICc	HQC
ML	0.7832	0.3569	69.8834	72.1544	70.4834	70.4546
LS	0.7256	0.3889	70.1306	72.4016	70.7306	70.7017
WLS	0.7524	0.3718	69.9507	72.2217	70.5507	70.5219
PC	1.4535	0.8749	270.5572	272.8282	271.1572	271.1283
MPS	0.7045	0.3431	70.4238	72.6948	71.0238	70.9950
PWM	1.4028	0.5037	124.7133	126.9843	125.3133	125.2845
CM	0.7780	0.4023	70.0773	72.3483	70.6773	70.6485
AD	0.7457	0.3749	69.9831	72.2541	70.5831	70.5542

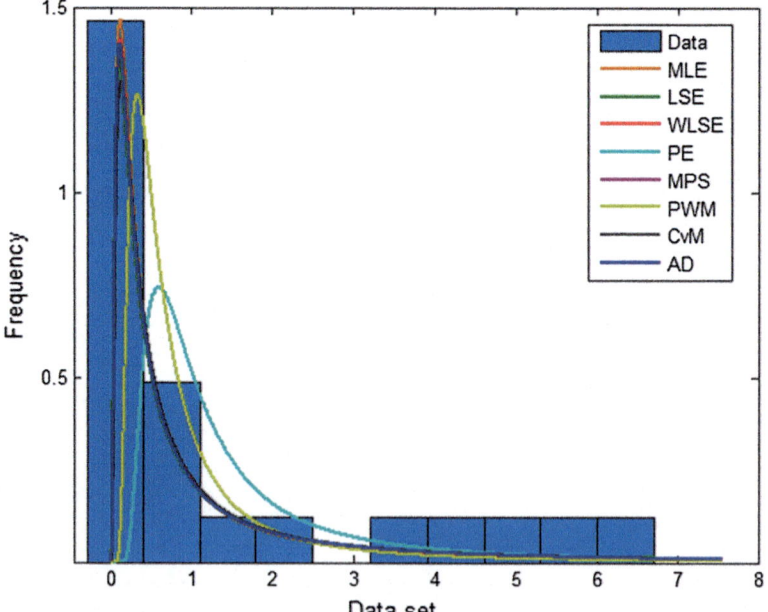

Fig. 7 Histogram of failure times data with fitted pdfs

PWM and PC estimators show the worst performance. These results are also in agreement with the simulation results.

7 Conclusion

In this chapter, we obtain the estimators of model parameters of IW distribution using the ML, LS, WLS, PC, MPS, PWM, AD, and CM methods. We perform an extensive Monte Carlo simulation study to compare the efficiencies of these estimators. It is concluded that the ML and MPS estimators show the best performance among the others. The PC and PWM estimators demonstrate the weakest performance. Robustness properties are also investigated. The PC and PWM estimators are found to be the most sensitive to the outliers. Moreover, it can be seen that the MPS estimators are affected badly by the outliers. In terms of minimum distance methods, the AD estimators perform well under the true model. However, the CM estimators are more resistant to outliers than AD estimators. At the end of the study, we use two different data sets to illustrate the modeling performance of IW distribution. It is concluded that IW distribution provides a good fit for modeling rainfall and failure times data sets.

References

Akgül, F.G., Şenoğlu, B., Arslan, T.: An alternative distribution to Weibull for modeling the wind speed data: Inverse Weibull distribution. Energy Convers. Manag. **114**, 234–240 (2016)

Asgharzadeh, A., Abdi, M., Nadarajah, S.: Interval estimation for gumbel distribution using climate records. Bull. Malays. Math. Sci. Soc. **39**(1), 257–270 (2016)

Ashoori, F., Ebrahimpour, M., Bozorgnia, A.: Modeling of maximum prediction using maximal generalized extreme value distribution. Commun. Stat. Theory Methods **46**(6), 3025–3033 (2017)

Bagheri, S.F., Alizadeh, M., Nadarajah, S.: Efficient estimation of the pdf and the cdf of the exponentiated gumbel distribution. Commun. Stat. Simul. Comput. **45**, 339–361 (2016)

Burnham, K.P., Anderson, D.R.: Multimodel inference: understanding AIC and BIC in model selection. Sociol. Methods Res. **33**, 261–304 (2004)

Calabria, R., Pulcini, G.: On the maximum likelihood and least-squares estimation in the inverse Weibull distribution. Stat. Appl. **2**(1), 53–66 (1990)

Cheng, R.C.H., Amin, N.A.K.: Estimating parameters in continuous univariate distributions with a shifted origin. Journal of Royal Statistical Society: Series B **45**, 394–403 (1983)

Drapella, A.: The complementary Weibull distribution: unknown or just forgotten. Qual Reliab Eng Inter **9**(4), 383–385 (1993)

Erto, P., Rapone, M.: Non-informative and practical Bayesian confidence bounds for reliable life in the Weibull model. Reliab. Eng. **7**, 181–191 (1984)

Fréchet, M.: Sur la loi de probabilite de lecart maximum. Ann. Soc. Polon. Math. **6**(93) (1927)

Greenwood, J.A., Landwehr, J.M., Matalas, N.C., Wallis, J.R.: Probability weighted moments: definition and relation to parameters of several distributions expressable in inverse form. Water Resour. Res. **15**(5), 1049–1054 (1979)

Hosking, J.R.M., Wallis, J.R., Wood, E.F.: Estimation of the generalized extreme value distribution by the method of probability weighted moment. J Technometr. **27**(3), 251–261 (1985)

Kao, J.H.K.: Computer methods for estimating Weibull parameters in reliability studies. IRE Trans. Reliab. Qual. Control **13**, 15–22 (1958)

Keller, A.Z., Kamath, R.R.: Alternative reliability models for mechanical systems. In: 3rd International Conference on Reliability and Maintainability, Toulouse (1982)

Landwehr, J.M., Matalas, N.C., Wallis, J.R.: Probability weighted moments compared with some traditional techniques in estimating gumbel parameters and quantiles. Water Resour. Res. **15**(5), 1055–1064 (1979)

Louzada, F., Ramos, P.L., Perdona, G.S.C.: Different estimation procedures for the parameters of the extended exponential geometric distribution for medical data. Comput. Math. Methods Med. (2016). https://doi.org/10.1155/2016/8727951

Luceño, A.: Fitting the generalized Pareto distribution to data using maximum goodness-of-fit estimators. Comput. Stat. Data Anal. **51**(2), 904–917 (2006)

Luceño, A.: Maximum likelihood vs. maximum goodness of fit estimation of the three-parameter Weibull distribution. J. Stat. Comput. Simul. **78**(10), 941–949 (2008)

Mudholkar, G.S., Kollia, G.D.: Generalized Weibull family: a structural analysis. Commun. Stat. Theory Methods **23**(4), 1149–1171 (1994)

Murthy, D.N., Bulmer, M., Eccleston, J.A.: Weibull model selection for reliability modelling. Reliab. Eng. Syst. **86**, 257–267 (2004)

Pasari, S., Dikshit, O.: Impact of three-parameter models in probabilistic assessment of earthquake hazards. Pure Appl. Geophys. **171**, 1251–1281 (2014)

Rinne, H.: The Weibull Distribution: A Handbook. CRC Press, Boca Raton (2009)

Santo, A.P.J.E., Mazucheli, J.: Comparison of estimation methods for the Marshall-Olkin expended Lindley distribution. J. Stat. Comput. Simul. **85**(17), 3437–3450 (2015)

Sharma, V.K., Singh, S.K., Singh, U.: A new upside-down bathtub shaped hazard rate model for survival data analysis. Appl. Math. Comput. **239**, 242–253 (2014)

Singh, S.K., Singh, U., Kumar, D.: Bayesian estimation of parameters of inverse Weibull distribution. J. Appl. Stat. **40**(7), 1597–1607 (2013)

Soukissian, T.H., Tsalis, C.: The effect of the generalized extreme value distribution parameter estimation methods in extreme wind speed prediction. Nat. Hazards **78**, 1777–1809 (2015)

Tiku, M.L., Akkaya, A.D.: Robust Estimation and Hypothesis Testing. New Age International (P) Limited, Publishers, New Delhi (2004)

Wolfowitz, J.: Estimation by the minimum distance method. Ann. Inst. Stat. Math. **5**, 9–23 (1953)

Wolfowitz, J.: The minimum distance methods. Ann. Math. Stat. **28**, 75–88 (1957)

Liu-Type Negative Binomial Regression: A Comparison of Recent Estimators and Applications

Yasin Asar

Abstract This chapter introduces a new biased estimator, that is a generalization of Liu-type estimator (Liu in Commun Stat Theory Methods 32:1009–1020 2003), for the negative binomial regression model. Since the variance of the maximum likelihood estimator (MLE) is inflated when there is multicollinearity between the explanatory variables, a new biased estimator is proposed to solve the problem and decrease the variance of MLE in order to make stable inferences. Moreover, we obtain some theoretical comparisons between the new estimator and some other existing estimators via matrix mean squared error criterion. Furthermore, a Monte Carlo simulation study is designed to evaluate performance of the estimators in the sense of mean squared error. Finally, real data applications are used to illustrate the benefits of new estimator.

Keywords Liu-type estimator · Negative binomial regression
Multicollinearity · MSE · MLE

Mathematics Subjects Classification: Primary 62J07 · Secondary 62J02

1 Introduction

In real life contexts, the observations are not independent and identically distributed (iid) all the time. The data often comes in the form of nonnegative integers or counts which are not iid. A count refers to the number of times an event occurs. Therefore, it is the realization of a nonnegative integer-valued random variable. The main interest of a researcher may depend on the covariates which are assumed to affect the parameters of the conditional distribution of events, given the covariates. This is generally achieved by a regression model of count (Cameron and Trivedi 2013). Thus, count

Y. Asar (✉)
Department of Mathematics-Computer Sciences,
Necmettin Erbakan University, 42090 Konya, Turkey
e-mail: yasar@konya.edu.tr; yasinasar@hotmail.com

© Springer International Publishing AG 2018   23
M. Tez and D. von Rosen (eds.), *Trends and Perspectives
in Linear Statistical Inference*, Contributions to Statistics,
https://doi.org/10.1007/978-3-319-73241-1_2

regression models such as Poisson regression or negative binomial (NB) regression is mostly used in the field of health, social, economic and physical sciences such that the nonnegative and an integer-valued aspect of the outcome play an important role in the analysis.

Although twenty-two different versions of NB model were mentioned by Hilbe (2011), the traditional NB model which was symbolized as NB2 by Cameron and Trivedi (1986) is the main topic of this chapter. When the dependent variable has a greater variance than the mean, the so-called over-dispersion occurs in Poisson models. If there is a positive correlation between the responses or an excess variation in the response counts, then over-dispersion may occur in data (Hilbe 2011, Ch. 7). In these situations, NB2 model is more useful than Poisson regression model since NB2 allows for random variation in the Poisson conditional mean, h_i, by letting $h_i = z_i \mu_i$ where $\mu_i = \exp(x_i \beta)$ such that x_i is the ith ow of the data matrix X of order $n \times (p+1)$ with p explanatory variables, β is the coefficient vector of order $(p+1) \times 1$ with intercept and z_i is a random variable following the gamma distribution such that $z_i \sim \Gamma(\delta, \delta)$, $i = 1, 2, ..., n$.

The density function of the dependent variable y_i is given by

$$\Pr\left(y = y_i | x_i\right) = \frac{\Gamma\left(\theta^{-1} + y_i\right)}{\Gamma\left(\theta^{-1}\right) \Gamma\left(1 + y_i\right)} \left(\frac{\theta^{-1}}{\theta^{-1} + \mu_i}\right)^{\theta^{-1}} \left(\frac{\mu_i}{\theta^{-1} + \mu_i}\right)^{y_i}$$

where the over-dispersion parameter θ is given as $\theta = 1/\delta$. The conditional mean and variance of the distribution are given respectively as follows:

$$E\left(y_i | x_i\right) = \mu_i,$$

$$Cov\left(y_i | x_i\right) = \mu_i \left(1 + \theta \mu_i\right).$$

The estimation of the coefficient vector β is usually obtained by maximizing the following log-likelihood function

$$L(\theta, \beta) = \sum_{i=1}^{n} \left\{ \left[\sum_{j=0}^{y_i - 1} \log\left(j + \theta^{-1}\right)\right] - \log\left(y_i!\right) - \left(y_i + \theta^{-1}\right) \log\left(1 + \theta \mu_i\right) + y_i \log\left(\theta\right) + y_i \log\left(\mu_i\right) \right\}$$

$$(1)$$

since $\log\left(\frac{\Gamma(\theta^{-1} + y_i)}{\Gamma(\theta^{-1})}\right) = \sum_{j=0}^{y_i - 1} \log\left(j + \theta^{-1}\right)$. The estimation of the parameter β is usually obtained by the method of maximum likelihood estimation (MLE) which can be obtained by maximizing the Eq. (1) with respect to β, namely, solving the following equations

$$S(\beta) = \frac{\partial L(\theta, \beta)}{\partial \beta} = \sum_{i=1}^{n} \frac{\left(y_i - \mu_i\right)}{1 + \theta \mu_i} x_i = 0. \tag{2}$$

Since the Eq. (2) is nonlinear in β, one should use the following scoring method

$$\beta^{(r)} = \beta^{(r-1)} + I^{-1}\left(\beta^{(r-1)}\right) S\left(\beta^{(r-1)}\right) \tag{3}$$

where $S\left(\beta^{(r-1)}\right)$ is the first derivative of the log-likelihood function evaluated at $\beta^{(r-1)}$ and

$$I^{-1}\left(\beta^{(r-1)}\right) = E\left(\frac{\partial^2 L\left(X;\beta\right)}{\partial\beta\,\partial\beta'}\right) = X'W\left(\beta^{(r-1)}\right)X,$$

$W\left(\beta^{(r-1)}\right) = \text{diag}\left(\frac{\mu_i\left(\beta^{(r-1)}\right)}{1+\theta\mu_i\left(\beta^{(r-1)}\right)}\right)$ evaluated at $\beta^{(r-1)}$. In the final step of the algorithm, MLE of β is obtained as follows:

$$\hat{\beta}_{MLE} = \left(X'\hat{W}X\right)^{-1}X'\hat{W}\hat{Z}$$

where $\hat{Z}$ is a vector with the ith element equal to $\log\left(\hat{\mu}_i\right) + \frac{y_i-\hat{\mu}_i}{\hat{\mu}_i}$, $\hat{W}$ and $\hat{\mu}_i$ are the values of $W\left(\beta^{(r-1)}\right)$ and $\mu_i\left(\beta^{(r-1)}\right)$ at the final step respectively, the hats show the iterative nature of the algorithm. This method is also known as the iteratively re-weighted least squares algorithm (IRLS), please see Månsson (2012), Hilbe (2011), Myers et al. (2010) for extra details. Here, we give the steps of the IRLS algorithm as follows:

(1) In the first step, one needs an initial vector of coefficients, say ordinary least squares (OLS) estimator.
(2) Calculate $\mu_i = \exp\left(x_i\beta\right)$ and than W.
(3) Get Z using μ and W obtained in the previous step.
(4) Calculate a new estimate of β via Eq. (3).
(5) Terminate the algorithm until the convergence is achieved.

However, when the matrix $X'\hat{W}X$ is ill-conditioned, i.e., the correlation between the explanatory variables are high, MLE becomes unstable and its variance is inflated. This problem is called multicollinearity. In this situation, some of the eigenvalues of the matrix $X'\hat{W}X$ become small (see Månsson (2012), MacKinnon and Puterman (1989)). Moreover, Lesaffre and Marx (1993) discussed the ill-conditioned nature of the weighted cross products matrix in generalized linear models in great details. Also, the condition number being the ratio of the largest eigenvalue to the smallest eigenvalue of the matrix $X'\hat{W}X$ is a measure of collinearity (see Smith and Marx (1990), Weissfeld and Sereika (1991)) and therefore can be used to check the existence of collinearity.

Although, applying shrinkage estimators is very popular in linear model to solve the multicollinearity problem (see Hoerl et al. (1975), Kibria (2003), Liu (2003), Lipovetsky and Conklin (2005) etc.), count models have not been investigated in the presence of multicollinearity. Therefore, as an exception, Månsson (2012) proposed to use the ridge regression (Hoerl and Kennard 1970) in negative binomial regression models. The negative binomial ridge regression estimator (RR) is obtained as follows:

$$\hat{\beta}_{RR} = \left(X'\hat{W}X + kI\right)^{-1}X'\hat{W}\hat{Z},\ k>0$$

where I is the $(p + 1) \times (p + 1)$ identity matrix. The author proposed to use some existing ridge estimators to estimate the ridge parameter k.

Moreover, Liu estimator (Liu 1993) is generalized to the negative binomial regression model by Månsson (2013) and the following negative binomial Liu estimator (LE) is obtained:

$$\hat{\beta}_{LE} = \left(X'\hat{W}X + I\right)^{-1}\left(X'\hat{W}X + dI\right)\hat{\beta}_{MLE}, \ 0 < d < 1.$$

Finally, motivated by the idea that combining the two estimators might inherit the advantages of both estimators, a two-parameter estimator which is a combination of RR and LE has been proposed in Huang and Yang (2014).

The purpose of this chapter is to generalize Liu-type estimator (Liu 2003) to the negative binomial regression and discusses some properties of the new estimator. The organization of the chapter is as follows: In Sect. 2, Liu-type negative binomial estimator (LT) is proposed, matrix mean squared error (MMSE) and mean squared error (MSE) properties are investigated and selection of the shrinkage parameters is discussed. In order to compare the performance of the estimators MLE, RR, LE, and LT, a Monte Carlo simulation is designed and its results are discussed in Sect. 3. Real data applications are demonstrated to illustrate the benefits of LT in Sect. 4. Finally, a brief summary and conclusion are provided.

2   New Estimator and MSE Properties

2.1   Construction of LT

Consider the linear regression model $Y = X\beta + \varepsilon$ where X is an $n \times p$ data matrix, β is the $p \times 1$ coefficient vector, ε is the $n \times 1$ random error vector satisfying $\varepsilon_i \sim N\left(0, \sigma^2\right)$ and Y is the $n \times 1$ dependent variable. When there is multicollinearity, the matrix $X'X$ becomes ill-conditioned and some of the eigenvalues of the matrix $X'X$ becomes close to zero and the condition number $\kappa = \left(\vartheta_{max}/\vartheta_{min}\right)$ becomes very high such that $\vartheta_j, j = 1, 2, ..., p$ are the eigenvalues of $X'X$. Thus, the ordinary least square estimator (OLS), $\hat{\beta}_{OLS} = \left(X'X\right)^{-1}X'Y$ becomes unstable. Therefore, Hoerl and Kennard (1970) proposed ridge estimator $\hat{\beta}_{Ridge} = \left(X'X + kI\right)^{-1}X'Y$ which is obtained by augmenting $0 = k^{1/2}\beta + \varepsilon'$ to the original equation. However, large values of k make the distance between $k^{1/2}\beta$ and 0 increase. One should use large values of k which impose more bias to the ridge estimator and to control the condition number. Therefore, Liu (2003) proposed to augment $\left(-d/k^{1/2}\right)\hat{\beta}_{OLS} = k^{1/2}\beta + \varepsilon'$ to the original equation and obtain Liu-type estimator $\hat{\beta}_{k,d} = \left(X'X + kI\right)^{-1}\left(X'X - dI\right)\hat{\beta}$ where $k > 0$, $-\infty < d < \infty$ and $\hat{\beta}$ is any estimator. The results showed that $\hat{\beta}_{k,d}$ has a better performance than OLS and ridge estimator in the sense of MSE. Therefore, we define a generalization of Liu-type estimator in negative binomial model to solve the problem of multicollinearity. In this study, following Månsson (2012), (Månsson, 2013) a generalization of Liu-type estimator (LT) to the negative binomial regression model is proposed as follows

$$\hat{\beta}_{LT} = \left(X'\hat{W}X + kI\right)^{-1} \left(X'\hat{W}X - dI\right) \hat{\beta}_{MLE}$$

where $k > 0$ and $-\infty < d < \infty$.

From the definition of LT, it is easy to see that LT is a general estimator including MLE, RR and LE as follows, if d equals $-k$, then we have $\hat{\beta}_{LT} = \hat{\beta}_{MLE}$ and if $k = 1$ with a minus sign in front of d, then $\hat{\beta}_{LT} = \hat{\beta}_{LE}$, and finally if $d = 0$ we have $\hat{\beta}_{LT} = \hat{\beta}_{RR}$.

In order to see the superiority of the estimator LT, MMSE containing all the relevant information regarding the estimators can be used as a comparison criterion. MMSE and MSE being the trace of MMSE of an estimator $\tilde{\beta}$ are respectively defined by

$$MMSE(\tilde{\beta}) = E\left[(\tilde{\beta} - \beta)(\tilde{\beta} - \beta)'\right] = \text{Cov}(\tilde{\beta}) + \text{bias}(\tilde{\beta})\text{bias}(\tilde{\beta})',$$

$$MSE(\tilde{\beta}) = \text{tr}\left(MMSE(\tilde{\beta})\right) = E\left[(\tilde{\beta} - \beta)'(\tilde{\beta} - \beta)\right]$$
$$= \text{tr}\left[\text{Cov}(\tilde{\beta})\right] + \text{bias}(\tilde{\beta})'\text{bias}(\tilde{\beta})$$

where $\text{Cov}(\tilde{\beta})$ is the covariance matrix and $\text{bias}(\tilde{\beta}) = E(\tilde{\beta}) - \beta$ is the bias vector of the estimator $\tilde{\beta}$.

Thus MSE and MMSE of MLE are given by the following equations respectively

$$MSE(MLE) = tr\left(X'WX\right)^{-1} = \sum_{j=1}^{p+1} \frac{1}{\lambda_j},$$

$$MMSE(MLE) = \left(X'WX\right)^{-1}$$

where λ_j is the jth eigenvalue of the matrix $X'WX$ (see Månsson (2012)).

A transformation is constructed in order to present the explicit form of the MMSE and MSE functions. Let $Q'X'WXQ = \Lambda = diag\left(\lambda_1, \lambda_2, ..., \lambda_{p+1}\right)$ and $\alpha = Q'\beta$ where $\lambda_1 \geq \lambda_2 \geq ... \geq \lambda_{p+1} > 0$ and Q is the matrix whose columns are the eigenvectors of the matrix $X'WX$.

The bias, covariance, MMSE and MSE functions of LT, RR, and LE are obtained respectively as follows:

(i) Characteristics of RR:

$$b_{RR} = \text{bias}\,(RR) = -kQ\Lambda_k^{-1}\alpha,$$

$$\text{Cov}\,(RR) = Q\Lambda_k^{-1}\Lambda\Lambda_k^{-1}Q',$$

$$MMSE(RR) = Q\Lambda_k^{-1}\Lambda\Lambda_k^{-1}Q' + k^2Q\Lambda_k^{-1}\alpha\alpha'\Lambda_k^{-1}Q',$$

$$MSE\,(RR) = \sum_{j=1}^{p+1} \frac{\lambda_j}{\left(\lambda_j+k\right)^2} + \frac{k^2\alpha_j^2}{\left(\lambda_j+k\right)^2},$$

$MSE\,(RR)$ was also computed in Månsson (2012).

(ii) Characteristics of LE:

$$b_{LE} = \text{bias}\,(LE) = (d-1)\,Q\Lambda_1^{-1}\alpha,$$

$$\text{Cov}\,(LE) = Q\Lambda_1^{-1}\Lambda_d\Lambda^{-1}\Lambda_d\Lambda_1^{-1}Q',$$

$$MMSE\,(LE) = Q\Lambda_1^{-1}\Lambda_d\Lambda^{-1}\Lambda_d\Lambda_1^{-1}Q' + (d-1)^2\,Q\Lambda_1^{-1}\alpha\alpha'\Lambda_1^{-1}Q',$$

$$MSE\,(LE) = \sum_{j=1}^{p+1} \frac{\left(\lambda_j+d\right)^2}{\lambda_j\left(\lambda_j+1\right)^2} + \frac{(d-1)^2\,\alpha_j^2}{\left(\lambda_j+1\right)^2}.$$

$MSE\,(LE)$ was also computed in Månsson (2013).

(iii) Characteristics of LT:

$$b_{LT} = \text{bias}\,(LT) = -\,(d+k)\,Q\Lambda_k^{-1}\alpha,$$

$$\text{Cov}\,(LT) = Q\Lambda_k^{-1}\Lambda_d^*\Lambda^{-1}\Lambda_d^*\Lambda_k^{-1}Q',$$

$$MMSE\,(LT) = Q\Lambda_k^{-1}\Lambda_d^*\Lambda^{-1}\Lambda_d^*\Lambda_k^{-1}Q' + (d+k)^2\,Q\Lambda_k^{-1}\alpha\alpha'\Lambda_k^{-1}Q',$$

$$MSE\,(LT) = \sum_{j=1}^{p+1} \frac{\left(\lambda_j-d\right)^2}{\lambda_j\left(\lambda_j+k\right)^2} + \frac{(d+k)^2\,\alpha_j^2}{\left(\lambda_j+k\right)^2}. \tag{4}$$

where $\Lambda_k = \Lambda + kI$, $\Lambda_d^* = \Lambda - dI$, $\Lambda_1 = \Lambda + I$ and $\Lambda_d = \Lambda + dI$.

After computing the MMSE and MSE functions, LT is compared to the other estimators in the sense of MMSE in the following theorems by using the following lemma:

Lemma 1 (Farebrother 1976) *Let M be a positive definite (p.d.) matrix, α be a vector of nonzero constants and c be a positive constant. Then $cM - \alpha\alpha' > 0$ if and only if $\alpha'M\alpha < c$.*

2.2 Comparison of LT Versus MLE

The following theorem presents the condition that LT is superior to MLE:

Theorem 2 Let $(d + k)(2\lambda_j + k - d) > 0$, $j = 1, 2, ..., p + 1$ and $b_{LT} = \text{bias}(\hat{\beta}_{LT})$. Then $MMSE(MLE) - MMSE(LT) > 0$ if and only if $b'_{LT} [\Lambda^{-1} - \Lambda_k^{-1}\Lambda_d^*\Lambda^{-1}\Lambda_d^*\Lambda_k^{-1}]^{-1} b_{LT} < 1$.

Proof The difference between the MMSE functions of MLE and LT is obtained by

$$MMSE(MLE) - MMSE(LT) = Q\left(\Lambda^{-1} - \Lambda_k^{-1}\Lambda_d^*\Lambda^{-1}\Lambda_d^*\Lambda_k^{-1}\right) Q' - b_{LT}b'_{LT}$$

$$= Q\text{diag}\left\{\frac{1}{\lambda_j} - \frac{(\lambda_j - d)^2}{\lambda_j(\lambda_j + k)^2}\right\}_{j=1}^{p+1} Q' - b_{LT}b'_{LT}$$

The matrix $\Lambda^{-1} - \Lambda_k^{-1}\Lambda_d^*\Lambda^{-1}\Lambda_d^*\Lambda_k^{-1}$ is p.d. if $(\lambda_j + k)^2 - (\lambda_j - d)^2 > 0$ which is equivalent to $[(\lambda_j + k) - (\lambda_j - d)][(\lambda_j + k) + (\lambda_j - d)] > 0$. Simplifying the last inequality, one gets $(d + k)(2\lambda_j + k - d) > 0$. The proof is finished by Lemma 1. $\square$

2.3 Comparison of LT Versus RR

The following theorem gives the condition that LT is superior to RR:

Theorem 3 Let $b = Q\Lambda_k^{-1}\alpha$ and $d < 2\min(\lambda_j)$. If $(d^2 + 2dk) b' [\Lambda_k^{-1}\Lambda\Lambda_k^{-1} - \Lambda_k^{-1}\Lambda_d^*\Lambda^{-1}\Lambda_d^*\Lambda_k^{-1}]^{-1} b < 1$, then $MMSE(RR) - MMSE(LT) > 0$.

Proof The difference between the MMSE functions of RR and LT is obtained by

$$MMSE(RR) - MMSE(LT) = Q\left(\Lambda_k^{-1}\Lambda\Lambda_k^{-1} - \Lambda_k^{-1}\Lambda_d^*\Lambda^{-1}\Lambda_d^*\Lambda_k^{-1}\right) Q' + b_{RR}b'_{RR} - b_{LT}b'_{LT}$$

$$= Q\text{diag}\left\{\frac{\lambda_j}{(\lambda_j + k)^2} - \frac{(\lambda_j - d)^2}{\lambda_j(\lambda_j + k)^2}\right\}_{j=1}^{p+1} Q' - (d^2 + 2dk) bb'$$

$$= Q\text{diag}\left\{\frac{2d\lambda_j - d^2}{\lambda_j(\lambda_j + k)^2}\right\}_{j=1}^{p+1} Q' - (d^2 + 2dk) bb'.$$

Since $(d^2 + 2dk) bb'$ is nonnegative definite, it is enough to prove that $Q\left(\Lambda_k^{-1}\Lambda\Lambda_k^{-1} - \Lambda_k^{-1}\Lambda_d^*\Lambda^{-1}\Lambda_d^*\Lambda_k^{-1}\right) Q' - (d^2 + 2dk) bb'$ is p.d. Now let $d < 2\min(\lambda_j)$, then using Lemma 1, the proof is finished. $\square$

2.4   Comparison of LT Versus Le

The following theorem presents the condition that LT is superior to LE:

Theorem 4   *If* $b'_{LT} \left[\Lambda_1^{-1} \Lambda_d \Lambda^{-1} \Lambda_d \Lambda_1^{-1} - \Lambda_k^{-1} \Lambda_d^* \Lambda^{-1} \Lambda_d^* \Lambda_k^{-1} \right]^{-1} b_{LT} < 1$ *and* $\lambda_j (k + 2d - 1) + d(k + 1) > 0$, $0 < d < 1$, *then* $MMSE (LE) - MMSE (LT) > 0$.

Proof   The difference between the MMSE functions of LE and LT is obtained by

$$MMSE (LE) - MMSE (LT) = Q \left(\Lambda_1^{-1} \Lambda_d \Lambda^{-1} \Lambda_d \Lambda_1^{-1} - \Lambda_k^{-1} \Lambda_d^* \Lambda^{-1} \Lambda_d^* \Lambda_k^{-1} \right) Q' + b_{LE} b'_{LE} - b_{LT} b'_{LT}$$

$$= Q \operatorname{diag} \left\{ \frac{(\lambda_j - d)^2}{\lambda_j (\lambda_j + 1)^2} - \frac{(\lambda_j - d)^2}{\lambda_j (\lambda_j + k)^2} \right\}_{j=1}^{p+1} Q' + b_{LE} b'_{LE} - b_{LT} b'_{LT}$$

$$= Q \operatorname{diag} \left\{ \frac{\lambda_j (k + 2d - 1) + d(k + 1)}{\lambda_j (\lambda_j + k)^2} \right\}_{j=1}^{p+1} Q' + b_{LE} b'_{LE} - b_{LT} b'_{LT}$$

Similarly, since $b_{LE} b'_{LE}$ is nonnegative definite, it is enough to prove that $Q \left(\Lambda_1^{-1} \Lambda_d \Lambda^{-1} \Lambda_d \Lambda_1^{-1} - \Lambda_k^{-1} \Lambda_d^* \Lambda^{-1} \Lambda_d^* \Lambda_k^{-1} \right) Q' - b_{LT} b'_{LT}$ is positive definite. Letting $\lambda_j (k + 2d - 1) + d(k + 1) > 0, 0 < d < 1$, it is easy to see that Lemma 1 leads to the desired result.   □

2.5   Estimating the Parameters k and d

The selection of shrinkage parameters in biased estimators has always been an important issue. There are different types of estimation techniques of the ridge parameter and Liu parameter in the literature (see Månsson (2012, 2013)). In this study, motivated by the works of Hoerl and Kennard (1970); Kibria (2003); Månsson and Shukur (2011), some methods to select the values of the parameters *k* and *d* are proposed.

Following Hoerl and Kennard (1970), differentiating the Eq. (4) with respect to the parameter *k*, it is easy to obtain the following equation:

$$\frac{\partial MSE(\hat{\beta}_{LT})}{\partial k} = \sum_{j=1}^{p+1} \left(\frac{-2\lambda_j (\lambda_j + k) (\lambda_j - d)^2}{\lambda_j^2 (\lambda_j + k)^4} + \frac{2 (k + d) (\lambda_j + k) \hat{\alpha}_j^2 - 2 (k + d)^2 (\lambda_j + k) \hat{\alpha}_j^2}{(\lambda_j + k)^4} \right).$$
$$(5)$$

Simplifying the numerator of the Eq. (5) and solving for *k*, one can get the following individual estimators

$$\hat{k}_{LT}^j = \frac{\lambda_j - d \left(1 + \lambda_j \hat{\alpha}_j^2 \right)}{\lambda_j \hat{\alpha}_j^2}, \quad j = 1, 2, ..., p + 1.$$

The condition $\lambda_j - d\left(1 + \lambda_j\hat{\alpha}_j^2\right) > 0$ should hold to get a positive value of $\hat{k}_{LT}^j$. Thus, the following restriction

$$d < \frac{\lambda_j}{1 + \lambda_j\hat{\alpha}_j^2} \tag{6}$$

should be satisfied.

Now, following Kibria (2003), the following method is proposed to estimate the parameter k using the mean function:

$$\hat{k}_{AM} = \frac{1}{p+1} \sum_{j=1}^{p+1} \left(\frac{\lambda_j - d\left(1 + \lambda_j\hat{\alpha}_j^2\right)}{\lambda_j\hat{\alpha}_j^2} \right)$$

which is the arithmetic mean of $\hat{k}_{LT}^j$.

Moreover, following Alkhamisi et al. (2006), the maximum function is used to obtain the following estimator:

$$\hat{k}_{MAX} = \max \left(\frac{\lambda_j - d\left(1 + \lambda_j\hat{\alpha}_j^2\right)}{\lambda_j\hat{\alpha}_j^2} \right).$$

After choosing the parameter d using Eq. (6), one can estimate the value of k using one of the methods proposed. By plugging-in these estimates in LT, a better performance may be observed.

In the following section, a Monte Carlo simulation is designed to compare the performance of the estimators for different scenarios.

To estimate the ridge parameter to be used in RR, Månsson (2012) proposed different methods. In this study, $K5 = \max\left(\sqrt{\frac{\hat{\alpha}_j^2}{\hat{\sigma}_j^2}}\right)$ was used in the simulation since the author reported that $K5$ is the best estimator in most of the situations investigated.

Moreover, some methods were proposed by Månsson (2013) to choose the shrinkage parameter d to be used in LE. However, $D5 = \max\left(0, \ \min\left(\frac{\hat{\alpha}_j^2}{1/\lambda_j + \hat{\alpha}_j^2}\right)\right)$ had the lowest MSE value in most of the situations. Therefore $D5$ is used to estimate d in LE in the simulation study.

3   Monte Carlo Simulation Study

3.1   Design of the Simulation

In the previous section, some theoretical comparisons are provided. In this section, an extensive Monte Carlo simulation study is designed to evaluate the performances of the estimators. Here is the description of the simulation.

Firstly, the observations of the explanatory variables are generated using the following equation

$$x_{ij} = \left(1 - \rho^2\right)^{1/2} z_{ij} + \rho z_{ip}$$

where $i = 1, 2, \ldots, n, j = 1, 2, \ldots p$, and ρ^2 represents the correlation between the explanatory variables and z_{ij}'s which are independent random numbers obtained from the standard normal distribution.

The dependent variable of the NB regression model is generated using random numbers following the negative binomial distribution $NB\left(\mu_i, \mu_i + \theta\mu_i^2\right)$ where $\mu_i = \exp\left(x_i\beta\right)$, $i = 1, 2, \ldots, n$. The slope parameters are decided such that $\sum_{j=}^{p} \beta_j^2 = 1$, which is a commonly used restriction in the field (see Kibria (2003)).

In the design of simulation, three different values of ρ corresponding to 0.90, 0.95, 0.99 are considered. The value of θ is taken to be 1.0 and 2.0 due to Månsson (2012). Moreover, the following small, moderate and large sample size values are considered: 50, 100 and 200. The numbers of explanatory variables are taken to be 4 and 6.

The simulation is repeated 2000 times, convergence tolerance is taken to be 10^{-8} and the estimated MSE values of the estimators are computed as follows:

$$MSE(\tilde{\beta}) = \frac{\sum_{r=1}^{2000} (\tilde{\beta} - \beta)'_r (\tilde{\beta} - \beta)_r}{2000}, \tag{7}$$

where $\tilde{\beta}_r$ is an estimator of β at the rth replication.

3.2   Results of the Simulation

The estimated MSE values obtained from the Monte Carlo simulation are presented in Tables 1 and 2. It is observed from tables that the factors affecting the performance of the estimators are the value of θ, the sample size n, the number of explanatory variables p and the degree of correlation ρ.

According to Tables 1 and 2, when the value of θ increases, the estimated MSE values increase. As the degree of correlation increases, MSE of MLE is inflated and MSE of RR is affected negatively. LT with k_{AM} and k_{MAX} show better performance than MLE and RR since an increase in the degree of correlation affects LT with

Table 1 Estimated MSE values when $p = 4$

$\theta = 1$				$\theta = 2$		
n	50	100	200	50	100	200
$\rho = 0.90$						
$LT(k_{AM})$	0.2483	0.2095	0.1816	0.3782	0.2714	0.2446
$LT(k_{MAX})$	0.2632	0.2503	0.2444	0.3409	0.3242	0.3023
RR	1.3117	0.4823	0.2606	2.2441	0.8058	0.4131
LE	0.5761	0.3537	0.2164	0.7115	0.4952	0.3125
MLE	1.3640	0.4901	0.2629	2.4722	0.8298	0.4168
$\rho = 0.95$						
$LT(k_{AM})$	0.3517	0.2468	0.1925	0.5141	0.3377	0.2435
$LT(k_{MAX})$	0.3025	0.2956	0.2581	0.3892	0.3611	0.3178
RR	2.1664	1.2658	0.5033	3.3184	1.8066	0.8421
LE	0.7008	0.5668	0.3534	0.7977	0.6843	0.4894
MLE	2.2329	1.6571	0.5176	3.7717	2.8076	0.8691
$\rho = 0.99$						
$LT(k_{AM})$	0.7346	0.4317	0.2866	1.0059	0.7081	0.4235
$LT(k_{MAX})$	0.3556	0.3018	0.2635	0.3948	0.3743	0.3335
RR	4.9866	3.2202	2.5914	5.5368	3.5991	3.3719
LE	0.7128	0.7048	0.6975	0.8280	0.7363	0.7144
MLE	10.0215	6.1448	2.8633	18.1536	10.0081	4.7138

k_{MAX} slightly, i.e., LT with k_{MAX} is the most stable estimator in the study. LE has also better performance than MLE and RR, however, LT with k_{AM} and k_{MAX} has the best performance in most of the situations considered.

Moreover, increasing the number of explanatory variables also affects the estimators negatively, i.e., their estimated MSE increases. Although high correlation makes an increase in the MSE of LT with k_{MAX} when $p = 4$, it becomes robust to the correlation when $p = 6$. According to the results of the simulation, LT with k_{MAX} has the best performance among the estimators.

4   Real Data Applications

4.1   Sweden Traffic Data

In this subsection, we illustrate the benefits of the new estimator LT using a real dataset. The dataset is taken from the official website of the Department of Transport

Table 2 Estimated MSE values when $p = 6$

$\theta = 1$				$\theta = 2$		
n	50	100	200	50	100	200
$\rho = 0.90$						
LT(k_{AM})	0.3374	0.2339	0.1961	0.4678	0.2971	0.2497
LT(k_{MAX})	0.4309	0.3663	0.3056	0.4800	0.4148	0.3522
RR	2.1804	0.9271	0.4479	3.3428	1.4560	0.7134
LE	0.9627	0.5997	0.3586	1.2341	0.7702	0.5113
MLE	2.4655	0.9289	0.4501	4.3043	1.4674	0.7199
$\rho = 0.95$						
LT(k_{AM})	0.3933	0.2514	0.2020	0.5808	0.3451	0.2470
LT(k_{MAX})	0.3792	0.3271	0.2730	0.4477	0.3953	0.3411
RR	3.8504	1.8333	0.8282	5.2209	3.0453	1.5874
LE	1.0414	0.8734	0.5688	1.2917	1.1277	0.8297
MLE	5.1537	1.9955	0.8319	8.9456	3.5382	1.6650
$\rho = 0.99$						
LT(k_{AM})	0.7928	0.4866	0.2976	1.5414	0.7228	0.4800
LT(k_{MAX})	0.3701	0.3170	0.2655	0.5113	0.3851	0.3402
RR	10.3993	8.9291	3.9404	12.3918	11.8867	6.7377
LE	1.0980	1.0196	1.0905	1.5306	1.2021	1.1914
MLE	19.1744	10.6314	5.4518	36.6638	17.7556	8.4051

Analysis in Sweden.[1] A similar dataset is used by Månsson (2013). The dependent variable is the number of pedestrians killed and the explanatory variables are the number of kilometers driven by cars X_1 and trucks X_2. In this application, we try to investigate the effect of changing the usage of cars and trucks on the number of pedestrians killed. There are 21 different counties in Sweden and the data are pooled during the year 2013 for different counties. The condition number is approximately 210.9146 showing that there is a moderate multicollinearity. The negative binomial regression model with intercept is estimated using IRLS algorithm for different estimators considered in this study. The results are reported in Table 3.

According to Table 3, considering the coefficients of MLE, the effect of increasing X_1 has a negative impact on the number of pedestrian killed which is not expected. It is known that the signs of coefficients may be wrong when there is multicollinearity. Moreover, the effect of increasing X_1 is low while the effect of increasing X_2 is high.

If we use biased estimators, the effect of increasing X_1 becomes positive which is expected and the effect of increasing is lower when compared to MLE. When we compare the standard errors of estimators, it is observed that LT with k_{AM} and k_{MAX} have lower standard errors than other estimators which makes them more

---

[1] www.trafa.se, accessed January 10, 2016.

Table 3 Coefficients, standard errors and MSE values of estimators for Sweden traffic data

	$LT(k_{AM})$	$LT(k_{MAX})$	RR	LE	MLE
MSE					
	0.9493	0.6002	1.8975	1.2069	22.3520
Coefficients					
β_0	2.3135	2.2615	2.3352	2.1943	2.3731
β_1	0.6031	0.5395	0.4826	0.2423	−0.6003
β_2	0.9121	0.7123	1.1791	0.8522	2.5690
Standard errors					
β_0	0.3081	0.3010	0.3111	0.2920	0.3165
β_1	0.6515	0.5017	0.9493	0.7479	3.3505
β_2	0.6551	0.5055	0.9483	0.7451	3.3205

stable. Thus, the estimator LT should be preferred since it has lower standard errors compared to other estimators and meaningful coefficients compared to MLE.

Moreover, LT with k_{AM} and k_{MAX} have less MSE values than the other estimators. We also plot the MSE values of the estimators LT and RR for $0 < k < 1$ and LE for $0 < d < 1$. We estimate the parameter d using (6) for LT. According to Fig. 1, we observe that when $0 < d < 0.16$ MSE of LE is smaller than MSE of LT. Otherwise LT has the least MSE value.

Finally, we provide some information to justify the theorems given in Sect. 2. The estimated parameter values of LT are as follows: $k_{AM} = 0.2489$, $k_{MAX} = 0.4901$ and $d = 0.0192$. To justify Theorem 2, we consider the followings: $\min\left[(d + k)\left(2\lambda_j + k - d\right)\right] = 0.2881 > 0$ and $b'_{LT}\left[\Lambda^{-1} - \Lambda_k^{-1}\Lambda_d^*\Lambda^{-1}\Lambda_d^*\Lambda_k^{-1}\right]^{-1}$ $b_{LT} = 2.8459 \times 10^{-5} < 1$ and the eigenvalues of the difference matrix $MMSE\,(MLE) - MMSE\,(LT)$ are 0.0071, 0.6927 and 21.0520 which are positive. Hence $MMSE\,(MLE) - MMSE\,(LT)$ is positive definite. Thus, Theorem 2 is satisfied.

Similarly, we compute the followings to justify Theorem 3, $2\min\left(\lambda_j\right) = 0.0947 > d = 0.0192$ and

$$\left(d^2 + 2dk\right) b'\left[\Lambda_k^{-1}\Lambda\Lambda_k^{-1} - \Lambda_k^{-1}\Lambda_d^*\Lambda^{-1}\Lambda_d^*\Lambda_k^{-1}\right]^{-1} b = 7.2232 \times 10^{-8} < 1$$

using $k_{MAX} = 0.4901$ for both RR and LT. The eigenvalues of the difference $MMSE\,(RR) - MMSE\,(LT)$ are 0.0002, 0.0203 and 0.1058 which are all positive, showing that the difference is positive definite.

$$\left(d^2 + 2dk\right) b'\left[\Lambda_k^{-1}\Lambda\Lambda_k^{-1} - \Lambda_k^{-1}\Lambda_d^*\Lambda^{-1}\Lambda_d^*\Lambda_k^{-1}\right]^{-1} b = 4.0861 \times 10^{-8} < 1$$

for both RR and LT and again the difference is positive definite (the eigenvalues are 0.0003, 0.0346 and 0.6454). Thus Theorem 3 is satisfied.

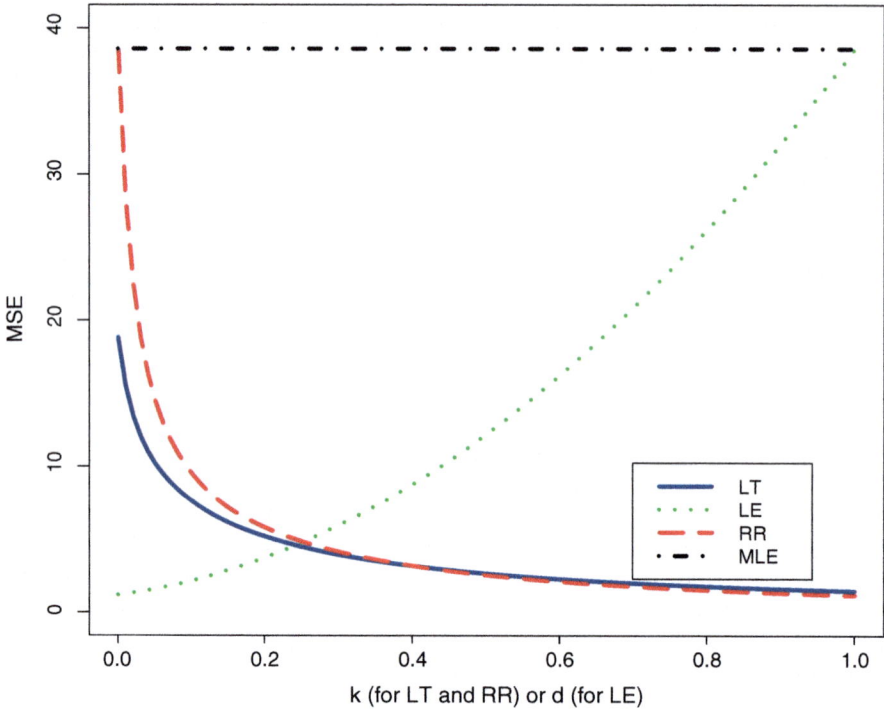

Fig. 1 Plot of MSE versus the parameters for Sweden traffic data

Again, we consider the following computations to justify Theorem 4. We let $k = k_{AM} = 0.2489$ and d are computed using (6) as 0.0192 for both LE and LT. However, $\min\left[\lambda_j(k + 2d - 1) + d(k + 1)\right]$ becomes negative and does not satisfy the precondition of Theorem 4. Thus, we try using $D5$ in both LE and LT to estimate the parameter d which is computed as 0.1528 and set $k = 1.1$. Now, $\min\left[\lambda_j(k + 2d - 1) + d(k + 1)\right] = 0.0192 > 0$ which satisfies the precondition of Theorem 4.

$$b'_{LT}\left[\Lambda_1^{-1}\Lambda_d\Lambda^{-1}\Lambda_d\Lambda_1^{-1} - \Lambda_k^{-1}\Lambda_d^*\Lambda^{-1}\Lambda_d^*\Lambda_k^{-1}\right]^{-1}b_{LT} = 8.7066 \times 10^{-5} < 1.$$

The eigenvalues of the difference matrix $MMSE(LE) - MMSE(LT)$ are 0.0007, 0.1905 and 0.5929 which are all positive. Hence the difference matrix is positive definite.

Thus, we observe that Theorems given in Sect. 2 are satisfied.

Table 4 The correlation matrix of the data

	NWM	NRC	NS	NOG	NCG	NGR1	NGR2
NWM	1.0000	0.0337	0.1362	−0.2257	0.5069	0.6062	0.6987
NRC	0.0337	1.0000	0.1695	−0.0580	0.2391	0.2801	−0.1531
NS	0.1362	0.1695	1.0000	−0.3151	−0.2254	−0.0989	0.0048
NOG	−0.2257	−0.0580	−0.3151	1.0000	0.3382	0.4597	−0.0653
NCG	0.5069	0.2391	−0.2254	0.3382	1.0000	0.7620	0.2858
NGR1	0.6062	0.2801	−0.0989	0.4597	0.7620	1.0000	0.5579
NGR2	0.6987	−0.1531	0.0048	−0.0653	0.2858	0.5579	1.0000

4.2 Football Teams Data

In this subsection, another data set[2] regarding the football teams competing in the 2014–2015 Super League Season in Turkey is considered. A similar data set is also analyzed by Türkan and Özel (2016) for the season 2012–2013. According to Türkan and Özel (2016), the data is appropriate for the Poisson regression model. However, we try to fit a negative binomial regression model because the variance (9.76) of the dependent variable is larger than the mean (7.33). Similar to their study, we have selected the number of won matches (NWM) as the dependent variable and the followings are the explanatory variables: the number of red cards (NRC), the number of substitutions (NS), the number of matches ending over 2.5 goals (NOG), the number of matches completed with goals (NCG), the ratio of the goals scored in number of matches [NGR1 = NGS/NM], and the ratio of goals scored in the sum of goals conceded and goals scored [NGR2 = NGS/(NGC + NGS)].

The correlation matrix of the data is given in Table 4. It is seen from Table 4 that, the dependent variable NWM has a negative correlation with NOG. It seems that the only high correlation between the variables NRG1 and NCG is 0.76 and the other bivariate correlations are not that high. However, the bivariate correlations may not have higher correlations in the case of collinearity, see Montgomery et al. (2015). Moreover, the eigenvalues of the matrix $X'\hat{W}X$ are 8.3623, 1.1082, 0.6132, 0.5473, 0.3134, 0.1856 and 0.0391. The condition number is computed as 213.6088 which shows that there is a moderate multicollinearity problem.

In Table 5 we present the MSE values, coefficients and the standard errors of estimators. According to Table 5, it is observed that the estimated MSE value of LT with k_{AM} and k_{MAX} are smaller than the others. Although, the variables NRC and NOG have negative impacts on NWM when MLE is used, only NOG has a negative coefficient when the biased estimators LT, RR, and LE are used. This may happen when there is collinearity problem (Montgomery et al. (2015)).

Moreover, the estimator LT has the least standard error values among others which also shows the superiority of LT over the others.

---

[2]Please see http://www.tff.org and http://www.sahadan.com, accessed November 15, 2016.

Table 5 Coefficients, standard errors and MSE values of estimators for football teams data

MSE	LT(k_{AM})	LT(k_{MAX})	RR	LE	MLE
	0.2768	0.4396	1.0537	1.1735	38.6024
Coefficients					
Intercept	1.3052	0.6908	1.7109	1.7275	1.9292
NRC	0.0114	0.0042	0.0107	0.0097	−0.3409
NS	0.0370	0.0110	0.1069	0.1133	0.2597
NOG	−0.0472	−0.0133	−0.1642	−0.1760	−1.0072
NCG	0.0740	0.0239	0.1788	0.1875	0.2244
NGR1	0.0901	0.0292	0.2318	0.2433	1.2479
NGR2	0.1136	0.0357	0.2850	0.2992	0.2431
Standard errors					
Intercept	0.2331	0.1233	0.3059	0.3089	0.3459
NRC	0.1445	0.0424	0.4226	0.4484	1.8995
NS	0.1430	0.0418	0.4244	0.4510	1.6376
NOG	0.1400	0.0416	0.4071	0.4320	2.2249
NCG	0.1350	0.0413	0.3785	0.4012	2.4855
NGR1	0.1304	0.0410	0.3380	0.3556	3.8594
NGR2	0.1419	0.0422	0.4059	0.4301	2.4839

5 Conclusion

In this study, a new biased estimator which is a generalization of Liu-type estimator is proposed for the negative binomial regression models. We also review some existing estimators namely, negative binomial Liu estimator and negative binomial ridge estimator. We obtain some theoretical comparisons between the estimators using MMSE and some conditions showing that LT is superior to the others.

Moreover, we design a Monte Carlo simulation to understand the effects of the degree of correlation among the explanatory variables, the sample size and the number of explanatory variables. LT has a better performance than the others in the sense of MSE criterion in most of the cases considered in the simulation. Finally, we show that LT is a better choice and all the theoretical derivations are satisfied in real data applications and it is recommended to the researchers.

References

Alkhamisi, M., Khalaf, G., Shukur, G.: Some modifications for choosing ridge parameters. Commun. Stat. Theory Methods **35**, 2005–2020 (2006)

Cameron, A.C., Trivedi, P.K.: Econometric models based on count data. Comparisons and applications of some estimators and tests. J. Appl. Econom. **1**, 29–53 (1986)

Cameron, A.C., Trivedi, P.K.: Regression Analysis of Count Data, vol. 53. Cambridge University Press (2013)

Farebrother, R.: Further results on the mean square error of ridge regression. J. R. Stat. Soc. Ser. B (Methodol.) 248–250 (1976)

Hilbe, J.: Negative Binomial Regression. Cambridge University Press (2011)

Hoerl, A.E., Kennard, R.W.: Ridge regression: biased estimation for nonorthogonal problems. Technometrics **12**, 55–67 (1970)

Hoerl, A.E., Kennard, R.W., Baldwin, K.F.: Ridge regression: some simulations. Commun. Stat. Theory Methods **4**, 105–123 (1975)

Huang, J., Yang, H.: A two-parameter estimator in the negative binomial regression model. J. Stat. Comput. Simul. **84**, 124–134 (2014)

Kibria, B.M.G.: Performance of some new ridge regression estimators. Commun. Stat. Simul. Comput. **32**, 419–435 (2003)

Lesaffre, E., Marx, B.D.: Collinearity in generalized linear regression. Commun. Stat. Theory Methods **22**(7), 1933–1952 (1993)

Lipovetsky, S., Conklin, W.M.: Ridge regression in two-parameter solution. Appl. Stoch. Models Bus. Ind. **21**, 525–540 (2005)

Liu, K.: A new class of biased estimate in linear regression. Commun. Stat. Theory Methods **22**, 393–402 (1993)

Liu, K.: Using Liu-type estimator to combat collinearity. Commun. Stat. Theory Methods **32**, 1009–1020 (2003)

MacKinnon, M.J., Puterman, M.L.: Collinearity in generalized linear models. Commun. Stat. Theory Methods **18**, 3463–3472 (1989)

Månsson, K.: On ridge estimators for the negative binomial regression model. Econ. Model. **29**, 178–184 (2012)

Månsson, K.: Developing a Liu estimator for the negative binomial regression model: method and application. J. Stat. Comput. Simul. **83**, 1773–1780 (2013)

Månsson, K., Shukur, G.: A Poisson ridge regression estimator. Econ. Model. **28**, 1475–1481 (2011)

Montgomery, D.C., Peck, E.A., Vining, G.G.: Introduction to Linear Regression Analysis. Wiley (2015)

Myers, R.H., Montgomery, D.C., Vining, G.G., Robinson, T.J.: Generalized Linear Models with Applications in Engineering and the Sciences. Wiley, New Jersey (2010)

Smith, E.P., Marx, B.D.: Ill-conditioned information matrices, generalized linear models and estimation of the effects of acid rain. Environmetrics **1**(1), 57–71 (1990)

Türkan, S., Özel, G.: A new modified Jackknifed estimator for the Poisson regression model. J. Appl. Stat. **43**(10), 1892–1905 (2016)

Weissfeld, L.A., Sereika, S.M.: A multicollinearity diagnostic for generalized linear models. Commun. Stat. Theory Methods **20**(4), 1183–1198 (1991)

Appraisal of Performance of Three Tree-Based Classification Methods

Huruy Debessay Asfha and Betul Kan Kilinc

Abstract Classification methods use different algorithms to get better performance in research fields such as statistics, machine learning and computational analysis. This study reviews the traditional method, recursive partitioning, as well as newer classification algorithms, conditional inference tree and evolutionary tree. Variations and improvements in algorithms, data types with or without missing values, and special applications are widely used in this field. Although classification algorithms have been studied often and performed reasonably well, there is no existing one that performs best among the others. Using a real dataset, the classification methods under consideration are applied and the results are compared.

Keywords Classification · Decision tree · Missing values · Tree-based

1 Introduction

Building accurate and efficient classifiers for large data are one of the essential tasks of data mining and machine learning researches. Classification and regression trees (CART) by Breiman et al. (1984) are among the common data mining techniques used to explore and model complex datasets with strong nonlinear associations and different variable types. CART uses recursive partitioning method to build classification and regression trees for predicting categorical and continuous dependent variables respectively. Therneau (1983) first used the recursive partitioning method, which implements many of the ideas found in Breiman et al. (1984) while writing a Stanford University technical report which was later modified in a report of Mayo Clinic Division of Biostatistics in 1997 (Therneau and Atkinson 1997). Recursive

H. D. Asfha
Department of Statistics, Graduate School of Sciences, Anadolu University,
Eskisehir, Turkey
e-mail: asfhah32@gmail.com

B. Kan Kilinc (✉)
Faculty of Science, Department of Statistics, Anadolu University, Eskisehir, Turkey
e-mail: bkan@anadolu.edu.tr

© Springer International Publishing AG 2018   41
M. Tez and D. von Rosen (eds.), *Trends and Perspectives
in Linear Statistical Inference*, Contributions to Statistics,
https://doi.org/10.1007/978-3-319-73241-1_3

partitioning method performs an exhaustive search over all possible splits in maximizing an information gain of nodes and selects the covariate showing the best split. Entropy-based measures of information gain such as Gini gain or information gain are used as a criterion to select the best split. Variable selection is however, biased in favor of variables with more potential splits when these criteria are used (Breiman et al. 1984; Hothorn et al. 2014). Categorical covariate variables with the higher number of categories are more likely to be selected. Moreover, variable selection bias can occur due to the difference in the number of missing values in covariate variables (Strobl et al. 2007).

To deal with the variable selection problem of the classical recursive methods, Hothorn et al. (2006) introduced an unbiased recursive partitioning method, the conditional inference trees, which is based on conditional distributions. The basic idea of this approach is to measure the association between response and covariate variables and then determine how significant this relationship is. The algorithm selects the variable with the highest association for splitting, however, if the association is found to be not significant, the process of splitting stops (Hothorn et al. 2006, 2014).

In recursive partitioning, the split at each internal node is selected to maximize the homogeneity of the next step (homogeneity of daughter nodes) without any consideration of nodes further down the tree. This yields to only locally optimal trees which could be far from a globally optimal solution. Furthermore, a tree fitted by recursive partitioning tends to overfit the data which, however, can be addressed by pruning or cross-validation mechanisms in order to obtain an optimal tree (Breiman et al. 1984; Grubinger et al. 2015; Therneau and Atkinson 1997).

As an alternative to the classical classification and regression methods, Grubinger et al. (2014) proposed an evolutionary method which uses a global search over the tree parameters in order to obtain a globally optimal tree rather than locally optimal one. This approach has a good balance between predictive accuracy and complexity of the tree, but nevertheless it is computationally demanding (Grubinger et al. 2014, 2015).

The aim of this study is to evaluate the performance of three tree-based classification methods: recursive partitioning, conditional inference tree, and evolutionary learning tree in determining outcomes (injured or uninjured) of vehicle accidents. Different performance indicators of classification methods are used for comparison of the classifiers. Furthermore, the effect of the missing values in covariates in the evaluated models are discussed.

1.1 Recursive Partitioning

Recursive partitioning (rpart) is a classification and/or regression method developed by Therneau (1983), which basically implements the concepts in Breiman et al. (1984); Therneau (1983). It constructs a decision tree by splitting each node on the tree into two daughter nodes. Rpart uses an impurity measure to choose the variable, which best splits the data into two groups.

Suppose, the response variable to be classified has a total of C categories, which are represented by the indexes $1, 2, \ldots, C$. The impurity measure of a given node, say A, is then given by

$$Q(A) = \sum_{i=1}^{C} f(p_{iA}), \tag{1}$$

where f is impurity function and p_{iA} is the proportion of observations in node A which belong to category i. $Q(A) = 0$ indicates that node A contains observations of only one category, hence, A is pure.

The impurity function, f, can either be Gini index or entropy (information index) which are given as follows:

$$Gini - index : f(p) = p \times (1 - p), \tag{2}$$

$$Information - index : f(p) = -p \times log(p). \tag{3}$$

By default Rpart employs the Gini index impurity function (Venables and Ripley 2002). For a two-class problem, the measures differ only slightly and will nearly always choose the same split point (Therneau and Atkinson 1997). To split node A into its two daughter nodes, A_L and A_R (left and right daughter nodes), the split with maximal impurity reduction, ΔQ is used.

$$\Delta Q = p(A) \times Q(A) - [p(A_L) \times Q(A_L) + p(A_R) \times Q(A_R)]. \tag{4}$$

After finding the variable which best splits the data into two groups, the data will be separated accordingly. The same process will be applied to each subgroup and will continue recursively until the subgroups either reach a minimum size or until no improvement can be made (Therneau and Atkinson 1997). Generally, the recursively built model using rpart is more complex than needed and the resulting tree overfits the data. Thereafter, the second stage of the procedure in rpart is then trimming back the constructed tree using cross-validation rule in order to avoid complexity, and hence finds the optimal tree.

1.2 Conditional Inference Trees

Conditional inference trees (ctree) modify the exhaustive search of split variable and split point in rpart by using statistical significance tests in selecting the variable and subsequently selecting the optimal split point for that particular variable. To accomplish this, Hothorn et al. (2006), constructed unified tests for the independence by means of the conditional distribution of linear statistics in the permutation test framework (Hothorn et al. 2006).

Let $D(Y/X^j)$ denote the conditional distribution of Y given the jth covariate, X^j. A partial hypothesis for independence of Y on X^j can then be written as

$H_0^j: D(Y \mid X^j) = D(Y)$, where $D(Y)$ is the distribution of Y. A global null hypothesis of independence between the covariates and the response variable formulated using the partial hypotheses is then given by

$$H_0 : \cap_{j=1}^m H_0^j, \tag{5}$$

where m is the number of covariates and H_0^j, are the partial hypotheses.

In order to find the optimal split variable, the global null hypothesis is first tested. Failure to reject H_0 stops the recursion indicating that the response variable, Y, is independent of the covariates. However, if H_0 is rejected, the association between the response and each of the covariates is measured by test statistics or p-values indicating the deviation from the partial hypotheses (H_0^j).

The algorithm will choose the covariate which has the strongest association with the response and a split point will be chosen from that variable in order to split the space into two disjoint groups. Recursively, these procedures will be repeated. While selecting a variable for classifying the dataset into the daughter branches, the approach of ctree algorithm avoids the selection of variables with many potential splits more often than those with fewer potential splits. Furthermore, a statistically motivated stopping criteria, not rejecting the global null hypothesis of independence is implemented (Hothorn et al. 2014).

1.3 Evolutionary Learning Trees

As presented in Breiman et al. (1984), if the complexity of a tree is measured by a function of the number of terminal nodes, without further considering the depth or the shape of the trees, then the goal of the classifiers is to find that classification and regression tree which optimizes some trade-off between prediction performance and complexity which is given by

$$\hat{\theta} = argmin_{\theta \in \Theta} loss\{Y, f(X, \theta)\} + comp(\theta), \tag{6}$$

where Y and $f(X, \theta)$ represents the actual and predicted values of the response variable respectively. The function $loss(...)$ is equivalent to misclassification rate (MC) in the case of classification and mean square error (MSE) for regression.

The function $comp(.)$ is monotonically non-decreasing in the number of terminal nodes of the tree θ, thus penalizing more complex models in the tree selection process (Grubinger et al. 2014, 2015). Since θ is subset of Θ, the overall parameter space, finding $\hat{\theta}$ requires a search over all Θ which is computationally unmanageable. Forward search recursive partitioning methods only search each pair of split variable and split point once, and independently of the subsequent split rules, hence typically leading to a globally sub-optimal solution $\hat{\theta}$.

In evolutionary learning trees (evtree), a split rule is randomly generated to initialize each of the individual trees of the population in the root node. For numeric or ordinal split variables, uniform probability is used to select split point, whereas for nominal, each of the categories has 50% chance to be assigned to the left or the right daughter nodes. However, considering that nodes should not be empty, if all the categories are assigned to one side, at least one category has to be assigned to the other side (Grubinger et al. 2014, 2015).

In every iteration, each tree is selected once to be modified by one of the variation operators, which will randomly be selected from the four types of mutation operators and one crossover operator. For the next generation (iteration), survivor trees are selected based on the evaluation function defined in Grubinger et al. (2014). The modification procedure of the trees continues until some defined termination condition is fulfilled. The tree with the minimum evaluation function will then be returned as optimal tree (Grubinger et al. 2014, 2015).

2 Comparison Measures

In this study, the accuracy measures used to compare the performance of the models were predictive accuracy, misclassification rate, evaluation function, kappa, sensitivity, specificity, positive prediction value, and negative prediction value.

Misclassification rate (MC) for a classification model is defined as the proportion of misclassified observations by the model during prediction process (Hastie et al. 2009) and it is given by

$$MC = \frac{1}{N} \sum_{i=1}^{N} I(Y_i \neq f(X, \theta)), \tag{7}$$

where N is the total number of observations and $I(w)$, for an arbitrary logical argument w is an indicator function which takes a value 1 when w is true and 0 otherwise. From Table 1, the total number of misclassified observations is obtained by summing up the number of false positives (B) and false negatives (C), hence, the misclassi-

Table 1 Relationship between observed accident outcome and predicted values

Accident outcome				
		injured (+)	Non-injured (−)	Total
Predicted values	Injured (+)	A	B	E = A + B
	Non-injured (−)	C	D	F = C + D
	Total	G = A + C	H = B + D	T = A + B + C + D

fication rate is computed as $MC = \frac{B+C}{T}$. Furthermore, accuracy, the proportion of correctly predicted values by the model, is given by $1 - MC = \frac{A+D}{T}$.

An evaluation function, a measure of the quality of a classification tree, is commonly expressed as a function of the misclassification rate (MC) and the complexity of the tree (Grubinger et al. 2014, 2015). The evaluation function used and given by Eq. 6 is adopted from Grubinger et al. (2015). For a classification tree with M terminal nodes and a misclassification rate, MC, the $loss(.)$ function is a function of the total number of observations, N and MC and is written as

$$loss(Y, f(X, \theta)) = 2N \times MC(Y, f(X, \theta)) = 2 \sum_{i=1}^{N} I(Y_i \neq f(X, \theta)) \qquad (8)$$

whereas, the $comp(.)$ function which is a measure of the complexity of the tree, is expressed in terms of the number of terminal nodes weighted by a user-defined constant, α, and $log\ N$ (Grubinger et al. 2014, 2015).

$$comp(\theta) = \alpha \times M \times log\ N \qquad (9)$$

Cohen's Kappa coefficient (Kappa) is a statistic which measures the inter-rater agreement or concordance between the predicted and the observed counts of the categories and is given by

$$k = \frac{P_0 - P_e}{1 - P_e} = 1 - \frac{1 - P_0}{1 - P_e}, \qquad (10)$$

where $P_0 = \frac{A+D}{T}$ is the relative observed agreement between the predicted and actual, and $P_e = \frac{G}{T} \times \frac{E}{T} + \frac{H}{T} \times \frac{F}{T} = \frac{G.E+H.F}{T^2}$ is the hypothetical probability of chance agreement (Hoehler 2000; Nelson 2015; Tang et al. 2015). $k = 1$ indicates that the classified values are in complete agreement with the actual responses, on the other hand, if there is no agreement other than what would be expected by chance, $k \leq 0$.

Sensitivity and specificity are ways of measuring how good a classification model is. Sensitivity is defined as the proportion of true positives that are correctly classified by the model and is given by $\frac{A}{G}$, and, specificity which is given by $\frac{D}{H}$ is the proportion of true negatives which are correctly predicted. The terms positive and negative are used to refer to presence or absence of a condition of interest, here injury of a person (Altman and Bland 1994).

Another approach of diagnosing model performance is using predictive values which help to know the probabilities that the model will give correct identification of the observations (Altman and Bland 1994). Positive predictive value (PPV) is the proportion of accident outcomes predicted as injured by the model which are actually injured, and negative predictive value (NPV) is the proportion of accident outcomes predicted as uninjured by the model who are actually noninjured (Altman and Bland 1994; Kuhn et al. 2016).

$$PPV = \frac{Sens \times Prev}{(Sens \times Prev) + (1 - Spec) \times (1 - Prev)} = \frac{A}{E} \quad (11)$$

$$NPV = \frac{Spec \times (1 - Prev)}{(1 - Sens) \times Prev + Spec \times (1 - Prev)} = \frac{D}{F} \quad (12)$$

where $Sens$ = Sensitivity, $Spec$ = Specificity and $Prev$ = Prevalence which is given by $\frac{G}{T}$.

3 Data Description

The data set used for comparison is obtained from a district police department from 2014 to 2015 for rural highways for the city of Aydin in Turkey. Traffic accident data contain 195 records with several missing values among variables. The names and the labels of the variables are summarized in Table 2. The missing values existing in covariates are listed in Table 3. As the dependent variable that is the accident outcome has only 3 invalid observations, totally 192 observations are taken into

Table 2 Variable description

Variable name	Labels	Variable name	Labels
Road type	Dual carriageway		Motorbikes and motorcycles
	One-way road	Vehicle type	Cars
	Two-way road		Bus and Minibus
			Trucks and Others
Road paving	Asphalt		
	Surface treatment	Driver's	Primary level
		Education level	Middle level
Place characteristics	Urban		High school
	Interurban		University
			Unknown
Time of day	Daytime		
	Nighttime	Traffic lights	Exist
			Out of order
Atmospheric factors	Dry		Non exist
	Wet		
		Lamps	Exist
Road characteristics	Dry		Out of order
	Wet		Non exist
Gender	Male	Driver's age	12–81 years old
	Female		

Table 3 Number of missing values in each covariate variable (only covariates which contain missing value are included)

Variable name	n	p
Accident outcome	3	0.016
Place characteristics	1	0.005
Traffic lights	14	0.073
Lamps	20	0.104
Age	2	0.010
Vehicle type	1	0.005
Driver's education level	15	0.078
Gender	2	0.010

n: number of missing values; p: proportion of missing values

consideration for analysis. Both categorical and numerical variables are used to build a classification model for the response variable. The response variable, *Accident outcome*, is binary with 0 representing for 'injured' and 1 for 'uninjured'. The only continuous variable used is 'Driver's Age'.

Variables which contain missing values are given in Table 3 along with the number and proportion of missing values exist in each variable. It should be noted that covariates which do not contain missing values are not included in the table.

4   Results and Discussion

Model performance was assessed in two different scenarios depending on whether there exists a missing value in covariate variables or not. In the first scenario, data which contain missing values in the covariate variables were used, whereas, in the second, missing values in all the covariates were assigned a value using simple imputation method. In both scenarios, however, since ctree algorithm does not support missing value (it stops execution) in response variable, a simple imputation method was used to replenish those in the 'Accident outcome' variable. Moreover, 70% of the data were used for training and the remaining 30% were used for validating or testing the model.

The classification methods considered in this study vary the way they handle missing values in the covariate variables. In rpart and ctree, surrogate splits are established if the chosen splitting variable contains missing values (Hothorn et al. 2014; Therneau and Atkinson 1997). An evtree model however, removes them if there are any for it does not support missing values (Grubinger et al. 2015).

A cross-tabulation of model outputs and the observed results for the two scenarios, using data with missing values and data free of missing values, are given in Table 4. The predictions are made based on the testing set which is made up of 40 injured and 19 uninjured observations. It can be seen that evtree models show greater

Table 4 Cross-tabulation of the actual data and predicted values

Using data comprising missing values in covariates

rpart					ctree			evtree		
		I	NI	Total	I	NI	Total	I	NI	Total
Predicted	I	35	12	47	40	19	59	29	12	41
Values	NI	5	7	12	0	0	0	11	7	18

Using data free of missing values

rpart					ctree			evtree		
		I	NI	Total	I	NI	Total	I	NI	Total
Predicted	I	36	13	49	40	19	59	36	12	48
Values	NI	4	6	10	0	0	0	4	7	11

I: injured(+); NI: noninjured(−)

improvement in classifying testing set correctly when complete data is used. Ctree algorithm however, is found to classify the observations in the testing set to the group of injured which is the majority of the two. It produced the same results regardless of the existence or nonexistence of missing values in the covariate variable.

Rpart trees are pruned based on their respective optimal complexity parameter (cp) and the optimal trees are shown in Figs. 1 and 2. It is important to note that in both scenarios, rpart algorithm picked vehicle type and lamp as splitting variables. As provided in Table 3, both splitting variables comprise missing values which means that in node 1 and node 3, a surrogate split was established on the process of discriminating the data. Even though it is not a guarantee that useful surrogates can always be found, in this application, it was effective for it resulted to a tree with a more or less the same to that with no missing values (Figs. 3 and 4).

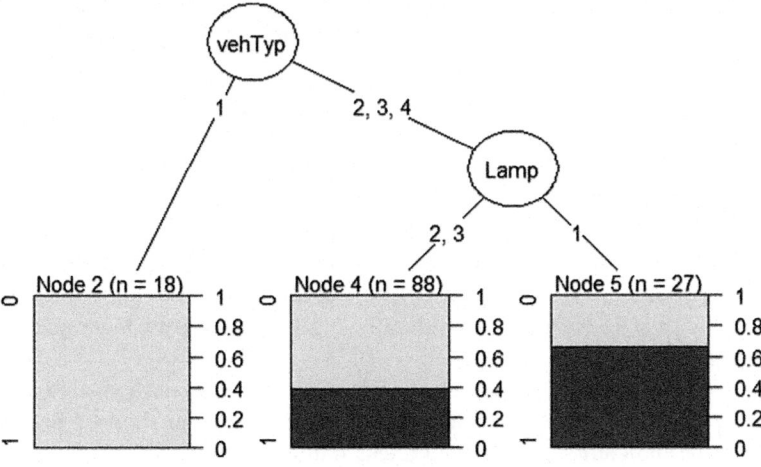

Fig. 1 Rpart tree for vehicle accident data which comprises missing values

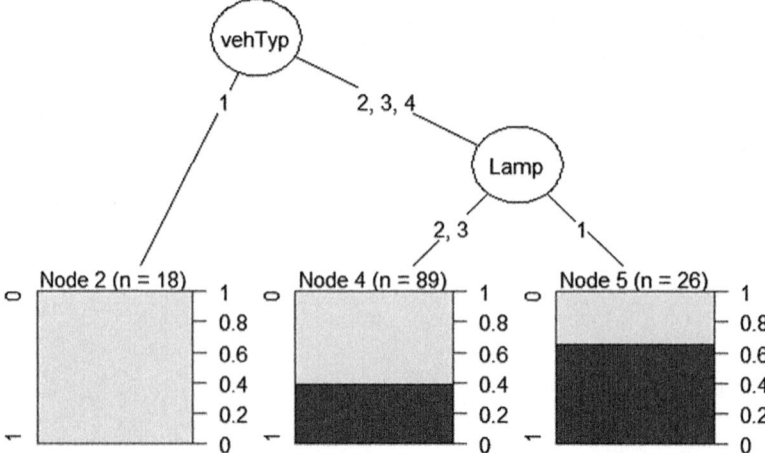

Fig. 2   Rpart tree for vehicle accident data which is free of missing values

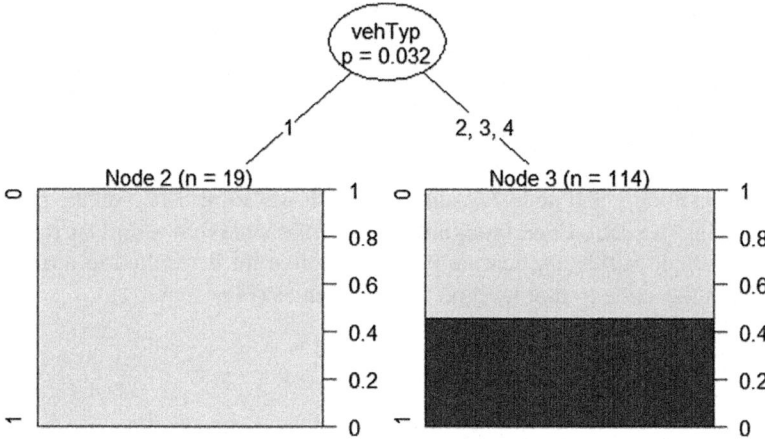

Fig. 3   Ctree tree for vehicle accident data which has missing values

The effect of deletion of incomplete observations in the case of evtree algorithm can be seen in Figs. 5 and 6, in which 'Lamp' was the split variable in the first, whereas in the later, the variable 'age' was selected prior to 'Lamp'. In addition to this, it can be seen in Fig. 5 that only 113 observations were used by the algorithm instead of the total 133 which shows that the algorithm removes incomplete entries in the training stage.

Table 5 shows the performance measures of the models when applied to the incomplete data. It can be noted that the pruned rpart model shows a better result in most of the indicators; it produced a less complex model with better prediction accuracy. The most complex tree was achieved by evtree algorithm which had also

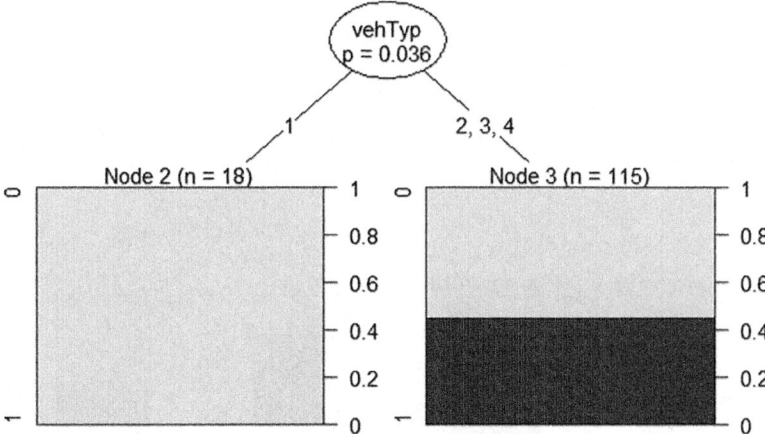

Fig. 4 Ctree tree for vehicle accident data which is free of missing values

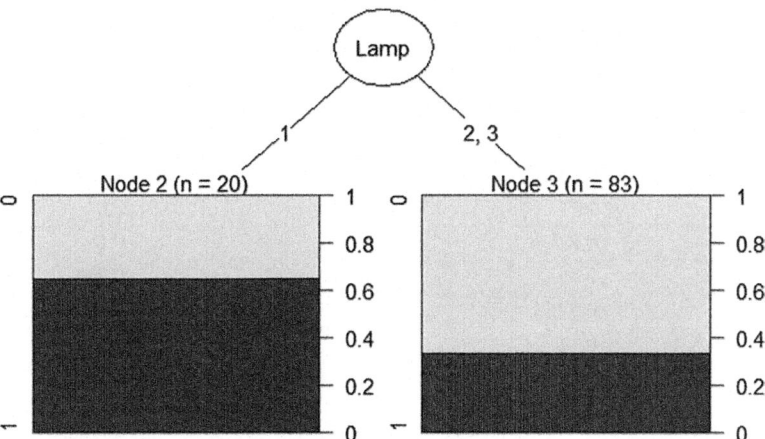

Fig. 5 Evtree tree for data which comprises missing values

the least accurate. Referring to the interpretations of Cohen's Kappa c oefficient presented in (Mary 2012), there exists a fair agreement between the actual data and the prediction s made by rpart whereas that of evtree have got a slight agreement. For ctree however, this indicator is zero which indicates an absence of agreement.

In addition to comparing classifiers based on their prediction performance, it is also important to look at the type of errors made by each model. In classifying observations like accidents, models which commit less 'type-II error' are more preferable for they are minimizing the proportion of wrongly classifying 'injured' as 'uninjured'. While ctree was found to have a zero type-II error (1- sensitivity), the 100% type-I error (1- specificity) committed is of great concern in which the model might be regarded as inadequate classifier in this case. Rpart has the upper hand when sensi-

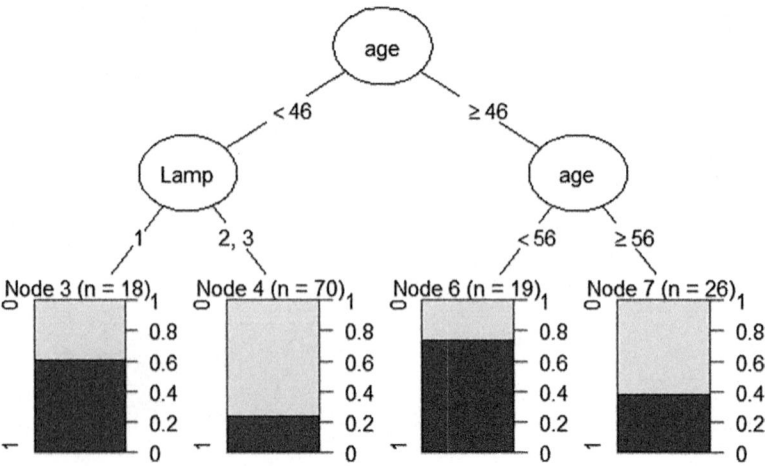

Fig. 6 Evtree tree for data which is free of missing values

Table 5 Performance of tree-based models using data which comprises missing values in the covariate variables

Performance measures	Classification methods		
	rpart	ctree	evtree
Evaluation function	91.32	95.44	113.48
Accuracy	0.71	0.68	0.61
Kappa	0.27	0	0.1
Sensitivity	0.88	1	0.73
Specificity	0.37	0	0.37
PPV	0.74	0.68	0.71
NPV	0.58	–	0.39
Prevalence	0.68	0.68	0.68

tivity is taken into consideration. Nonetheless, evtree made a slight balance between the two types of errors when compared to that of rpart.

Performance indicators of the three models when the classification methods were applied to data without missing values are provided in Table 6. Compared to the results in Table 5, the performance of evtree has improved greatly. On the contrast, rpart showed slight decrement in most of the performance measures, whereas, ctree has showed no difference. Furthermore, a better agreement between the actual and predicted values was obtained using the evtree model unlike scenario one where a better agreement was obtained from the rpart model. The improvement in the performance of evtree could be due to the increase in sample size used in the evtree model construction when incomplete observations are replenished.

Table 6 Performance of tree-based models using data which is free of missing values

Performance measures	Classification methods		
	rpart	ctree	evtree
Evaluation function	91.31	95.44	91.7
Accuracy	0.71	0.68	0.73
Kappa	0.27	0	0.30
Sensitivity	0.90	1	0.90
Specificity	0.32	0	0.37
PPV	0.73	0.68	0.75
NPV	0.60	–	0.64
Prevalence	0.68	0.68	0.68

Fig. 7 ROC curve for tree-based classifiers in the presence of missing values

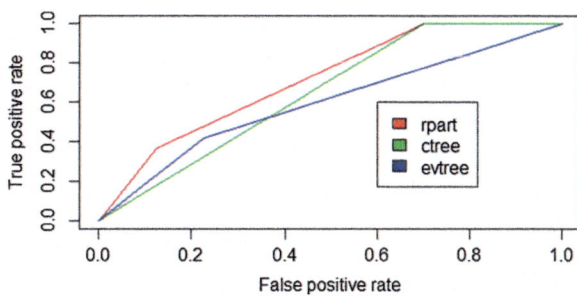

Fig. 8 ROC curve for tree-based classifiers in the absense of missing values

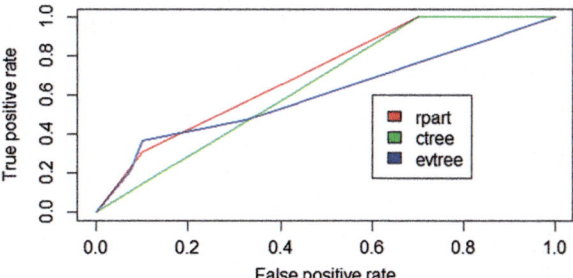

Below are the ROC curves with missing values and without missing values presented by Figs. 7 and 8, respectively. Furthermore, the corresponding area under the curve (AUC) value was tabulated in Table 7. In Figs.7 and 8, the y-axis denotes the true positive rate (or sensitivity) and x-axis denotes the true negative rate (or specificity). The better the classification models, the more quickly the true positive rate nears 1 (or 100%). This curve and the AUC show that rpart has predictive ability to discriminate the accident outcome.

Table 7 AUC areas of the ROC curves for the tree-based classifiers using data with and without missing values

AUC area		
	WM	WO
rpart	0.724	0.711
ctree	0.650	0.650
evtree	0.598	0.607

WM: with missing values WO: without missing values

5 Conclusion

In this study, three machine learning models; rpart, ctree, and evtree were used to classify a data of vehicle accident outcomes and compare their performance. The comparison of the algorithms was taken under two different scenarios. In the first scenario, data which includes missing values in covariate variables was used and in the second, the missing values were replenished using imputation method. In this study, two interesting points can be mentioned;

(i) Ctree method was found to be the least best performer in both the scenarios when overall performance measures are considered. Moreover, it was found to stop classifying the dataset at the second stage of the tree-building process, which could somehow be because of the structure of the dataset.

(ii) Evtree algorithm looks sensitive to missing values. When data free of missing values was used, the algorithm was found to outperform the others.

With the quick-emergence of a number of new algorithms to the family of classifiers, their performance in various applications should be assessed well. For a particular application like the 'Vehicle accident', only when a more comprehensive data is used can we make more appropriate selection of algorithms. Therefore, such kind of studies is recommended to figure out why the above-mentioned points came to happen.

References

Altman, D.G., Bland, J.M.: Diagnostic tests 1: sensitivity and specificity. Br. Med. J. **308**, 1552 (1994)

Altman, D.G., Bland, J.M.: Diagnostic tests 2: predictive values. Br. Med. J. **309**, 102 (1994)

Breiman, L., Friedman, J., Olshen, Stone, C.: Classification and Regression Trees, Wadsworth, Inc. Monterey, California, U.S.A. (1984)

Grubinger, T., Zeileis, A., Pfeiffer, K.P. : evtree: evolutionary learning of globally optimal classification and regression trees in R. J. Stat. Softw. **61** (2014)

Grubinger, T., Zeileis, A., Pfeiffer, K.P., (2015). evtree: evolutionary learning of globally optimal classification and regression trees in R. https://cran.r-project.org/web/packages/evtree/vignettes/evtree.pdf. Accessed 27 Oct 2017

Hastie, T., Tibshirani, R., Friedman, J.: The Elements of Statistics Learning; Data Mining, Inference, and Prediction, 2nd edn, pp. 305–311. Springer Science+Business Media, LLC (2009)

Hoehler, F.K.: Bias and prevalence effects in kappa viewed in terms of sensitivity and specificity. J. Clin. Epidemiol. **53** (2000)

Hothorn, T., Hornik, K., Zeileis, A.: Unbiased recursive partitioning: a conditional inference framework. J. Comput. Graph. Stat. **15**(3), 651674 (2006)

Hothorn, T., Hornik, K., Zeileis, A.: ctree: conditional inference trees. https://cran.r-project.org/web/packages/partykit/vignettes/ctree.pdf (2014). Accessed 16 Oct 2017

Kuhn, M., Contributions from Wing, J., Weston, S., Williams, A., Keefer, C., Engelhardt, A., Cooper, T., Mayer, Z., Kenkel, B., the R Core Team, Benestry, M., Lescarbeau, R., Ziem, A., Scrucca, L., Tang, Y., and Candan, C.: Caret: classification and regression training. https://CRAN.R-project.org/package=caret (2016). Accessed 13 Oct 2017

McHugh, Mary, L.: Biochemia Medica **22**(3), 276–282 (2012)

Nelson, K.P., Edwards, D.: Measures of agreement between many raters for ordinal classifications. Stat. Med. **34**(23), 3116–3132 (2015)

Strobl, C., Boulesteix, A., Augustin, T.: Unbiased split selection for classification trees based on the Gini index. Comput. Stat. Data Anal. **52**(1), 483–501 (2007)

Tang, W., Hu, J., Zhang, H., Wu, P., He, H.: Kappa coefficient: a popular measure of rater agreement. Shanghai Arch. Psychiatry **27**(1), 62–67 (2015)

Therneau, T.M.: A short introduction to recursive partitioning. Orion Technical Report 21, Stanford University, Statistics Department (1983)

Therneau, T.M., Atkinson, E.J.: An introduction to recursive partitioning using the rpart routines. Technical Report 61 (1997). http://www.mayo.edu/research/documents/biostat-61pdf/doc-10026699. Accessed 16 Oct 2017

Therneau, T. M., Atkinson, B., Ripley, M. B.: rpart: recursive partitioning and regression trees. (2010). Accessed 13 Oct 2017. http://CRAN.R-project.org/package/rpart

Venables, W.N., Ripley, B.D.: Modern Applied Statistics with S, vol. 9, 4th edn, pp. 251–269. Springer (2002)

High-Dimensional CLTs for Individual Mahalanobis Distances

D. Dai and T. Holgersson

Abstract Statistical analysis frequently involves methods for reducing high-dimensional data to new variates of lower dimension for the purpose of assessing distributional properties, identification of hidden patterns, for discriminant analysis, etc. In classical multivariate analysis such matters are usually analysed by either using principal components (PC) or the Mahalanobis distance (MD). While the distributional properties of PC's are fairly well established in high-dimensional cases, no explicit results appear to be available for the MD under such cases. The purpose of this chapter is to bridge that gap by deriving weak limits for the MD in cases where the dimension of the random vector of interest is proportional to the sample size (n, p-asymptotics). The limiting distributions allow for normality-based inference in cases when the traditional low-dimensional approximations do not apply.

Keywords Mahalanobis distance · Increasing dimension · Weak convergence · Marchenko-Pastur distribution · Outliers · Pearson family distributions

1 Mahalanobis Distances

Mahalanobis distance is considered to be one of the most fundamental concepts in multivariate analysis since its introduction in 1930 by Mahalanobis (1930), Anderson 2003, (Mardia 1977). It is used in a wide range of applications, including graphical analysis (Healy 1968; Andrews et al. 1973), outlier detection (Wilks 1963), discriminant analysis (Pavlenko 2003), multivariate calibration (De Maesschalck et al. 2000), non-normality testing (Mardia 1974), covariate matching (Frölich 2012), chemometrics (Todeschini et al. 2013) and construction of multivariate process control charts (Montgomery and Woodall 1999) to mention a few. These methods have mainly been developed in a finite dimension framework (i.e. when the dimension p of the random vector of interest is a fixed number, independent of the sample size n),

D. Dai · T. Holgersson (✉)
Department of Economics and Statistics, Linnaeus University, 351 95 Växjö, Sweden
e-mail: thomas.holgersson@lnu.se

© Springer International Publishing AG 2018
M. Tez and D. von Rosen (eds.), *Trends and Perspectives in Linear Statistical Inference*, Contributions to Statistics,
https://doi.org/10.1007/978-3-319-73241-1_4

and are usually not directly applicable under more general asymptotics. During the last decades, much of the research in multivariate analysis has shifted focus from classical fixed-dimension asymptotics where $p/n \to 0$ as $n \to \infty$ to settings where $p/n \to c > 0$ as $n, p \to \infty$ since many real-world applications are argued to be better described by this case. Bai and Silverstein (2010) report an almost exponential growth in the number of publications concerning analysis of high-dimensional data. While most of this research has been concerned with core matters such as development of efficient point estimators or tests of covariance structures etc., few results are available for diagnostic analysis or distance measures. In particular, the asymptotic distributions of individual MD distances appear to be unknown under a general high-dimensional setting, and plain applications of standard methods, such as those described in Wilks (1963), may lead to erratic conclusions unless p/n is very close to zero. Moreover, asymptotic normality is important not only in empirical applications, but also in theoretical perspectives since normality implies many other important properties such as uniform integrability, stability, unimodality, finiteness of exponential moments, etc. This chapter derives weak limits of two common MD estimates under increasing dimension asymptotics. It is shown that the so-called leave-one-out version of the MD, which is commonly used in outlier analysis, limits a different point than the traditional MD estimate and has a different convergence rate. Hence, while the choice between the two MD estimates is unimportant in low dimensions, some care needs to be taken otherwise. We will start by introducing some definitions and symbolism, state some finite-dimensional distributions and then pass onto increasing dimension settings.

Definition 1 Let $\mathbf{X}_i : p \times 1$ be a random vector such that $E\left[\mathbf{X}_i\right] = \boldsymbol{\mu}$ and $\boldsymbol{\Sigma} = E\left[\left(\mathbf{X}_i - \boldsymbol{\mu}\right)\left(\mathbf{X}_i - \boldsymbol{\mu}\right)'\right]$ for $i = 1, \ldots, n$. Then,

$$D_{ii} := \left(\mathbf{X}_i - \boldsymbol{\mu}\right)'\boldsymbol{\Sigma}^{-1}\left(\mathbf{X}_i - \boldsymbol{\mu}\right).$$

The D_{ii} statistic is a measure of the scaled distance between an individual observation $\mathbf{X}_i$ and its expected value $\boldsymbol{\mu}$. Estimators of D_{ii} may be obtained by simply replacing the unknown parameters with appropriate estimators. In case both $\boldsymbol{\mu}$ and $\boldsymbol{\Sigma}$ are unknown and replaced by the standard estimators, we obtain the well-known sample MD defined below.

Definition 2 Let $\bar{\mathbf{X}} = n^{-1}\mathbf{X}'\mathbf{1}$ where $\mathbf{1}$ is a vector of ones of appropriate dimension and $\mathbf{S} = n^{-1}\sum_{i=1}^{n}\left(\mathbf{X}_i - \bar{\mathbf{X}}\right)\left(\mathbf{X}_i - \bar{\mathbf{X}}\right)'$. Then, we define an estimator of D_{ii} by

$$d_{ii} := \left(\mathbf{X}_i - \bar{\mathbf{X}}\right)'\mathbf{S}^{-1}\left(\mathbf{X}_i - \bar{\mathbf{X}}\right).$$

Further discussions of the aforementioned statistic d_{ii} are available in Mardia (1977) and Mardia et al. (1980). Note that although the inverse sample covariance matrix is sometimes expressed using a different divisor (for example, $n - p - 1$ instead of n), the statistic d_{ii} is unbiased in the sense that $E\left[d_{ii}\right] = E\left[D_{ii}\right] = p$. Alternative estimators are available as leave-one-out estimators, obtained by omitting a specific observation from the estimation of the sample mean vector and covariance matrix, thereby avoiding correlations among the components within the MD estimator. Formally, this is performed as follows:

Definition 3 Let $\mathbf{S}_{(i)} := (n-1)^{-1} \sum_{k=1, k \neq i}^{n} \left(\mathbf{X}_k - \bar{\mathbf{X}}_{(i)}\right) \left(\mathbf{X}_k - \bar{\mathbf{X}}_{(i)}\right)'$ where $\bar{\mathbf{X}}_{(i)} := (n-1)^{-1} \sum_{k=1, k \neq i}^{n} \mathbf{X}_k$. Then, an estimator of D_{ii} is defined as follows:

$$d_{(ii)} := \left(\mathbf{X}_i - \bar{\mathbf{X}}_{(i)}\right)' \mathbf{S}_{(i)}^{-1} \left(\mathbf{X}_i - \bar{\mathbf{X}}_{(i)}\right).$$

The sample mean vector and the sample covariance matrix are independent when sampling from a normal distribution; hence, all components within the estimator $d_{(ii)}$ will be mutually independent, resulting in estimators with distributional properties different from those of d_{ii}. Furthermore, these estimators are robust to the effect of a single outlier in the sense of not contaminating the estimators of μ and Σ^{-1}.

Few, if any, results are available regarding the limiting distributions of d_{ii} and $d_{(ii)}$ in high-dimensional settings. The remainder of this chapter is, therefore, concerned with the asymptotic distributions of d_{ii} and $d_{(ii)}$ in such cases.

Proposition 1 *Let* $\{X_{ij}\}$, $i = 1, \ldots, n, j = 1, \ldots, p$, *be a row-wise i.i.d. double array of normally distributed variables and* $p \leq n-1$. *Then, for any* i, *as* $p/n \rightarrow c \in (0,1)$, *the following holds:*

$$\frac{\sqrt{p}}{\sqrt{2}} \left(p^{-1} D_{ii} - 1\right) \overset{\ell}{\rightarrow} N(0,1) \text{ as } n \rightarrow \infty, p \rightarrow \infty.$$

Proof Since D_{ii} is invariant to affine transformations we may assume $X_{ij} \overset{i.i.d.}{\sim} N(0,1)$ and hence $D_{ii} \sim \chi_{(p)}^2$ for any i, independently of j. The asymptotic normality of chi-square variables with increasing degrees of freedom is well known Anderson (2003), and Proposition 1 is derived. ∎

The asymptotic distribution obtained in Proposition 1 may easily be derived under some general moment restrictions of the parent variable $\mathbf{X}_i$ and relaxation of the normality assumption. However, in real applications, the (inverse) covariance matrix is unknown and has to be estimated. As a consequence, the distribution of D_{ii} depends on that of $\mathbf{X}_i$. We will, therefore, retain the normality assumption of Proposition 1 in the derivation of the asymptotic properties of sample MDs to access known finite sample properties and subsequently relax this assumption.

Theorem 1 *Let* $\{X_{ij}\}$, $i = 1, \ldots, n, j = 1, \ldots, p$, *be distributed as in Proposition 1 and* $p \leq n-2$. *Then,*

$$\frac{\sqrt{p}}{\sqrt{2}} \frac{\left(p^{-1} d_{ii} - 1\right)}{\sqrt{(1-c)}} \overset{\ell}{\rightarrow} N(0,1) \text{ as } n \rightarrow \infty, p \rightarrow \infty.$$

Proof Let $\mathbf{G} = (d_{ij})$, $i, j = 1, \ldots, n$ be the matrix of sample MDs as specified in Definition 2 where $d_{ij} = (\mathbf{X}_i - \bar{\mathbf{X}})' \mathbf{S}^{-1} (\mathbf{X}_j - \bar{\mathbf{X}})$ and let $\mathbf{H} = \mathbf{I} - n^{-1} \mathbf{1} \mathbf{1}'$. Following Mardia (1977) we have, for any p and $p \leq n - 1$, that

$$n^{-1} \mathbf{H}_r^{-1/2} \mathbf{G}_r \mathbf{H}_r^{-1/2} \sim \mathbf{B}_r \left(\frac{1}{2} p, \frac{1}{2} (n - p - 1) \right),$$

where $\mathbf{G}_r$ and $\mathbf{H}_r$ are principal sub-matrices of order r from $\mathbf{G}$ and $\mathbf{H}$ respectively, $\mathbf{B}_r$ has a matrix $r \times r$ beta type I distribution. Then, $n^{-1} \mathbf{H}_1^{-1/2} \mathbf{G}_1 \mathbf{H}_1^{-1/2} = n^{-1} (1 - n^{-1})^{-1/2} g_{11} (1 - n^{-1})^{-1/2} = g_{11}/(n - 1) = M_{11}$, which is distributed as a univariate beta type I distribution. Hence, $M_{ii} \sim B_I (\alpha, \beta)$ where $\alpha = p/2$ and $\beta = (n - p - 1)/2$. It may be shown that the asymptotic skewness and kurtosis coefficients of $\sqrt{p} \left(p^{-1} d_{ii} - 1 \right) \big/ \sqrt{2 (1 - c)}$ are given by $\lim_{n,p \to \infty} \gamma_1 = 0$, $\lim_{n,p \to \infty} \gamma_2 = 0$ (details omitted) and hence the beta-distributed M_{ii} limits a normal distribution in the Pearson chart determined by the plane of γ_1^2 and γ_2^2 (Ord 1972), which completes the proof. ∎

Below we derive a CLT for the leave-one-out MD, which is seen to behave rather different than d_{ii}.

Theorem 2 *Let* $\{X_{ij}\}$ *$i = 1, \ldots, n$, $j = 1, \ldots, p$, be distributed as in Proposition 1 and $p \leq n - 1$. Then,*

$$\frac{\sqrt{p}}{\sqrt{2}} \sqrt{(1 - c)} \left(p^{-1} (1 - c) d_{(ii)} - 1 \right) \xrightarrow{\ell} N(0, 1) \text{ as } n \to \infty, p \to \infty.$$

Proof First, we note that $d_{(ii)}$ is invariant to affine transformations and we may hence assume that $X_{ij} \overset{i.i.d.}{\sim} N(0, 1)$ and therefore $(\mathbf{X}_i - \bar{\mathbf{X}}_{(i)}) \sim N \left(\mathbf{0}_{p \times 1}, \left(\frac{n}{n-1} \right) \mathbf{I}_{p \times p} \right)$ for any $p \leq n - 1$, and $(n - 1) \mathbf{S}_{(i)} \sim Wishart (n - 2, \mathbf{I})$. It then follows that $p^{-1} d_{(ii)} \sim \left[(n - 2)/(n - p - 1) \right] F_{(p, n-p-1)}$ (Mardia et al. 1980). The F-distribution is well known to be a particular parameterization of the beta prime distribution (Ord 1972), which, in turn, is a special case of the Pearson type VI distribution. It may be shown that the skewness and excess kurtosis both limit zero as $n \to \infty$, $p \to \infty$, and hence the standardized $d_{(ii)}$ asymptotically converges to a normal distribution within the Pearson plane, and Theorem 2 is established. ∎

Although Theorems 1 and 2 assume a Gaussian distribution of the parent variable X_{ij}, one may expect a limiting Gaussian distribution to be valid under more general distributions of X_{ij}. We will show that this is indeed the case.

Theorem 3 *Let* $\{X_{ij}\}$, *$i = 1, \ldots, n$, $j = 1, \ldots, p$, be a row-wise i.i.d. double array and* $E\left[\left((X_{1j} - E[X_{1j}]) \big/ \sigma_j \right) \right]^4 = 3$ *with* $p \leq n - 1$. *Then,*

$$\frac{\sqrt{p}}{\sqrt{2}}(1-c)^{3/2}\left(p^{-1}d_{(ii)} - (1-c)^{-1}\right) \overset{\ell}{\to} N(0,1) \text{ as } n \to \infty, p \to \infty.$$

Proof Let $\mathbf{S} = \{S_{jj'}\}, S_{jj'} = n^{-1}\sum_{i=1}^{n}\left(X_{ij} - \bar{X}_j\right)\left(X_{ij'} - \bar{X}_{j'}\right), \bar{X}_j = n^{-1}\sum_{i=1}^{n}X_{ij}, j, j' = 1\dots p$. From Jonsson (1982) and Bai et al. (2007) we then have that $\sqrt{p/2}\left(p^{-1}tr\left(\mathbf{S}_{(i)}^{-1}\right) - (1-c)^{-1}\right)$ converges to a zero mean normal distribution with variance $\int z^{-2}dF_c(z)$, where $dF_c(z) = (2\pi zc)^{-1}\sqrt{((1+\sqrt{c})^2 - z)(z - (1-\sqrt{c})^2)}$ is the standard Marchenko-Pastur distribution (1967), Bai et al. (2007), Arharov (1971). It may be shown that $\int z^{-2}dF_c(z) = (1-c)^{-3}$ (Arharov 1971; Serdobolskii 2000; Glombek 2014). By using Bai et al. (2007)'s Corollary 2, we have that $\left\{p^{-1}d_{(ii)} - p^{-1}tr\left(\mathbf{S}_{(i)}^{-1}\right)\right\} \overset{a.s.}{\to} 0$, which completes the proof. ∎

Before deriving the CLT for the estimator d_{ii}, we will establish an important identity.

Proposition 2 $\left(1 - n^{-1}d_{ii}\right)\left(\left(1 + d_{(ii)}n^{-1}\right) / \left(1 + d_{(ii)}n^{-2}\right)\right) = 1.$

Proof Set $\mathbf{W} = n\mathbf{S}$ and $\mathbf{W}_{-i} = (n-1)\mathbf{S}_{-i} = \sum_{k=1, k\neq i}^{n}\mathbf{Y}_k\mathbf{Y}_k'$ where $\mathbf{Y}_k = \left(\mathbf{X}_k - \bar{\mathbf{X}}\right)$. Let $d_{-ii} = y_i'\mathbf{S}_{-i}^{-1}y_i$. Then $\left|\mathbf{W}_{-i} + \mathbf{Y}_i\mathbf{Y}'_i\right| = \left|\mathbf{W}_{-i}\right|\left(1 + \mathbf{Y}'_i\mathbf{W}_{-i}^{-1}\mathbf{Y}_i\right)$ and $\left|\mathbf{W} - \mathbf{Y}_i\mathbf{Y}'_i\right| = \left|\mathbf{W}\right|\left(1 - \mathbf{Y}'_i\mathbf{W}^{-1}\mathbf{Y}_i\right)$ and so $\left|\mathbf{W}_{-i} + \mathbf{Y}_i\mathbf{Y}'_i\right| / \left|\mathbf{W}_{-i}\right| = 1 + (n-1)^{-1}d_{-ii}$ and $\left|\mathbf{W} - \mathbf{Y}_i\mathbf{Y}'_i\right| / \left|\mathbf{W}\right| = 1 - n^{-1}d_{ii}$. Hence $\left(1 - cp^{-1}d_{ii}\right)\left(1 + cp^{-1}d_{-ii}\right) = 1$. Now d_{-ii} is not independent of observation i, since $\mathbf{X}_i$ is included in the sample mean vector $\bar{\mathbf{X}}$, and does hence not fully represent the leave-one-out estimator of D_{ii}. However, using the identity $\left(\mathbf{X}_1 - \bar{\mathbf{X}}\right) = ((n-1)/n)\left(\mathbf{X}_1 - \bar{\mathbf{X}}_{(i)}\right)$ and substituting it in the above expressions we find that $\left(1 - n^{-1}d_{ii}\right)\frac{(1 + d_{(ii)}n^{-1})}{(1 + d_{(ii)}n^{-2})} = 1$. This is an exact identity, independent of distributional properties of the parent variable $\mathbf{X}$, and can be used to derive properties of d_{ii} as a function of $d_{(ii)}$, and vice versa. ∎

Since $1 \leq \left\{\left(1 + d_{(ii)}n^{-1}\right) / \left(1 + d_{(ii)}n^{-2}\right)\right\}$, it follows that $0 \leq n^{-1}d_{ii} \leq 1$. For example, if $c = p/n$, then as $n, p \to \infty$, $0 \leq p^{-1}d_{ii} \leq c^{-1}$, which explicitly shows how the range of $p^{-1}d_{ii}$ is restricted by c. In particular, $p^{-1}d_{ii}$ will have too low a range relative to $p^{-1}D_{ii}$ for large values of c (for example, if $\mathbf{X}_i$ is normally distributed, then $p^{-1}D_{ii} \sim p^{-1}\chi^2_{(p)}$, which is not bounded).

Theorem 4 *Let $\{X_{ij}\}, i = 1, \dots, n, j = 1, \dots, p$, be distributed as in Theorem 3. Then*

$$\frac{\sqrt{p}}{\sqrt{2}}\frac{1}{\sqrt{1-c}}\left(p^{-1}d_{ii} - 1\right) \overset{\ell}{\to} N(0,1) \text{ as } n \to \infty, p \to \infty.$$

Proof From Theorem 3, we know that $\frac{\sqrt{p}}{\sqrt{2}}(1-c)^{3/2}\left(p^{-1}d_{(ii)} - (1-c)^{-1}\right) \overset{\ell}{\to} N(0,1)$ as $n, p \to \infty$. Using the identity from Proposition 2, we may write $p^{-1}d_{ii} = k_1 p^{-1}d_{(ii)} /$

$\left(1 + pn^{-1}p^{-1}d_{(ii)}\right)$ where $k_1 = (n-1)/n$. Since d_{ii} is a measurable function of $d_{(ii)}$, we can apply Cramer's theorem (Ferguson 1996; Birke and Dette 2005) to obtain the asymptotic distribution of d_{ii} as $\left(g\left(p^{-1}d_{(ii)}\right) - g\left(\mu\right)\right) \overset{\ell}{\to} N\left(0, \dot{g}^2\left(\mu\right)\right)$. Since $g\left(\mu\right) = k_1\mu/\left(1 + pn^{-1}\mu\right)$, its first derivative is $\dot{g}\left(\mu\right) = k_1/\left(1 + pn^{-1}\mu\right)^2$ where $\mu = (1-c)^{-1}$. Thus, it follows that $\sqrt{p}/\sqrt{2(1-c)}\left(p^{-1}d_{ii} - 1\right) \overset{\ell}{\to} N\left(0, 1\right)$ as $np \to \infty$.   ∎

Note that, although Theorems 3 and 4 are more general in the sense that they do not require a normal distribution of the parent variable $\mathbf{X}_i$, or even the assumption of all moments to be finite, Theorems 1 and 2 are still relevant as they provide a little more information than Theorems 3 and 4, namely that the normalized MD distribution limits the normal distribution *within* the Pearson (γ_1, γ_2) plane, and hence possesses all Pearson family properties in its path to the normal distribution.

2 Finite Sample Distributions

Proposition 1 and Theorems 1–4 above establish asymptotic normality of D_{ii}, d_{ii} and $d_{(ii)}$ as $p \to \infty$. In this section, we pay some attention to the rates of convergence and also to the relations between D_{ii}, d_{ii} and $d_{(ii)}$. In particular, since the two MD estimates d_{ii} and $d_{(ii)}$ differ only in that observation X_i is void in the covariance matrix used in $d_{(ii)}$, one may expect them to behave similarly when n is large. This is, however, not the case when p is proportional to n. Some demonstrations of this are given below.

Proposition 3 *Let the skewness coefficient of some random variable x be defined by $\gamma_1(x) := E[x - E[x]]^3/\left(E[x - E[x]]^2\right)^{3/2}$. Then the skewness of D_{ii}, d_{ii} and $d_{(ii)}$ are, under sampling from normal distributions, given as follows:*

$$(i) \quad \gamma_1\left(D_{ii}\right) = \frac{\sqrt{8}}{\sqrt{p}}.$$

$$(ii) \quad \gamma_1\left(d_{ii}\right) = \frac{4\left(n - 2p - 1\right)\sqrt{n+1}}{(n+3)\sqrt{2p\left(n - p - 1\right)}}.$$

$$(iii) \quad \gamma_1\left(d_{(ii)}\right) = \frac{(p + n - 3)\sqrt{8\left(n - p - 5\right)}}{(n - p - 7)\sqrt{p\left(n - 3\right)}}.$$

Proof When sampling from a normal distribution with i.i.d. variables $\mathbf{Z}'_i = \left(Z_{i1}, \ldots, Z_{ip}\right)$, the distributions of the sample MDs are, as explained on pages 3–4 above, determined by $D_{ii} \sim \chi^2_{(p)}$, $M_{11} = \frac{d_{11}}{(n-1)} \sim B\left(\alpha = p/2, \beta = (n-p-1)/2\right)$, and the leave-one-out MD $p^{-1}d_{(11)} \sim \{(n-2)/(n-p-1)\}F_{\left(d_1 = p, d_2 = n-p-1\right)}$.

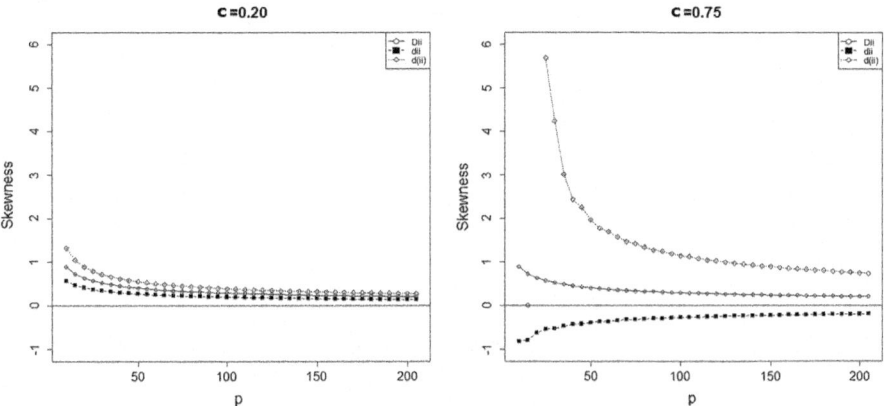

Fig. 1 Skewness of D_{ii}, d_{ii} and $d_{(ii)}$ w.r.t the ratio values $c = 0.2; 0.75$

The skewness coefficients of the chi-square, beta, and F distributions are available in Johnson et al. (1995), from where we have, using a slight abuse of notations,

$$\gamma_1\left(\chi^2_{(p)}\right) = \sqrt{8/p}, \quad \gamma_1\left(B_{(\alpha,\beta)}\right) = \frac{2(\beta-\alpha)\sqrt{\alpha+\beta+1}}{(\alpha+\beta+2)\sqrt{\alpha\beta}}, \quad \gamma_1\left(F_{(d_1,d_2)}\right) = \frac{(2d_1+d_2-2)\sqrt{8(d_2-4)}}{(d_2-6)\sqrt{d_1(d_1+d_2-2)}}.$$

Substituting for the appropriate degrees of freedom and simplifying, we reach (i–iii).

The denominator of $\gamma_1\left(d_{ii}\right)$ is always non-zero while the denominator of $\gamma_1\left(d_{(ii)}\right)$ is zero when $(n - p) = 7$. It may be shown in a similar manner that the denominator of the coefficient of kurtosis of $d_{(ii)}$ is zero for $(n - p) \leq 8$. Hence the first four moments are only finite when $n > 8$, and $d_{(ii)}$ may behave unexpectedly when $(n - p)$ assigns values in the neighbourhood of 8_+. Although the order of the skewness is O $\left(p^{-1/2}\right)$ for each of D_{ii}, d_{ii} and $d_{(ii)}$ the exact skewnesses are functions of different powers of (p/n), and hence the actual skewness for a given $\{n, p\}$ pairing may be quite different for the three MDs. This is demonstrated in Fig. 1, where the skewness is displayed for two values of (p/n) as p increases. When $(p/n) = 0.2$, the skewness coefficients almost coincide, contrary to the case when $(p/n) = 0.75$, where the skewness of $d_{(ii)}$ is seen to have a discontinuity point when p is low and then to assign a very high value and slowly approach zero. It is also noteworthy that d_{ii} and $d_{(ii)}$ are skewed in different directions. It is, in fact, possible to show that the distribution of $d_{(ii)}$ is in general very different from that of d_{ii}: suppose, with no loss of generality, that the mean of the parent variable $\mathbf{X}$ is known. Then the identity in Proposition 2 simplifies to $\left(1 + n^{-1}d_{(ii)}\right)\left(1 - n^{-1}d_{ii}\right) = 1$ and we may express the difference between the two estimators as $\delta_{ii} := \left\{n^{-1}d_{ii} - n^{-1}d_{(ii)}\right\} = -\frac{\left(n^{-1}d_{(ii)}\right)^2}{1+n^{-1}d_{(ii)}}$. Whenever $n^{-1}d_{(ii)}$ has a non-degenerate distribution then so has δ_{ii} by the continuous mapping theorem (Davidson 1994).

Another matter of concern is the rate of convergence to normality. For this purpose, a Monte Carlo simulation is conducted. The closeness between the actual

Table 1 Factors varied in the simulation

Factor	Symbol	Value
Distribution of parent variable	X	$\{N(0,1), t_{(6)}, K_3\}$
Sample-to-dimension quotient	c	$\{0.20, 0.75, 0.95, 0.98\}$
Dimension of random vector X	p	$\{10, 20, \ldots, 200\}$
Support of evaluated distributions	x	$\{-6, -5.98, \ldots, 5.98, 6\}$

distributions and the normal distribution is measured by the maximum discrepancy, also known as the Kolmogorov distance (Shiryayev 1992), while varying relevant quantities. Since each of the statistics D_{ii}, d_{ii} and $d_{(ii)}$ of Definitions 1–3 are invariant to affine transformations, the mean values and covariances need not concern us. On the other hand, the higher moments of the parent variable X_{ij} are likely to affect the convergence rates. Three different marginal distributions are, therefore, included: (i) the standard normal distribution, representing the ideal case, (ii) a Khintchine type of non-normal distribution with kurtosis equal to 3 (details are available in Appendix) and (iii) a t-distribution with six degrees of freedom, which violates the moment assumptions of Theorems 3–4. Further, two different values of the quotient $p/n = c$ are used; $c = 1/5$ represents an 'almost low-dimensional case', while $c = 0.75$ represents a genuine high-dimensional setting, where the dimension p is about the same as n. For the normal distribution the values $c = 0.95$ and $c = 0.98$ are included to investigate the behaviour close to singularity of **S**. The Kolmogorov distance is estimated by $\max_{x \in (-6,6)} |F_p(x) - F(x)|$ where $F_p(x)$ is the empirical CDF of the statistics in Theorems 3–4 and $F(x)$ is the CDF of the standard normal distribution, over increments $x, x + h$ with $h = 0.02$, using $r = 10^4$ replicates for each x. The factors varied in the simulation are summarized in Table 1.

From Figs. 2, 3, 4 and 5, some interesting properties are shown. It is seen that the convergence rate depends on the distribution of the parent variable in the way that MDs calculated from a normal distribution of the parent variable converge faster than those calculated from a Kinthchine distribution, even though they both satisfy the assumptions of Theorems 3–4. On the other hand, the $t_{(6)}$ distribution has a kurtosis equal to 5, which violate the assumptions of Theorems 3 and 4, and the limiting distributions of the MD estimates consequently do not limit a normal distribution. It is also seen that high values of the p/n ratio slow down the convergence rate. While this effect is in line with what may be expected, it is somewhat surprising to see that the leave-one-out estimator of Definition 3 converges more slowly than the more commonly used estimator d_{ii}, specified in Definition 2. Although the $d_{(ii)}$ estimator may still be preferred because of its robustness to a single influential value, it seems less favourable in analyses where data are expected to be i.i.d., due to the slower convergence rate.

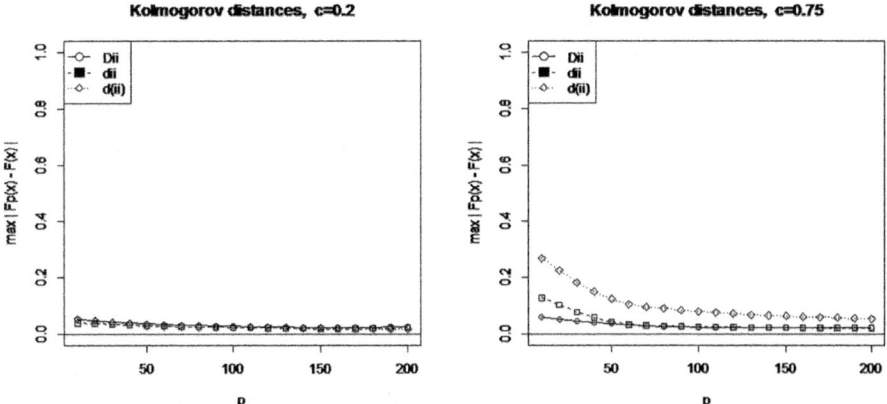

Fig. 2 Kolmogorov distances for MD when sampled from a normal distribution

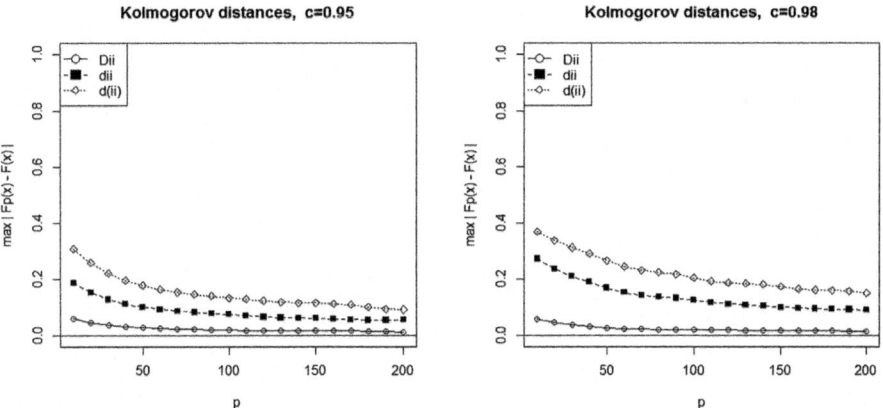

Fig. 3 Kolmogorov distances for MD when sampled from a normal distribution

3 Summary

In this chapter, central limit theorems are derived for two types of individual MDs in cases where the dimension of data increases proportionally with the sample size. These limiting distributions can be applied directly for inference of an individual Mahalanobis distance but also ensure asymptotic normality of functions of the MDs, which in turn are common in applications such as normality testing. They also serve as a basis for deriving asymptotic multivariate normality of vectors or even matrices of MD's. Furthermore, an explicit link between the leave-one-out estimator, obtained by omitting one observation from the estimation of mean vectors and the covariance matrix on which the MD depends, and the standard estimator is derived. This link is one-to-one and can be used to derive distributional properties of one type of

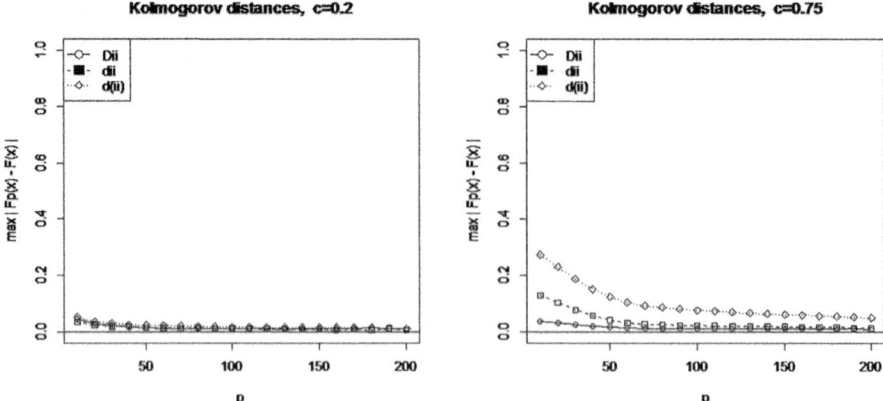

Fig. 4 Kolmogorov distances for MD when sampled from a Khintchine distribution

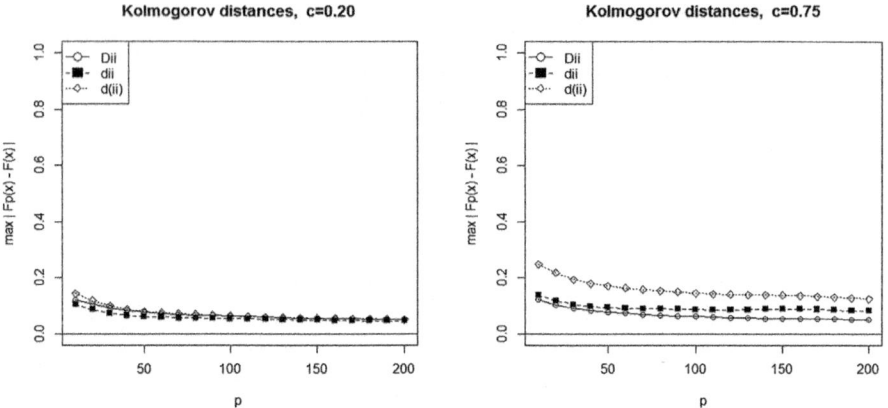

Fig. 5 Kolmogorov distances for MD when sampled from a t-distribution with df $= 6$

estimator as a function of the other. A Monte Carlo simulation is included, which shows the dependence of the convergence rates on the distribution of the population sampled from. It is also seen that the leave-one-out estimator converges more slowly than the traditional estimator, which is a somewhat unexpected finding. Although the proposed weak limits are useful in their own right, further research should involve uniform convergence in n, in order to explore properties of statistics based on the full set of n Mahalanobis distances resulting from a sample.

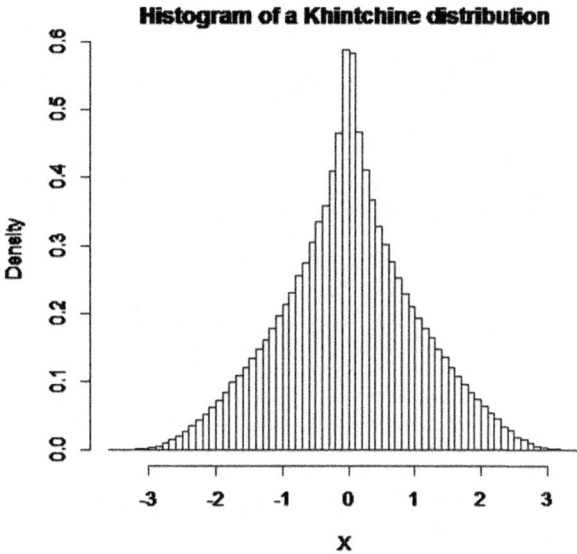

Fig. 6 Histogram of a Kinthchine distribution

Appendix

A family of distributions, discussed in Johnson (1987), is defined as $X_{ij} = \pi_{ij} R_{ij} U_{ij}$ where all components are individually and mutually i.i.d., where $p\left(\pi_{ij} = 1\right) = p\left(\pi_{ij} = -1\right) = 0.5$, $U_{ij} \sim U(0, 1)$, $R_{ij} \sim (Gamma(\lambda, q))^{\tau}$, where $U(0, 1)$ is the standard uniform distribution and $Gamma(\lambda, q)$ denotes the location-scale parametrization of the gamma distribution. The parameters are set to $\lambda = 0.000028$, $q = 012757$ and $\tau = 3.5$. This yields a symmetric distribution, labelled as K_3 in Sect. 2, with marginal kurtosis equal to 3. A histogram of 50 000 drawings from this distribution is displayed (Fig. 6).

References

Anderson, T.: An Introduction to Multivariate Statistical Analysis, 3 ed. Wiley, New York (2003)

Andrews, D., Gnanadesikan, R., Warner, J., Krishnaiah, P.R.: Methods for assessing multivariate normality. In: Krishnaiah, P.R. (ed.) Multivariate Analysis III, vol. 17, no. 1, pp. 95–116. Academic Press, New York (1973)

Arharov, L.: Limit theorems for the characteristic roots of a sample covariance matrix. Soviet Math. Dokl **12**, 1206–1209 (1971)

Bai, Z., Silverstein, J.W.: Spectral Analysis of Large Dimensional Random Matrices, vol. 20. Springer (2010)

Bai, Z., Miao, B., Pan, G.: On asymptotics of eigenvectors of large sample covariance matrix. Ann. Probab. **35**(4), 1532–1572 (2007)

Birke, M., Dette, H.: A note on testing the covariance matrix for large dimension. Stat. Probab. Lett. **74**, 281–289 (2005)

Davidson, J.: Stochastic Limit Theory: An Introduction for Econometricians. Oxford University Press, UK (1994)

De Maesschalck, R., Jouan-Rimbaud, D., Massart, D.L.: The Mahalanobis distance. Chemometr. Intell. Lab. **50**, 1–18 (2000)

Ferguson, T.S.: A Course in Large Sample Theory. Chapman & Hall, London (1996)

Frölich, M.: Programme Evaluation and Treatment Choice, vol. 524. Springer Science & Business Media (2012)

Glombek, K.: Statistical inference for high-dimensional global minimum variance portfolios. Scand. J. Stat. **41**, 845–865 (2014)

Healy, M.: Multivariate normal plotting. Appl. Stat. J. Roy. St. C. **17**, 157–161 (1968)

Johnson, M.E.: Multivariate Statistical Simulation: A Guide to Selecting and Generating Continuous Multivariate Distributions. Wiley, New York, NY (1987)

Johnson, N., Kotz, S., Balakrishnan, N.: Continuous Univariate Distributions. Wiley series in probability and mathematical statistics: Applied probability and statistics, vol. 2. Wiley (1995)

Jonsson, D.: Some limit theorems for the eigenvalues of a sample covariance matrix. J. Multivar. Anal. **12**, 1–38 (1982)

Mahalanobis, P.: On tests and measures of group divergence. J. Asiat. Soc. Beng. **26**, 541–588 (1930)

Marchenko, V.A., Pastur, L.A.: Distribution of eigenvalues for some sets of random matrices. Mat. Sb. **114**, 507–536 (1967)

Mardia, K.: Applications of some measures of multivariate skewness and kurtosis in testing normality and robustness studies. Sankhyā: Ind. J. Stat. Ser. B **36**(2), 115–128 (1974)

Mardia, K.: Mahalanobis distances and angles. Multivar. Analysis IV **4**(1), 495–511 (1977)

Mardia, K., Kent, J., Bibby, J.: Multivariate Analysis. Academic press, London (1980)

Montgomery, D.C., Woodall, W.: Research issues and ideas in statistical process control. J. Qual. Technol. **31**, 376–387 (1999)

Ord, J.K.: Families of Frequency Distributions. Griffin, London (1972)

Pavlenko, T.: On feature selection, curse-of-dimensionality and error probability in discriminant analysis. J. Stat. Plan. Infer. **115**, 565–584 (2003)

Serdobolskii, V.: Multivariate Statistical Analysis: A High-Dimensional Approach. Springer (2000)

Shiryayev, A.: On The Empirical Determination of A Distribution Law. Springer (1992)

Todeschini, R., Ballabio, D., Consonni, V., Sahigara, F., Filzmoser, P.: Locally centred Mahalanobis distance: a new distance measure with salient features towards outlier detection. Anal. Chim. Acta **787**, 1–9 (2013)

Wilks, S.: Multivariate statistical outliers. Sankhyā: Ind. J. Stat. Ser. A **25**(4), 407–426 (1963)

Bootstrap Type-1 Fuzzy Functions Approach for Time Series Forecasting

Ali Zafer Dalar and Erol Eğrioğlu

Abstract In this study, we proposed an alternative approach for time series forecasting. Many approaches have been developed and applied for forecasting in the literature. In the past years, most of these approaches are fuzzy system modelling approaches. Fuzzy functions approaches were proposed by Turksen (Appl Soft Comput 8:1178–1188 2008) because traditional fuzzy system modelling approaches are generally based on the fuzzy rule base. Fuzzy functions approaches do not need to use the rule base. Fuzzy functions approaches should employ randomness, and their values change randomly from sample to sample. Taking into consideration this change, researchers need to obtain estimators, but this process for nonlinear models is not an easy task to carry out. Thus, bootstrap methods can be used in order to overcome this problem. In this chapter, we proposed a new approach that uses fuzzy c-means techniques for clustering, type-1 fuzzy functions approach for fuzzy system modelling and subsampling bootstrap method for probabilistic inference. By means of the proposed method, researchers can obtain forecast distribution, forecasts can be obtained from the distribution of forecasts as a measure of central tendency, and combine many different forecast results. For experimental study, we used Istanbul Stock Exchange 100 indices as data sets. For comparison of the results obtained from the proposed method, some other methods that are well known in the literature are used.

Keywords Forecasting · Bootstrap methods · Type-1 fuzzy functions approach
Subsampling · Stock and bond exchange

A. Z. Dalar (✉) · E. Eğrioğlu
Department of Statistics, Forecast Research Laboratory,
Giresun University, 28200 Giresun, Turkey
e-mail: alizaferdalar@hotmail.com

E. Eğrioğlu
e-mail: erole1977@yahoo.com

© Springer International Publishing AG 2018   69
M. Tez and D. von Rosen (eds.), *Trends and Perspectives*
in Linear Statistical Inference, Contributions to Statistics,
https://doi.org/10.1007/978-3-319-73241-1_5

1 Introduction

In recent years, there has been growing interest in the development of new alternative methods that provide more accurate forecasts for time series. These methods are generally based on artificial neural networks (ANNs) or fuzzy set theory. While applying ANNs to the time series, they do not comprise any approximations to uncertainty. The methods based on fuzzy set theory comprise fuzzy approximation to uncertainty. These methods can be classified as fuzzy regression methods, fuzzy time series (FTS) methods, fuzzy inference systems (FIS) and fuzzy functions (FF) approaches for time series forecasting. The aims of these methods are to obtain more accurate interval and point predictions. In addition to this, these methods have not often been used in literature because of using linear models and requiring computations of complex mathematical programming problems.

FTS approach was initially proposed by Song and Chissom (1993a, b). FTS approaches that work with membership values are like FF approaches. FTS methods have populously been used for time series forecasting in recent years. These methods do not comprise any restrictions, unlike classical time series methods. The main problem of FIS is to take no account of membership values of FTS methods. Many of the FTS methods are rule based. Determining of the rules is a significant problem in FIS, and also this is one of the important factors which effect the performance of methods. The most common used FIS for time series forecasting is adaptive neuro-fuzzy inference system (ANFIS) that was proposed by Jang (1993). Traditional fuzzy system modelling methods are generally based on fuzzy rule bases. This is an important disadvantage situation, therefore, FF approaches were introduced.

Fuzzy functions approach was introduced by Turksen in 2005 (Turksen 2008). According to Turksen (2008), FF approaches have emerged from the idea of representing each unique rule of a fuzzy rule base system. After introducing FF approaches to the literature, these approaches are developed by using different kinds of artificial intelligent systems and fuzzy sets (Celikyilmaz and Turksen 2008a, b, 2009; Turksen 2009). Beyhan and Alci (2010) adapted FF to time series forecasting and used an embedded model. In Zarandi et al. (2013), a hybrid FF approach was proposed, and lagged variables were not used like in regression analysis. Aladag et al. (2016) proposed a type-1 fuzzy functions approach. In this approach, inputs of the system are lagged variables of time series, and these variables are determined by binary particle swarm optimisation. In Dalar et al. (2015), FF approach used for forecasting Turkey Electric Consumption time series data.

Randomness is a well-known expediency for uncertainty because of the human incapability of understanding. FF approaches should employ randomness, and their values change randomly from sample to sample. In order to consider this change, one needs to obtain estimators, but obtaining of estimators for nonlinear models is not easy. In order to overcome this problem, bootstrap methods can be used.

In linear statistical models, it is possible to obtain the probability distributions of estimators by means of an assumed probability distribution of error. However, the pre-assumed probability distribution of error term can cause problems. It is quite difficult to obtain the probability distribution of estimators with an assumption on error term in nonlinear and data-based methods like ANNs and FIS. Efron (1979) proposed an approach which is called as bootstrap. By means of Efron's method, one can obtain a sample for distribution of estimators, and also expected value, variance and confidence intervals of estimator through these samples.

Bootstrap methods for dependent data have been implemented in different ways. One can divide these methods in two, model-based and non-model-based methods. The non-model-based methods are moving block bootstrap method which was proposed by Künsch (1989) and Liu and Singh (1992), independently, and different versions of it. These methods are moving block bootstrap (MBB), non-overlapping block bootstrap (NBB), circular block bootstrap (CBB) which was proposed by Politis and Romano (1992), stationary block bootstrap (SBB) which was proposed by Politis and Romano (1994b), and subsampling moving block bootstrap (Sub-MBB) which discussed in Politis and Romano (1994a), Hall and Jing (1996) and Bickel et al. (1997). There exist several books about bootstrap and subsampling, e.g. Efron (1982), Hall (1992a, b), Efron and Tibshirani (1993), Shao and Tu (1995), Davison and Hinkley (1997), Politis et al. (1999), Lahiri (2003), Good et al. (2005), Chernick (2008) and Chernick and LaBudde (2011). Besides, several papers give overviews of various aspects of bootstrapping time series. Among them are Berkowitz and Kilian (2000), Bose and Politis (1995), Bühlmann (1997), Carey (2005), H rdle et al. (2003), Li and Maddala (1996), Politis (2003), Kreiss and Paparoditis (2011), Kreiss and Lahiri (2012), Jin et al. (2013), Hwang and Shin (2014), Cavaliere et al. (2015), Costa et al. (2015) and Pan and Politis (2016).

The chapter is organised as follows. The fundamental knowledge of T1FF approach and bootstrap methods are briefly summarised in the 2nd and 3rd sections. The proposed method is introduced in Sect. 4. In the 5th section, the results obtained from the implementation of the proposed method which was examined by ISE time series. Finally, in Sect. 6, conclusions are discussed.

2 Type-1 Fuzzy Functions Approach

Turksen (2008) proposed fuzzy functions approach instead of rule-based FIS. While a relation between inputs and output is established in rule-based FIS, a function is generated instead of a relation in FF approach. There is no need to determine any rules in FF approach. This is an important advantage of the T1FF approach.

Algorithm of Turksen (2008)'s T1FF approach is given below step by step.

Step 1. Inputs are lagged variables of time series. Matrix Z comprises of inputs and output of the system. Inputs and output of the system are clustered using fuzzy c-means (Bezdek 1981) clustering method.

Fuzzy c-means (FCM) clustering method can be applied by using the formulas given below.

$$v_i = \frac{\sum_{k=1}^{n}(\mu_{ik})^{fi}z_k}{\sum_{k=1}^{n}(\mu_{ik})^{fi}} , i = 1, 2, \ldots, c \qquad (2.1)$$

$$\mu_{ik} = \left[\sum_{j=1}^{c}\left(\frac{d(z_k, v_i)}{d(z_k, v_j)}\right)^{\frac{2}{fi-1}}\right]^{-1} , i = 1, 2, \ldots, c; k = 1, 2, \ldots, n \qquad (2.2)$$

where $d(z, v)$ is Euclidian distance and is computed by using the formula (2.3). Also, z_k is a vector whose elements are the elements compose of kth row of $\mathbf{Z}$. μ_{ik} is the degree of belongingness of kth observation to the ith cluster.

$$d(z_k, v_i) = \|z_k - v_i\| \qquad (2.3)$$

Step 2. Membership values of the input space are constituted as below.

$$\mu_{ik} = \left[\sum_{j=1}^{n}\left(\frac{d(x_k, v_i)}{d(x_k, v_j)}\right)^{\frac{2}{fi-1}}\right]^{-1} , i = 1, 2, \ldots, c; k = 1, 2, \ldots, n \qquad (2.4)$$

Step 3. For each cluster i, membership values of each input data sample, μ_{ik} and original inputs are gathered together, and ith fuzzy function is obtained from predicting $Y^{(i)} = X^{(i)}\beta^{(i)} + \varepsilon^{(i)}$ regression model. When the number of inputs is p, $X^{(i)}$ and $Y^{(i)}$ matrices are as follows:

$$X^{(i)} = \begin{bmatrix} \mu_{i1} & x_{11} & \cdots & x_{p1} \\ \mu_{i2} & x_{12} & \cdots & x_{p2} \\ \vdots & \vdots & \ddots & \vdots \\ \mu_{in} & x_{1n} & \cdots & x_{pn} \end{bmatrix}, Y^{(i)} = \begin{bmatrix} y_1 \\ y_2 \\ \vdots \\ y_n \end{bmatrix} \qquad (2.5)$$

Step 4. Output values are calculated by using the results obtained from fuzzy functions as follow:

$$\hat{y}_i = \frac{\sum_{i=1}^{c}\hat{y}_{ik}\mu_{ik}}{\sum_{i=1}^{c}\mu_{ik}}, k = 1, 2, \ldots, n \qquad (2.6)$$

Those who want more information can look the study of Turksen (2008).

3   Bootstrap Methods

The bootstrap method, initially introduced by Efron (1979) for independent variables and later extended to deal with more complex dependent variables by several authors, is a class of non-parametric methods that allow the statisticians to carry out statistical inference on a wide range of problems without imposing many structural assumptions on the underlying data-generating random process (Kreiss and Lahiri 2012). In this study, we used SBB in subsampling.

3.1   Subsampling

Before and after the development of non-parametric bootstrap methods, subsamples-based methods were developed to deal with special problems (Davison and Hinkley 1997). Subsampling has been applied for confidence intervals and variance estimates in both i.i.d. and dependent situations. Politis et al. (1999) summarise results on subsampling and compare it to the bootstrap. They include applications to i.i.d. samples, stationary and non-stationary time series (Chernick 2008).

3.2   Stationary Block Bootstrap

Stationary block bootstrap, which was proposed by Politis and Romano (1994b), is a special case of block bootstrap with random block length, and the length of each block is approximated by the geometric distribution. Instead of the assumption of the geometric distribution for i.i.d. variables, one can consider other forms of data distribution for the bootstrap procedure. In this study, we used the uniform distribution for this procedure. When applying SBB to the time series, unlike other block bootstrap methods (MBB, NBB and CBB), the new samples of time series are stationary.

4   The Proposed Method

In this study, subsampling stationary block bootstrap fuzzy functions approach (SSBFF) which use type-1 fuzzy functions (T1FF) based on subsampling bootstrap method is proposed. Algorithm of the proposed method is given below, step by step.

Algorithm. T1FF based on subsampling bootstrap method

Step 1. The lengths of the training and test sets (*ntrain* and *ntest*), the number of fuzzy clusters (*cn*) which will be used in fuzzy functions approach, fuzzy index value (f_i) used in fuzzy c-means method, the number of lagged variables (*p*) and the number of bootstrap repetitions (*nbst*) are determined.

Step 2. For each bootstrap sample, the steps from Step 2.1 to Step 2.6 are repeated. Bootstrap samples are taken from the training set. The forecast results for the test set are obtained from optimised fuzzy functions approach by using the bootstrap sample. As a result of this, the test set is fixed for all bootstrap repetitions.

Step 2.1. The inputs of the system are determined according to number of lagged variables. Matrix **Z**, composed of both inputs and output of the system, is generated. Then, elements of the matrix are clustered by using FCM technique. FCM can be applied using the formulas given below.

$$v_i = \frac{\sum_{k=1}^{n}(\mu_{ik})^{fi} z_k}{\sum_{k=1}^{n}(\mu_{ik})^{fi}}, i = 1, 2, \ldots, c \qquad (4.1)$$

$$\mu_{ik} = \left[\sum_{j=1}^{c}\left(\frac{d(z_k, v_i)}{d(z_k, v_j)}\right)^{\frac{2}{fi-1}}\right]^{-1}, i = 1, 2, \ldots, c; k = 1, 2, \ldots, ntrain$$

$$(4.2)$$

where $d(z, v)$ is Euclidian distance and is computed by using the formula (2.3).

Step 2.2. Membership values of the input space are found as follows:

$$\mu_{ik} = \left[\sum_{j=1}^{n}\left(\frac{d(x_k, v_i)}{d(x_k, v_j)}\right)^{\frac{2}{fi-1}}\right]^{-1}, i = 1, 2, \ldots, c; k = 1, 2, \ldots, ntrain$$

$$(4.3)$$

where x is an input matrix which is generated for lagged variables.

Celikyilmaz and Turksen (2009) used mathematical transformations of membership values. Their research indicated that the exponential and various logarithmic transformations of membership values can improve the performance of the system models.

In this study, for each cluster i by using membership values, μ_{i1}^2 and $\exp(\mu_{i1})$ are taken.

Step 2.3. For each cluster i, membership values of each input data sample, μ_{ik} and original inputs are gathered together, and ith fuzzy function is obtained from predicting $Y^{(i)} = X^{(i)} \boldsymbol{\beta}^{(i)} + \boldsymbol{\varepsilon}^{(i)}$ regression model. When the number of inputs is p, $X^{(i)}$ and $Y^{(i)}$ matrices are as follows:

$$X^{(i)} = \begin{bmatrix} 1 & \mu_{i1} & \mu_{i1}^2 & exp(\mu_{i1}) & x_{11} & \cdots & x_{p1} \\ 1 & \mu_{i2} & \mu_{i2}^2 & exp(\mu_{i2}) & x_{12} & \cdots & x_{p2} \\ \vdots & \vdots & \vdots & \vdots & \vdots & \ddots & \vdots \\ 1 & \mu_{i,ntrain} & \mu_{i,ntrain}^2 & exp(\mu_{i,ntrain}) & x_{1,ntrain} & \cdots & x_{p,ntrain} \end{bmatrix}, Y^{(i)} = \begin{bmatrix} y_1 \\ y_2 \\ \vdots \\ y_{ntrain} \end{bmatrix}$$

$$(4.4)$$

The parameters and predictions of the linear model are estimated by using following formulas:

$$\boldsymbol{\beta}^{(i)} = (X^{(i)\prime} X^{(i)})^{-1} X^{(i)\prime} Y^{(i)} \tag{4.5}$$

$$\boldsymbol{Y}^{(i)} = X^{(i)} \boldsymbol{\beta}^{(i)} \tag{4.6}$$

Step 2.4.   By using the results obtained from fuzzy functions, prediction values for training set are calculated as follow:

$$\hat{y}_i = \frac{\sum_{k=1}^{c} \hat{y}_{ik} \mu_{ik}}{\sum_{k=1}^{c} \mu_{ik}}, i = 1, 2, \ldots, ntrain \tag{4.7}$$

Step 2.5.   The membership values are obtained for test set inputs according to precalculated cluster centres. $X^{(i)}$ and $Y^{(i)}$ matrices are constituted for the test set as follows:

$$X^{(i)} = \begin{bmatrix} 1 & \mu_{i,ntrain+1} & \mu_{i,ntrain+1}^2 & exp(\mu_{i,ntrain+1}) & x_{i,ntrain+1} & \cdots & x_{p,ntrain+1} \\ 1 & \mu_{i,ntrain+2} & \mu_{i,ntrain+2}^2 & exp(\mu_{i,ntrain+2}) & x_{i,ntrain+2} & \cdots & x_{p,ntrain+2} \\ \vdots & \vdots & \vdots & \vdots & \vdots & \ddots & \vdots \\ 1 & \mu_{i,ntrain+ntest} & \mu_{i,ntrain+ntest}^2 & exp(\mu_{i,ntrain+ntest}) & x_{i,ntrain+ntest} & \cdots & x_{p,ntrain+ntest} \end{bmatrix}$$

$$Y^{(i)} = \begin{bmatrix} y_{ntrain+1} \\ y_{ntrain+2} \\ \vdots \\ y_{ntrain+ntest} \end{bmatrix}$$

$$\boldsymbol{Y}^{(i)} = X^{(i)} \boldsymbol{\beta}^{(i)}$$

$$(4.8)$$

Step 2.6.   Forecasts of fuzzy functions for the test set are obtained as follow:

$$\hat{y}_i^j = \frac{\sum_{k=1}^{c} \hat{y}_{ik} \mu_{ik}}{\sum_{k=1}^{c} \mu_{ik}}, i = ntrain+1, ntrain+2, \ldots, ntrain+ntest, j = 1, 2, \ldots, nbst \tag{4.9}$$

Table 1 The bootstrap forecasts

$\hat{y}_1^j$	$\hat{y}_2^j$	...	$\hat{y}_{ntest}^j$
$\hat{y}_1^1$	$\hat{y}_2^1$	...	$\hat{y}_{ntest}^1$
$\hat{y}_1^2$	$\hat{y}_2^2$	...	$\hat{y}_{ntest}^2$
⋮	⋮	...	⋮
$\hat{y}_1^{nbst}$	$\hat{y}_2^{nbst}$	...	$\hat{y}_{ntest}^{nbst}$
$\hat{y}_1$	$\hat{y}_2$	...	$\hat{y}_{ntest}$

Step 3. The bootstrap forecasts are calculated by using median statistic as follow, and template of the bootstrap forecasts is given in Table 1.

$$\hat{y}_i = median\left(\hat{y}_i^j\right), j = 1, 2, \ldots, nbst, i = 1, 2, \ldots, ntest \qquad (4.10)$$

Step 4. Calculate root mean square error (RMSE) and mean absolute percentage error (MAPE) values for test set.

$$RMSE = \sqrt{\frac{1}{T}\sum_{t=1}^{T}(y_t - \hat{y}_t)^2} \qquad (4.11)$$

$$MAPE = \frac{1}{T}\sum_{t=1}^{T}\left|\frac{y_t - \hat{y}_t}{y_t}\right| \qquad (4.12)$$

5   Implementation

In the implementation, the proposed method was applied to Istanbul Stock Exchange (ISE) data sets. Details of data sets are given below:

Set 1. It is daily observed between 02/01/2009 and 29/05/2009 dates, and consist of 103 observations.

Set 2. It is daily observed between 04/01/2010 and 31/05/2010 dates, and consist of 104 observations.

Set 3. It is daily observed between 03/01/2011 and 31/05/2011 dates, and consist of 106 observations.

Set 4. It is daily observed between 02/01/2012 and 31/05/2012 dates, and consist of 106 observations.

Set 5. It is daily observed between 02/01/2013 and 29/05/2013 dates, and consist of 106 observations.

The test sets are taken as the last 7 observations of each ISE data sets. All ISE data sets are forecasted by using the methods are listed below.

ARIMA: Autoregressive Integrated Moving Average Model, The best model was determined Box–Jenkins Procedure.

ES: Exponential Smoothing, Simple, Holt and winters exponential smoothing methods were applied and the best model was selected.

Song–Chissom: Song and Chissom time-invariant fuzzy time series method (Song and Chissom 1993a, b), the numbers of fuzzy sets were changed from 5 to 15 and the best numbers of fuzzy sets were selected.

T1FF: Type-1 fuzzy functions approach (Turksen 2008), the model order and the number of fuzzy sets were changed from 1 to 5 and from 5 to 15, respectively, and α-cut was taken 0 and 0.1.

In the proposed approach, we determined the number of lagged variables (m) between 1 and 5, with an increment of 1. The number of fuzzy clusters (cn) experienced between 3 and 7, with an increment of 1. The number of bootstrap repetitions is taken as 50, 75 and 100. RMSE and MAPE performance measures, whose formulas given in 4.11 and 4.12, are calculated for each method.

5.1 Analysis of Set 1 Time Series

The graph and analysis result of Set 1 time series is given in Fig. 1 and in Table 2, respectively.

As seen in Table 2, the best RMSE and MAPE values are obtained from the SSBFF approach. The best model for ARIMA was obtained as ARIMA(0, 1, 0). Alpha value of Simple-ES was obtained as 0.99993. The result of T1FF was obtained when cn, m and α-cut are taken as 7, 5 and 0, respectively. In the SSBFF approach, the best result was obtained when cn, m and $nbst$ are taken as 4, 2 and 50, respectively.

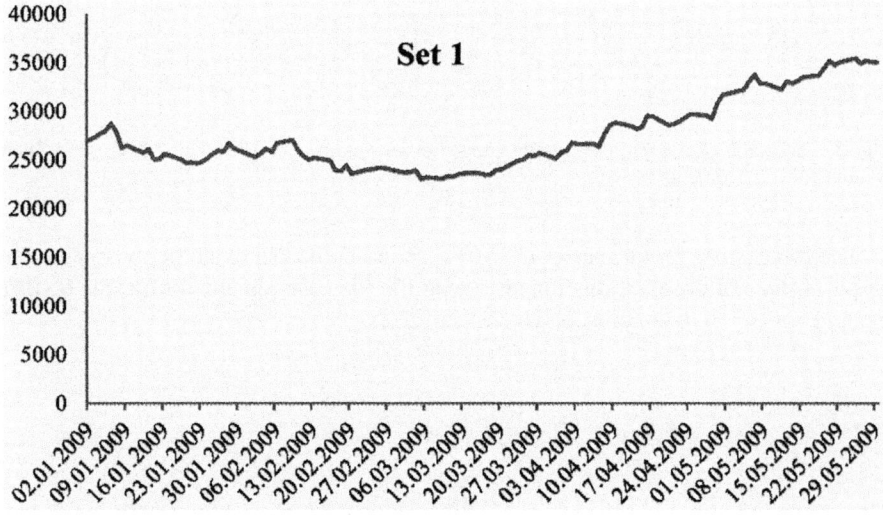

Fig. 1 The graph of Set 1 time series

Table 2 RMSE and MAPE values for test data of Set 1 time series

Date	Test data	ARIMA	ES	Song–Chissom	ANFIS	T1FF	SSBFF
21.05.2009	34721	35140	35140	33641	34981	35218	35067
22.05.2009	35015	34721	34721	33641	34501	35039	34693
25.05.2009	35408	35015	35014	33641	34932	35127	34957
26.05.2009	34861	35408	35408	33641	35363	35733	35303
27.05.2009	35169	34861	34861	33641	34707	35191	34820
28.05.2009	35021	35169	35169	33641	35127	35401	35092
29.05.2009	35003	35021	35021	33641	34929	35399	34962
	RMSE	344,910	344,930	1402,400	385,639	445,515	327,425*
	MAPE	0,0087	0,0087	0,0396	0,0169	0,0101	0,0082*

The best RMSE and MAPE values are marked with '*'

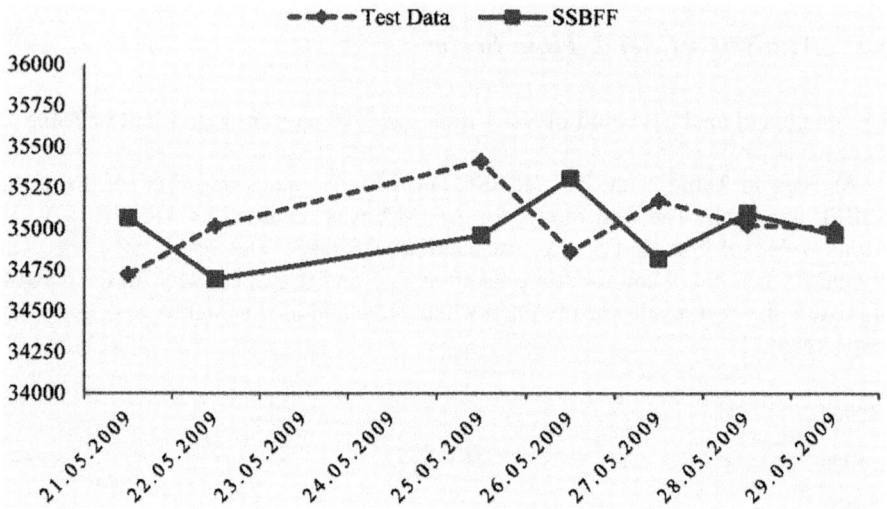

Fig. 2 The graph of the real observations together with the forecasts obtained from the SSBFF method for Set 1 time series

The forecasting performance of SSBFF approach is also examined visually. The graph of the real observations together with the forecasts obtained from the SSBFF method for Set 1 time series is given in Fig. 2.

5.2 Analysis of Set 2 Time Series

The graph and analysis result of Set 2 time series is given in Fig. 3 and in Table 3, respectively.

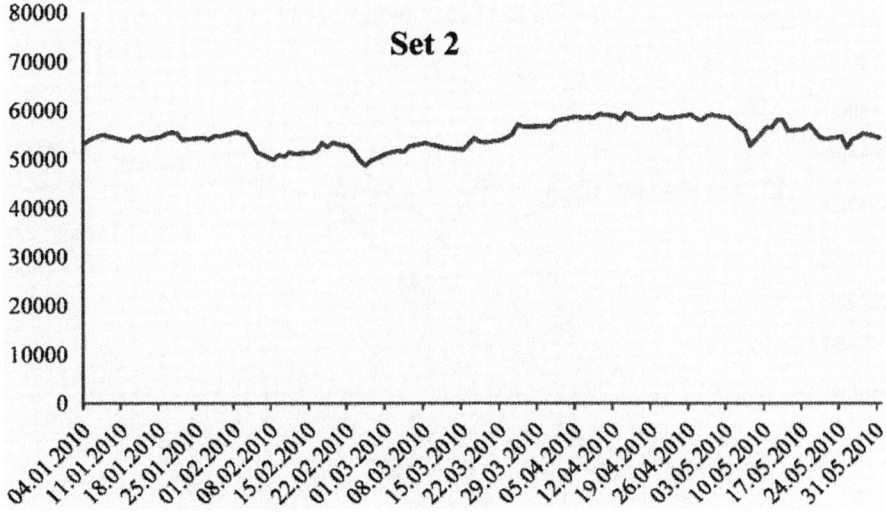

Fig. 3 The graph of Set 2 time series

Table 3 RMSE and MAPE values for test data of Set 2 time series

Date	Test data	ARIMA	ES	Song–Chissom	ANFIS	T1FF	SSBFF
21.05.2010	54112	54450	54520	53278	53924	54349	54424
24.05.2010	54558	54112	54123	53278	54901	54032	54101
25.05.2010	52257	54558	54546	53278	54519	54602	54528
26.05.2010	54104	52257	52321	53897	52037	52430	52410
27.05.2010	54498	54104	54054	53278	55253	54111	54094
28.05.2010	55234	54498	54486	53278	53616	54561	54470
31.05.2010	54385	55234	55213	54791	55403	55100	55198
RMSE		1221,000	1208,100	1127,500*	1402,219	1179,900	1179,634
MAPE		0,0183	0,0185	0,0182	0,0145*	0,0179	0,0178

The best RMSE and MAPE values are marked with '*'

As seen in Table 3, the best RMSE and MAPE values are obtained from the Song–Chissom and ANFIS methods, respectively. The best model for ARIMA was obtained as ARIMA(0, 1, 0). Alpha value of Simple-ES was obtained as 0.971991. The result of T1FF was obtained when cn, m and α-cut are taken as 6, 3 and 0, respectively. In the SSBFF approach, the best result was obtained when cn, m and $nbst$ are taken as 3, 2 and 50, respectively.

The forecasting performance of SSBFF approach is also examined visually. The graph of the real observations together with the forecasts obtained from the SSBFF method for Set 2 time series is given in Fig. 4.

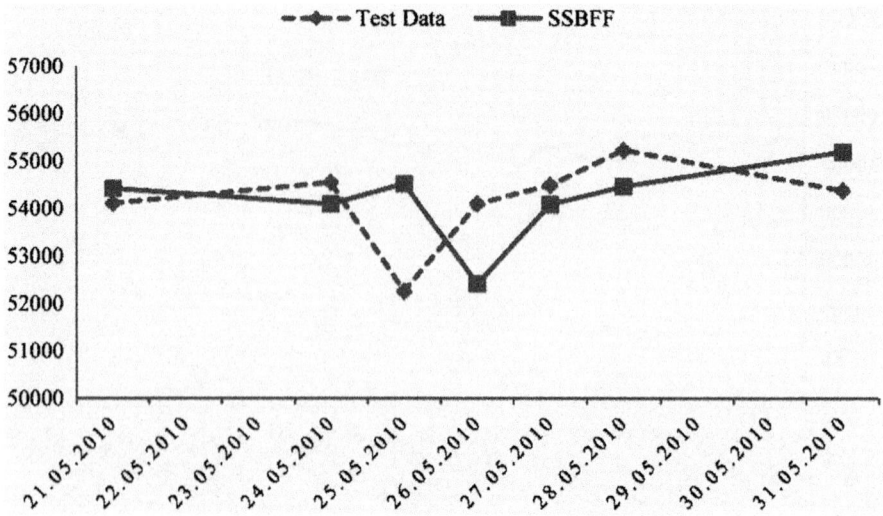

Fig. 4 The graph of the real observations together with the forecasts obtained from the SSBFF method for Set 2 time series

5.3 Analysis of Set 3 Time Series

The graph and analysis result of Set 3 time series is given in Fig. 5 and in Table 4, respectively.

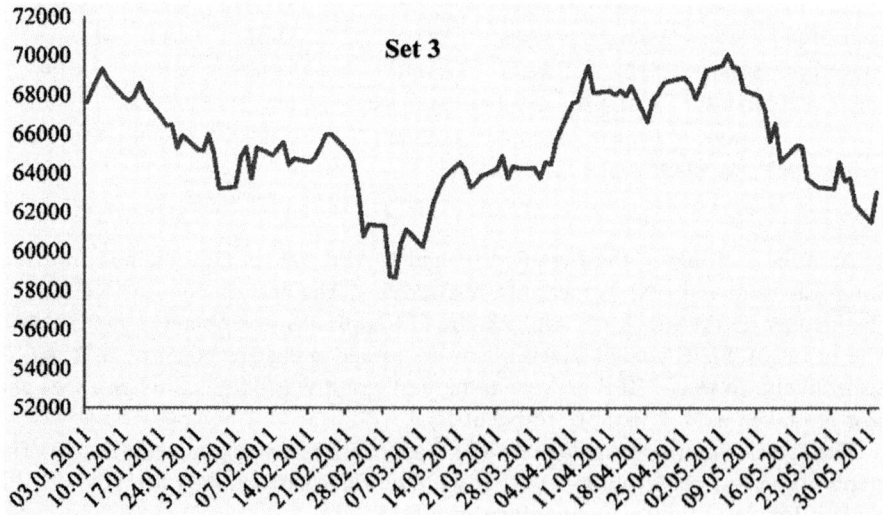

Fig. 5 The graph of Set 3 time series

Table 4 RMSE and MAPE values for test data of Set 3 time series

Date	Test data	ARIMA	ES	Song–Chissom	ANFIS	T1FF	SSBFF
23.05.2011	63210	63299	63300	62738	63192	63467	63504
24.05.2011	64561	63210	63210	62738	63897	63120	63354
25.05.2011	63609	64561	64557	64368	64553	64592	64662
26.05.2011	63755	63609	63613	64368	65336	63495	63539
27.05.2011	62407	63755	63755	64368	63462	64038	63898
30.05.2011	61492	62407	62412	62738	63685	62458	62476
31.05.2011	63046	61492	61495	61109	62184	61845	61799
	RMSE	1057,600	1057,000	1396,400	1224,672	1083,200	1031,546*
	MAPE	0,0144*	0,0144*	0,0200	0,0162	0,0153	0,0147

The best RMSE and MAPE values are marked with '*'

As seen in Table 4, the best RMSE value is obtained from the SSBFF. The best model for ARIMA was obtained as ARIMA(0, 1, 0). Alpha value of Simple-ES was obtained as 0.99999. The result of T1FF was obtained when cn, m and α-cut are taken as 7, 5 and 0, respectively. In the SSBFF approach, the best result was obtained when cn, m and $nbst$ are taken as 3, 4 and 50, respectively.

The forecasting performance of SSBFF approach is also examined visually. The graph of the real observations together with the forecasts obtained from the SSBFF method for Set 3 time series is given in Fig. 6.

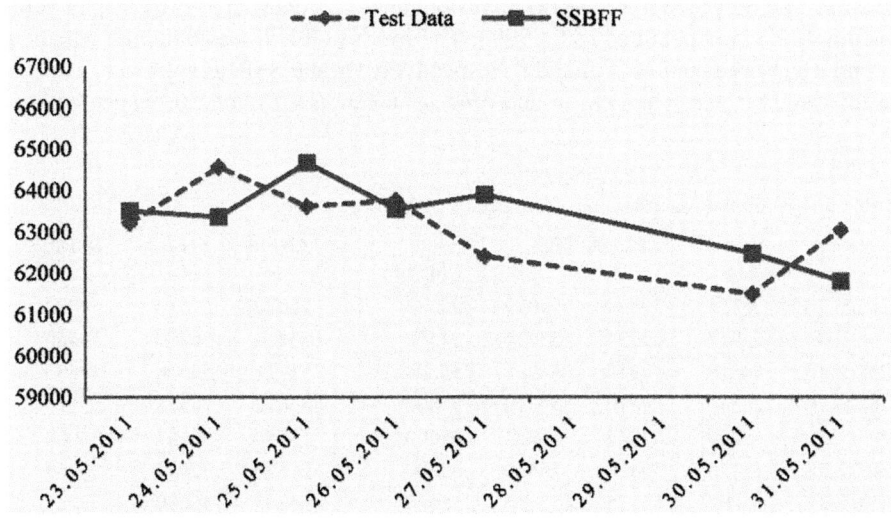

Fig. 6 The graph of the real observations together with the forecasts obtained from the SSBFF method for Set 3 time series

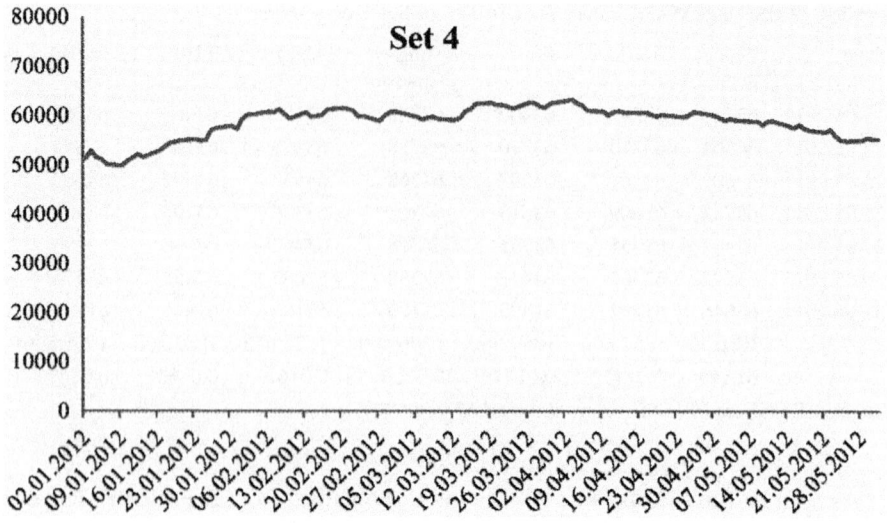

Fig. 7 The graph of Set 4 time series

5.4 Analysis of Set 4 Time Series

The graph and analysis result of Set 4 time series is given in Fig. 7 and in Table 5, respectively.

As seen in Table 5, the best RMSE and MAPE values are obtained from the SSBFF. The best model for ARIMA was obtained as ARIMA(0, 1, 0). Alpha value of Simple-ES was obtained as 0.99993. The result of T1FF was obtained when cn, m and α-cut are taken as 7, 5 and 0, respectively. In the SSBFF approach, the best result was obtained when cn, m and $nbst$ are taken as 4, 5 and 50, respectively.

Table 5 RMSE and MAPE values for test data of Set 4 time series

Date	Test data	ARIMA	ES	Song–Chissom	ANFIS	T1FF	SSBFF
23.05.2012	55734	57079	57079	57522	56020	57614	57004
24.05.2012	54917	55734	55734	57522	55434	56209	55456
25.05.2012	54810	54917	54917	55600	55271	55546	54855
28.05.2012	54844	54810	54810	55600	53612	55578	55110
29.05.2012	55450	54844	54844	55600	53540	55557	54960
30.05.2012	55125	55450	55449	55600	54550	56044	55465
31.05.2012	55099	55125	55125	55600	54805	55690	55027
	RMSE	650,560	650,739	1291,500	937,030	1034,200	577,970*
	MAPE	0,0084	0,0084	0,0183	0,0148	0,0162	0,0078*

The best RMSE and MAPE values are marked with '*'

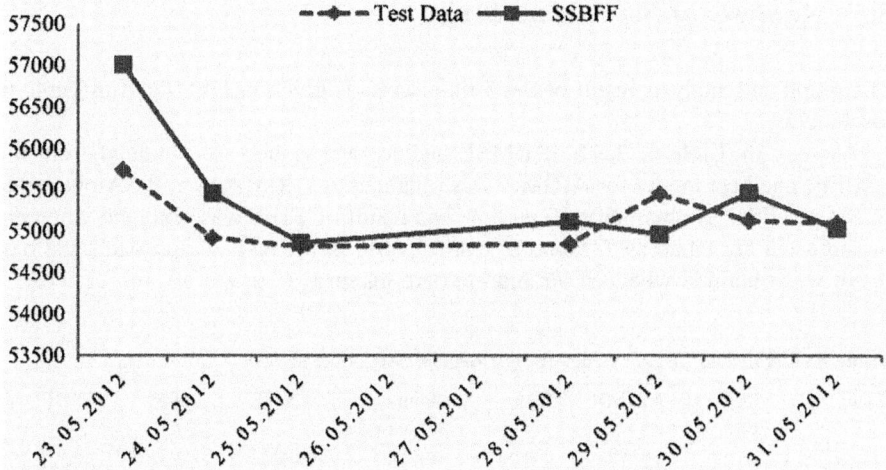

Fig. 8 The graph of the real observations together with the forecasts obtained from the SSBFF method for Set 4 time series

The forecasting performance of SSBFF approach is also examined visually. The graph of the real observations together with the forecasts obtained from the SSBFF method for Set 4 time series is given in Fig. 8.

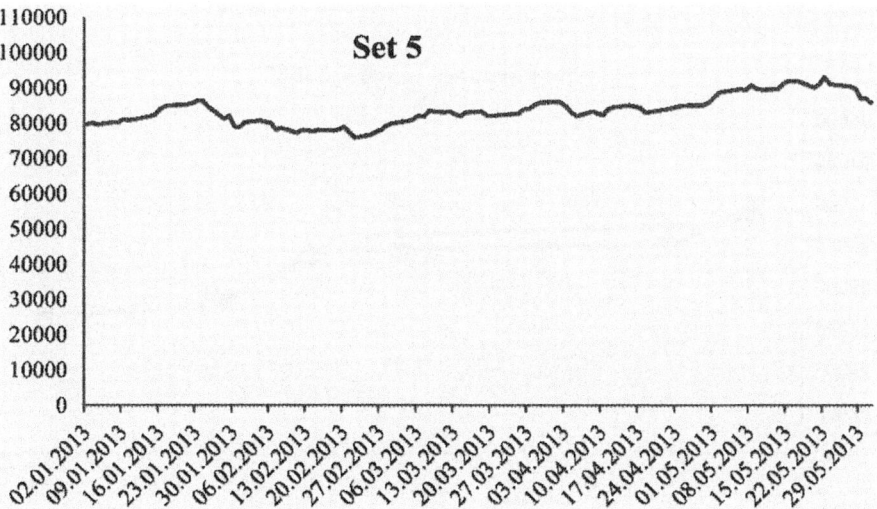

Fig. 9 The graph of Set 5 time series

5.5 Analysis of Set 5 Time Series

The graph and analysis result of Set 5 time series is given in Fig. 9 and in Table 6, respectively.

As seen in Table 6, the best RMSE and MAPE values are obtained from the SSBFF. The best model for ARIMA was obtained as ARIMA(0, 1, 0). Alpha value of Simple-ES was obtained as 0.99993. The result of T1FF was obtained when cn, m and α-cut are taken as 7, 5 and 0, respectively. In the SSBFF approach, the best result was obtained when cn, m and $nbst$ are taken as 6, 2 and 50, respectively.

Table 6 RMSE and MAPE values for test data of Set 5 time series

Date	Test data	ARIMA	ES	Song–Chissom	ANFIS	T1FF	SSBFF
23.05.2013	91351	93179	93179	90710	94507	93691	92021
24.05.2013	91016	91351	91351	90710	90138	91551	90007
27.05.2013	90547	91016	91016	90710	89376	91026	89851
28.05.2013	89916	90547	90547	90710	91672	90740	89557
29.05.2013	87175	89916	89916	90710	88091	90092	87515
30.05.2013	87170	87175	87175	87008	85637	86997	86728
31.05.2013	85990	87170	87170	87008	85162	86876	86725
	RMSE	1361,600	1361,600	1450,300	1649,960	1511,600	647,300*
	MAPE	0,0116	0,0116	0,0108	0,0124	0,0131	0,0068*

The best RMSE and MAPE values are marked with '*'

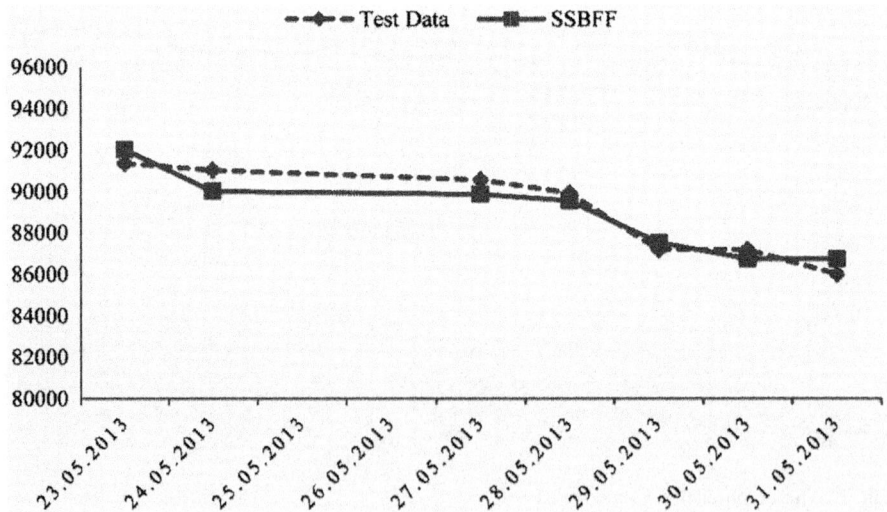

Fig. 10 The graph of the real observations together with the forecasts obtained from the SSBFF method for Set 5 time series

The forecasting performance of SSBFF approach is also examined visually. The graph of the real observations together with the forecasts obtained from the SSBFF method for Set 5 time series is given in Fig. 10.

It can be seen from the graphs of real observations together with forecasts, forecasts of the proposed method for all data sets are very close to the original observations. Although the result for Set 1 time series is the second best, forecasts of the proposed methods are good.

6 Conclusions

In this study, a subsampling stationary block bootstrap fuzzy functions approach which uses T1FF approach based on subsampling bootstrap method is proposed. And besides, the proposed approach is implemented to five different ISE time series. The performance of proposed method has been compared with some recent methods such as fuzzy functions approach and fuzzy time series methods available in the literature. As a result of the comparison, the proposed method obtained better forecasts compare to the other methods, for the time series that are used in this study except for Set 2 time series. By means of the proposed method, one can obtain forecast distribution, and forecasts can be obtained from the distribution of forecasts as a measure of central tendency, and combine many different forecast results.

References

Aladag, C.H., Yolcu, U., Egrioglu, E., Turksen, I.B.: Type-1 fuzzy time series function method based on binary particle swarm optimisation. Int. J. Data Anal. Tech. Strat. **8**(1), 2–13 (2016)

Berkowitz, J., Kilian, L.: Recent developments in bootstrapping time series. Econom. Rev. **19**, 1–48 (2000)

Beyhan, S., Alci, M.: Fuzzy functions based ARX model and new fuzzy basis function models for nonlinear system identification. Appl. Soft Comput. **10**, 439–444 (2010)

Bezdek, J.C.: Pattern Recognition with Fuzzy Objective Function Algorithms. Plenum Press, New York, USA (1981)

Bickel, P.J., Götze, F., Zwet, W.: Resampling fewer than n observations: Gains, loses and remedies for losses. Stat. Sinica **7**, 1–31 (1997)

Bose, A., Politis, D.N.: A review of the bootstrap for dependent samples. In: Bhat, B.R., Prakasa Rao, B.L.S. (eds.) Stochastic Processes and Statistical Inference, pp. 39–51. New Age International Publishers, New Delhi (1995)

Bühlmann, P.: Sieve bootstrap for time series. Bernoulli **3**, 123–148 (1997)

Carey, V.J.: Resampling methods for dependent data. J. Am. Statist. Assoc. **100**, 712–713 (2005)

Cavaliere, G., Politis, D.N., Rahbek, A.: Recent developments in bootstrap methods for dependent data. J. Time Ser. Anal. **36**(3), 269–271 (2015)

Celikyilmaz, A., Turksen, I.B.: Enhanced fuzzy system models with improved fuzzy clustering algorithm. IEEE Trans. Fuzzy Syst. **16**(3), 779–794 (2008a)

Celikyilmaz, A., Turksen, I.B.: Uncertainty modeling of improved fuzzy functions with evolutionary systems. IEEE Trans. Syst. Man Cybern. **38**(4), 1098–1110 (2008b)

Celikyilmaz, A., Turksen, I.B.: Modeling Uncertainty with Fuzzy Logic. Studies in Fuzziness and Soft Computing, pp. 240. Springer (2009)

Chernick, M.R.: Bootstrap Methods: a Guide for Practitioners and Researchers, 2nd edn. Wiley-Interscience, Hoboken, N.J. (2008)

Chernick, M.R., LaBuddle, R.A.: An Introduction to Bootstrap Methods with Applications to R. Wiley Publishing (2011)

Costa, M., Gonçalves, A.M., Silva, J.: Forecasting time series combining Holt-Winters and bootstrap approaches. In: AIP Conference Proceedings, vol. 1648, p. 110004 (2015)

Dalar, A.Z., Yolcu, U., Egrioglu, E., Aladag, C.H.: Forecasting turkey electric consumption by using fuzzy function approach. In: Proceeding ITISE 2015, International Work Conference on Time Series Analysis, Granada, Spain, July 1–3, vol. 543 (2015)

Davison, A.C., Hinkley, D.V.: Bootstrap Methods and their Application. Cambridge University Press, USA (1997)

Efron, B.: Bootstrap Methods: Another Look at the Jackknife. Ann. Stat. 7(1), 1–26 (1979)

Efron, B.: The Jackknife, the Bootstrap, and Other Resampling Plans. Society of Industrial and Applied Mathematics CBMS-NSF Monographs, vol. 38 (1982)

Efron, B., Tibshirani, R.: An Introduction to the Bootstrap. Chapman and Hall, New York (1993)

Good, P.I.: Resampling Methods: A Practical Guide to Data Analysis, 3rd edn. Birkhäuser (2005)

Hall, P.: On bootstrap confidence intervals in nonparametric regression. Ann. Statist. 20(2), 695–711 (1992a)

Hall, P.: The Bootstrap and Edgeworth Expansion. Springer, New York (1992b)

Hall, P., Jing, B.-Y.: On sample re-use methods for dependent data. J. R. Stat. Soc. Ser. B 58, 727–738 (1996)

Hwang, E., Shin, D.W.: A bootstrap test for jumps in financial economics. Econ. Lett. 125(1), 74–78 (2014)

Härdle, W., Horowitz, J., Kreiss, J.-P.: Bootstrap for time series. Int. Stat. Rev. 71, 435–459 (2003)

Jang, J.S.R.: ANFIS: Adaptive network based fuzzy inference system. IEEE Trans. Syst. Man Cybern. 23(3), 665–685 (1993)

Jin, S., Su, L., Ullah, A.: Robustify financial time series forecasting with bagging. Econom. Rev. 33, 575–605 (2013)

Kreiss, J.-P., Paparoditis, E.: Bootstrapping locally stationary time series. Technical Report (2011)

Kreiss, J.-P., Lahiri, S.N.: Bootstrap methods for time series. Time Ser. Anal. Methods Appl. 30, 3–26 (2012)

Künsch, H.R.: The jackknife and the bootstrap for general stationary observations. Ann. Stat. 17(3), 1217–1241 (1989)

Lahiri, S.N.: Resampling methods for dependent data. Springer, New York (2003)

Li, H., Maddala, G.S.: Bootstrapping time series models. Econom. Rev. 15, 115–158 (1996)

Liu, R.Y., Singh, K.: Moving blocks jackknife and bootstrap capture weak dependence. In: LePage, R., Billard, L. (eds.) Exploring the Limits of Bootstrap. Wiley, New York (1992)

Pan, L., Politis, D.N.: Bootstrap prediction intervals for linear, nonlinear and nonparametric autoregressions. J. Stat. Plann. Inference 177, 1–27 (2016)

Politis, D.N.: The impact of bootstrap methods on time series analysis. Stat. Sci. 18, 219–230 (2003)

Politis, D.N., Romano, J.P.: A circular block resampling procedure for stationary data. In: Lepage, R., Billard, L. (eds.) Exploring the Limits of Bootstrap, pp. 263–270. Wiley, New York (1992)

Politis, D.N., Romano, J.P.: The stationary bootstrap. J. Am. Stat. Assoc. 89, 1303–1313 (1994a)

Politis, D.N., Romano, J.P.: Large sample confidence regions based on subsamples under minimal assumptions. Ann. Stat. 22, 2031–2050 (1994b)

Politis, D.N., Romano, J.P., Wolf, M.: Subsampling. Springer, Berlin; New York (1999)

Shao, J., Tu, D.: The Jackknife and Bootstrap. Springer, New York (1995)

Song, Q., Chissom, B.S.: Forecasting enrollments with fuzzy time series-Part I. Fuzzy Sets Syst. 54, 1–10 (1993a)

Song, Q., Chissom, B.S.: Fuzzy time series and its models. Fuzzy Sets Syst. 54, 269–277 (1993b)

Turksen, I.B.: Fuzzy function with LSE. Appl. Soft Comput. 8, 1178–1188 (2008)

Turksen, I.B.: Fuzzy system models. Encyclopedia Complex. Syst. Sci., 4080–4094 (2009)

Zarandi, M.H.F., Zarinbal, M., Ghanbari, N., Turksen, I.B.: A new fuzzy functions model tuned by hybridizing imperialist competitive algorithm and simulated annealing. Application: Stock price prediction. Inf. Sci. **222**(10), 213–228 (2013)

A Weighted Ensemble Learning by SVM for Longitudinal Data: Turkish Bank Bankruptcy

Birsen Eygi Erdogan and Süreyya Özöğür Akyüz

Abstract Support Vector Machines (SVMs) are one of the most popular classification methods developed in recent years which have various application fields of research with diverse types of data sets. Longitudinal type of data is one of these types where a great deal of attention should be paid before applying SVMs. In this study, we modeled the decision maker with the idea of ensemble learning on longitudinal financial ratios to discriminate between weak and strong banks validated on Turkish commercial banks data where SVM is considered to be the base learner. We used the success status of the banks as the dependent variable and the financial ratios as independent variables. The results are compared in terms of the modeling performances and sensitivity measures which show the robustness of the model for finding positive instances, i.e. weak banks. The results show that ensemble learning performs better than a single learner. Moreover, we also validated that applying an appropriate normalization technique has strong effects on the performance of the learning step, especially when dealing with longitudinal data.

Keywords Bankruptcy · Kernel learning · Ensemble learning
Longitudinal data · Normalization · Support vector machines (SVM)

1 Introduction

One of the main goals of the classification techniques is to build a mathematical model which predicts the classes of the objects with a higher accuracy than a random guess. It has been shown that the decision of the community of learners

B. E. Erdogan (✉)
Department of Statistics, Marmara University, Istanbul, Turkey
e-mail: birsene@marmara.edu.tr

S. Ö. Akyüz
Faculty of Engineering and Natural Sciences, Bahçeşehir University, Istanbul, Turkey
e-mail: sureyya.akyuz@eng.bau.edu.tr

© Springer International Publishing AG 2018
M. Tez and D. von Rosen (eds.), *Trends and Perspectives in Linear Statistical Inference*, Contributions to Statistics,
https://doi.org/10.1007/978-3-319-73241-1_6

gives better results than a single one (Hansen and Salomon 1990; Utami et al. 2014). The community of learners, namely ensembles, can be constructed in different ways, such as bagging (bootstrap aggregating) or boosting which use a resampling approach to minimize the classification error.

The main idea of bagging is based on bootstrap and aggregation (Breiman 1996). In the bootstrap stage, a bootstrap sample is uniformly drawn with a replacement and the base learner (decision trees, neural networks, support vector machines (SVM) etc.) is trained for each learner. In the aggregation stage, a final classifier is built up using majority voting.

In boosting, the selection of the bootstrap samples is different where the sample is drawn based on the information obtained from the previous sample drawn. In a boosted sample, incorrectly predicted observations are chosen more often than the correctly predicted ones by former classifiers. This approach is especially useful for unbalanced data, where samples in one or more group of data are rare in the rest of the groups. Different boosting algorithms have been developed for binary classification problems subsequent to the most well-known boosting algorithm AdaBoost (Freund and Schapire 1997).

In this study, we have developed an ensemble learning model which aggregates the decision of multiple classifiers by a weighted combination. One of the novelties of the proposed model in this study is constructed on the idea of an ensemble of learners arising from the *annual* information of each bank. In other words, each classifier in the ensemble is trained for different time intervals so that the annual information is preserved by its decision maker for future prediction. As it is stated before that SVM is chosen as a base classifier when constructing the set of ensemble and the results of each classifier are aggregated by the proposed optimization model in Sect. 3.2

2 Bankruptcy Prediction

The banking sector is one the most important sectors to create more wealth and it has a major share in the financial field which needs to be tracked constantly. Sometimes the bankruptcy of a single bank may collapse the world's financial system. For this reason, besides the investors, ordinary people need to know the performance of the banks for their future plans. Therefore the prediction of the bankruptcy probability of a bank is a very important task.

Researchers from different disciplines have begun to focus their work on financial failure prediction models since the mid-1970s. In order to see the diversity and the development of the models made between 1970 and 2005 one can refer to Balcaen and Ooghe (2006). It is concluded that for the time period included in the study of Balcaen and Ooghe (2006), the researchers focused only on the classical statistical models (discriminant analysis, logit, probit, etc.) and it is stated that alternative methods (survival analysis, machine learning decision trees, expert system, and neural networks) must be compared with the classic statistical models. Hossari (2006) reviewed the latest studies focusing on corporate collapse over the period 1966–2004 where he concluded that in the last few decades beside the various statistical techniques several new methods have been proposed (ID3, rough sets analysis, tabu search, neural networks, etc.) on developing failure models to compare predictive power with those classical statistical models. Kumar and Ravi (2007) did comprehensive research with overlapping times between 1968 and 2005 and ended up almost in the same conclusion with Hossari stating that intelligent techniques such as neural networks, support vector machines and fuzzy logic can be utilized. In the study of Demyanyk and Hasan (2010) which reviews bank bankruptcy starting from 1968 up to date concludes that operations research techniques can be exploited in financial failure studies. For instance, in another bank bankruptcy study by Fethi and Pasiouras (2010) both operational research and artificial intelligence techniques are included over the period of 1998 to 2009.

There exist few types of research dealing with the prediction of financial failure focusing on the bankruptcy of Turkish banks. These studies are important for understanding the fatal consequences of the global financial crises in the case of Turkish financial failures, which were easily triggered by the global financial crises that peaked both in 2001 and in 2008. Important contributions to this literature include Canbas et al. (2005), Celik and Karatepe (2007), Boyacıoğlu et al. (2009), Erdogan (2013, 2016).

It can be seen from the literature mentioned above that in the last few decades artificial intelligence techniques have become more prevalent because of their advantages of not having strict theoretical statistical assumptions such as normality, linearity, homoscedasticity etc. Neural networks were the most preferred tools until a more advanced method SVMs have been reported as being superior among all. The advantages of SVMs are listed by Härdle and Moro (2004), Auria and Moro (2008) as well as by Vapnik (1998) who pioneered the construction of SVMs. In a comparison of Artificial Neural Network, the main advantages of SVMs are having a global solution and the opportunity of using a kernel function in case of nonlinear data sets.

In this study, our aim is to build a model that maps financial ratios (input) to the position of the bank (output variable) by labeling the output values either "failed" or "successful". The bank is defined to be *failed* if it is transferred to Savings and Deposits Insurance Fund (SDIF) which is a governmental body works for fund management and insurance in the Turkish banking system.

The next section summarizes theoretical aspects of SVMs and ensemble models which are followed by Sect. 4 with the details of the implementation of the proposed method and finally, Sect. 5 presents the conclusion and discussion.

3 Theoretical Background

3.1 Support Vector Machines

Support Vector Machines which have been proven for their success in classification problems have begun to be used in a variety of fields such as text categorization, face recognition, image processing, loan deficiency and bankruptcy prediction since the early 1990s.

SVM is a supervised learning which learns the rule of classification for a given input–output pairs (x_i, y_i). Mathematically speaking, it constructs a hyperplane $f = \langle w, x \rangle + b$ that maps the input (sample points), $x_i \in R^n$ to the output (classes/labels), $y_i \in \{+1, -1\}$ in case of a binary classification problem. Here, $\langle ., . \rangle$ stands for dot product, $w \in R^n$ refers to the normal vector of the hyperplane and b refers to the bias term, Cristianini and Shawe-Taylor (2000).

The hyperplane defined above is a linear decision surface which separates the class of objects as best as possible by using the idea of maximum margin principle, Cristianini and Shawe-Taylor (2000). The maximum margin principle is defined by finding the maximum distance between the class boundaries. This approach reduces the probability of misclassification, Hamel (2009). The solution of the problem referring to finding a separating hyperplane is sparse i.e. depends only on the training points that have heavier constraints on the class boundaries so-called *support vectors*.

Mathematically, finding the maximum margin is calculated by maximizing the distance between two supporting hyperplanes determined by a formula $\frac{2}{|w|}$. As the norm of w becomes smaller, the margin becomes larger. When fitting a model usually a regularization parameter is added to the model as a penalty term to avoid overfitting. In SVM this is achieved by using a regularization parameter C corresponding to the error term ξ_i. A large C produces a small margin and a small C produces a large margin. The objective function can be expressed as a minimization problem given by Eq. (1), Hamel (2009),

$$\min_{w, \xi} w \cdot w + C \sum_{i=1}^{l} \xi_i \qquad (1)$$

For a given training set, $D = \{(x_1, y_1), (x_2, y_2), \ldots, (x_l, y_l)\} \subseteq \mathbb{R}^n \times \{+1, -1\}$, the primal optimization problem is formulated with the objective function given by Eq. (1) and the constraints constructed by supporting hyperplanes written in Eq. (2) below

$$y_i(w \cdot x_i - b) + \xi_i - 1 \geq 0 \quad \text{for all } (x_i, y_i) \in D \text{ s.t. } y_i \in \{+1, -1\}, \qquad (2)$$

$\xi_i \geq 0, \ C > 0.$

By using the theory of constrained optimization, the minimization problem given by Eqs. (1) and (2) is equivalent to maximize the Lagrangian function in (3):

$$L(\boldsymbol{\alpha}, \mathbf{w}, \boldsymbol{\gamma}, \boldsymbol{\xi}, b) = \frac{1}{2} \mathbf{w} \cdot \mathbf{w} + C \sum_{i=1}^{l} \xi_i - \sum_{i=1}^{l} \alpha_i (y_i (\mathbf{w} \cdot \mathbf{x}_i - b) + \xi_i - 1) - \sum_{i=1}^{l} \gamma \xi_i \quad (3)$$

Therefore the primal problem of SVM is turned into the following dual problem

$$\max_{\boldsymbol{\alpha}, \boldsymbol{\gamma}} \min_{\mathbf{w}, \boldsymbol{\xi}, b} L(\boldsymbol{\alpha}, \mathbf{w}, \boldsymbol{\gamma}, \boldsymbol{\xi}, b),$$

$$\text{subject to } \alpha_i \geq 0, \gamma_i \geq 0, i = 1, \ldots, l \quad (4)$$

where $\boldsymbol{\alpha}$ and $\boldsymbol{\gamma}$ are $l \times 1$ vectors corresponding Lagrangian multipliers for the constraints.

Based on the requirements of Necessary and Optimality Conditions of the constrained optimization theory when Karush Kuhn Tucker (KKT) conditions are applied to the problem (4), the dual problem can be written as:

$$\alpha^* = arg \max_{\alpha} \left(\sum_{i=1}^{l} \alpha_i - \frac{1}{2} \sum_{i=1}^{l} \sum_{j=1}^{l} \alpha_i \alpha_j y_i y_j \mathbf{x}_i \cdot \mathbf{x}_j \right)$$

$$\text{subject to } \sum_{i=1}^{l} \alpha_i y_i = 0, \quad (5)$$

$$C \geq \alpha_i \geq 0 \ i = 1, \ldots, l.$$

The Solution of the dual problem above are the indicators of support vectors which allows to write the equation of optimum separating hyperplane with the normal vector below:

$$\mathbf{w}^* = \sum_{i \in SV} \alpha_i^* y_i x_i, \quad (6)$$

where the offset term b^* of the hyperplane is computed by:

$$b^* = \frac{1}{|N_{SV}|} \sum_{i \in SV} \left(y_i - \sum_{j=1}^{N} \alpha_j^* y_j x_i \cdot \mathbf{x}_j \right). \quad (7)$$

Here, SV represents the set of support vectors. By using Eqs. (6) and (7), the linear decision function can be written by:

$$\hat{f}(\mathbf{x}) = \text{sign} \left(\sum_{i \in SV} \alpha_i^* y_i x_i \cdot \mathbf{x} + b^* \right). \quad (8)$$

The dot product of the inputs in the Eq. (8) leads to a very nice property of SVMs, called the **kernel trick**. In real-world problems, data can not generally be separated linearly. In this case, these data points can be transformed into a higher dimensional feature space by a nonlinear mapping $\phi(\mathbf{x})$ where the new data points become linearly separable in this new space so-called *feature space*. The new data points $\phi(\mathbf{x})$ in the feature space can be infinite dimensional in some cases where the computational cost becomes large and finding the nonlinear mapping $\phi(\mathbf{x})$ explicitly can be difficult. Kernel functions associated with these transformations by dot products in the feature space can be easily calculated in the input space.

The kernel trick is formulated by a dot product between the two data points in feature space F where the data points are transformed from the input space X to the feature space F by a mapping function $\varphi\colon X \to F$. Then the dot product in the input space in Eq. (8) is replaced by the dot product in the feature space F.

A kernel function $K\colon X \times X \to \mathbb{R}$ which measures the similarity between two points is defined by a dot product: $K(\mathbf{x_i}, \mathbf{x_j}) = \varphi(\mathbf{x_i}) \cdot \varphi(\mathbf{x_j})$. If the dot product in Eq. (8) is replaced by $K(\mathbf{x_i}, \mathbf{x_j})$, w^* and b^* can be written as:

$$w^* \cdot \varphi(\mathbf{x}) = \sum_{i \in SV} \alpha_i^* y_i K(\mathbf{x}_i, \mathbf{x}), \tag{9}$$

$$b^* = \frac{1}{|N_{SV}|} \sum_{i \in SV} \left(y_i - \sum_{j=1}^{N} \alpha_j^* y_j K(\mathbf{x}_i, \mathbf{x}) \right), \tag{10}$$

and hence the classification rule becomes:

$$\hat{f}(\mathbf{x}) = \mathrm{sign}\left(w^* \cdot \varphi(\mathbf{x}) + b^* \right). \tag{11}$$

Among all of the proposed functions in the literature, the most common kernel functions used for SVMs are polynomial, Gaussian, and sigmoid functions (Motai 2015). In this study, all of the three kernels are tested and the best training accuracy is obtained by the combination of Gaussian kernel and the linear kernel. In the rest of the chapter, all calculations and the models are based on the Gaussian and the linear kernel functions which are given in the following formulas respectively:

Gaussian	$K(\mathbf{x_i}, \mathbf{x_j}) = \exp\left(-\frac{1}{2\sigma^2} \left| \mathbf{x_i^T} - \mathbf{x_j} \right|^2 \right)$; σ: kernel width
Linear	$K(\mathbf{x_i}, \mathbf{x_j}) = \mathbf{x_i^T} \mathbf{x_j}$

3.2 Weighted Ensemble Learning for Longitudinal Data

Ensemble learning is a method of combining decisions of different learning models which has been widely used in machine learning because of its performance in classification problems (Zhang et al. 2006). In the literature, it has been empirically shown that the decision of the community of classifiers gives better accuracy than the decision of a single classifier (Kuncheva 2004). As it is stated before, a number of methods have been proposed over the decades such as bagging (Breiman 1996) and boosting (Freund and Schapire 1996). It has not gone beyond the empirical validation of ensemble learners since there is not any theoretically consistent explanation in the literature. Despite these negative expressions, there exist three theories which explain the success of ensemble learners. The first theory is founded on large margin classifiers (Mason et al. 2000) where ensemble classifiers enlarge the margins and enhance the generalization performance of output coding (Allwein et al. 2000). The second theory is based on the bias-variance decomposition of the error where ensemble classifiers reduce the variance or both bias and variance (Breiman 2008; Kong and Dietterich 1995; Schapire 1999). The last theory is developed by a set theoretical point of view in which the classifiers are considered as a set of points to remove all algorithmic details of classifiers and training procedures (Kleinberg 1996, 2000).

The effectiveness of ensemble methods depends highly on the diversity and the accuracy of the learning models within the ensemble set. The set of ensembles can be generated in different ways such as including different learning algorithms, changing the parameters of the base learner or changing the features of the data set in order to satisfy the accuracy and diversity dilemma. Following the selection of the best candidate solutions (classifier models), the ensemble enters into the process of aggregation of these solutions. The aggregation of the various classifier functions is combined by so-called *consensus function* which can be written in two different ways; by using optimization models and cluster-based representation. Developing optimization models for the aggregation of classifiers is based on minimizing the error within the training error by means of a loss function. On the other hand, cluster-based representation is founded on graph theoretical approach which takes into account of the similarity of points defined by a similarity matrix in terms of so-called co-association matrices.

In this study, we implemented a consensus function by using a weighted convex combination for longitudinal data. Each classifier is trained by SVM within a predefined time interval and tested on the data of the consecutive years. The performance of each classifier is then recorded to be used as corresponding weights. The weight vector $\boldsymbol{\beta} = (\beta_1, \beta_2, \beta_3)^{\mathbf{T}}$ is normalized with respect to the Euclidean norm. The consensus function of the model is formulated by

$$F(x) = \sum_{i=1}^{3} \beta_i \hat{f}_i(x) \tag{12}$$

where $\hat{f}_i(x)$ is the classifier function that is obtained from each year. The sign of the consensus function which is tested on data of year 2001 will give the aggregated decision of each classifier.

As it is stated before, the selection of the best candidates of learners among the ensemble is another problem in this field of research (Özöğür-Akyüz 2015). In fact, we do not need such selection in our problem since the ensemble is constructed by different time intervals where each classifier represents the corresponding year and therefore none of them needs to be eliminated from the ensemble. The implementation of our ensemble model is introduced in the next section.

4   Implementation and Results

4.1   Datasets

As mentioned earlier, in this research, a longitudinal classification study will be carried out with the help of long-term financial ratios of the banks with respect to the success status of the banks. For this purpose, we have used annual financial data of the Turkish commercial banks. In this study, the data set includes 41 banks operating in Turkey which is collected from The Banks Association of Turkey (BAT) web site (http://www.tbb.org.tr/english/) (Banks Association of Turkey). *Banks in Turkey (1998–2002).*

According to the Banks Association of Turkey's records, 4 of 41 operating banks went bankrupt in 1998, 9 in 1999, 14 in 2000 and 20 in 2001. The banks which were not transferred to SDIF are defined as "successful". As a training set, we have used the longitudinal data set between 1998 and 2000. With the aim of making a priori bankruptcy prediction, the test set is chosen as the data belonging to the year 2001. Therefore, there are 27 failed banks in the training set and 20 failed banks in the test set. In this study, the financial status of the bank is considered as the dependent variable with the labels −1 for successful and 1 for the failed banks. Annual financial ratios are used as independent variables. For more explanation about the data set, one can refer to Erdogan (2013). The failed banks list of Erdogan (2013) study has been updated using the following idea: if a bank was taken by SDIF in the first six month its failure comes from the previous year. For example, Ulusal Bank was taken over by SDIF at February 28, 2001, so the bank is coded as failed since 2000.

4.2   Normalization

As it is discussed by Moeller (2015), there are some risks of normalization in longitudinal studies. To examine the possible risks closely, we considered the original pooled data without normalization where the results show that whichever method is used the process failed to find the positive instances. In our second approach, vertical normalization is performed by omitting the panel structure of the data which also failed to catch the banks that are bankrupted. Therefore we carried out double normalization in order to cope with this problem.

The first normalization is performed on the same bank block individually (for example Akbank data from 1998 to 2001 is scaled separately, the same thing is done for every bank) and the second normalization is performed within the pooled data to bring all the data to the same metric reflecting the panel structure. For the same bank block, we used the Proportion of Maximum Scaling ("POMS") method (Moeller 2015), which transforms each scale to a metric from 0 to 1, by first making the scale range from 0 to the highest value, and then dividing the scores by the highest value (x = (observed-minimum)/(maximum-minimum)). We did not prefer to use the standardizing (Z score) repeated measures within individuals since this approach prevents examining mean-level differences between individuals since each individual's mean score becomes zero. For the normalization within pooled data, the input vector $\mathbf{x}$ is divided with its norm $\|\mathbf{x}\|_2$.

4.3 Parameter Estimation and Classification Performance Measures

The power of the SVM technique depends on the selection of a suitable kernel and the kernel parameters. Moreover, it is well known that the choice of regularization parameter C which controls the misclassification error, affects the classification performance. If C is too large, we have a high penalty for non-separable points and we may store many support vectors which cause an overfitting. If C is too small, we may have an underfitting.

In order to find the best parameters, 5 fold cross validation is performed on the training set. The regularization term C and Gaussian kernel width σ are chosen among the values given by: $C = [2^{-5}, 2^{-3}, 2^{-1}, 2, 2^3, 2^5, 2^7, 2^9, 2^{11}, 2^{13}, 2^{15}]$; $\sigma = [2^{-5}, 2^{-13}, 2^{-11}, 2^{-9}, 2^{-7}, 2^{-5}, 2^{-3}, 2^{-1}, 2, 2^3]$. The rest of the parameters are left as default values of libsvm (Chang 2011).

In this study, for the evaluation of the models, several classification performance measures such as Correct Classification Rate (CCR), Sensitivity (SEN) and Specificity (SPE) are used. All these measures are defined by using a confusion matrix. A confusion matrix of a standard classification problem is given in Table 1.

Here, "*True Positive*" is defined by the number of the models that classify a failed bank as failed and "*False Positive*" is given by the number of the banks classified as failed when, in fact, they are non-failed. "*False Negative*" occurs when the model classifies a failed bank as non-failed, and "*True Negative*" is the number of the banks classified correctly as non-failed.

Table 1 The results for the train/test set

Observed	Predicted	
	+1	−1
+1	*True Positive*	*False Negative*
−1	*False Positive*	*True Negative*

Generally, the most frequently used metric for model performance is CCR. It is defined by the percentage of correct classifications among the whole test samples. Specificity reflects how good a model is at finding negative instances and it is defined by the True Negative divided by the sum of all negatives. Sensitivity is another important metric for model performance, and it is determined by the ratio of True Positive among of all positives. When using unbalanced data, the value of sensitivity plays an important role in determining the robustness of the model since the number of correct prediction of the observations that are rare in a class reflects how model accurately predicts. For this reason, in this study, we preferred to focus on sensitivity. Furthermore, with a skeptical approach, we aim to find a model with a high sensitivity and a relatively low specificity. With this approach, the model correctly finds the failed banks, but mistakenly labels some banks as "failed". These mistakenly labeled banks may be monitored in the future to see if they will go to the bankrupt, even though they are not bankrupt at the time of the test year.

4.4   Implementation of Proposed Weighted Ensemble Method

The performance of the proposed model in this study is tested on Turkish Bank data which covers the time interval between 1998 and 2001. The set of the ensemble is created by including the longitudinal data of years 1998, 1999, and 2000 per classifier and tested on the data of the year 2001. Therefore, 3 classifiers exist in our ensemble.

The classifier construction process has two stages. In the first stage, a cross validation is performed within corresponding training sets to select the best regularization parameter using the Gaussian kernel based SVMs. There are similar studies in the literature which use separate kernels to determine regularization and kernel parameters (Hastie et al. 2009; Keerthi and Lin 2003). Once the optimum parameter C is determined, in the second stage, each year is trained on its own year using the linear kernel and tested on the following consecutive years from which the performances on each test are recorded to be weights for the overall ensemble model. If there is more than one test year and hence their corresponding weight, the mean of weights are considered as a single weight referring to that classifier. For instance, the first classifier trained in 1998 has two weights obtained from the test performances during 1999 and 2000. Therefore the weight of a model for 1998 data is the mean of these two weights. The second model which is trained on 1999 data has only one weight coming from the test performance on 2000 data. The third classifier is trained on 70% of the year 2000 and tested on 30% of the year 2000. The reason for the partitioning the data of the year 2000 is that it is the last year before the data to be predicted. In this study, data of the year 2001 is used only for testing performance of our model and is not considered in any of the training set of any classifier.

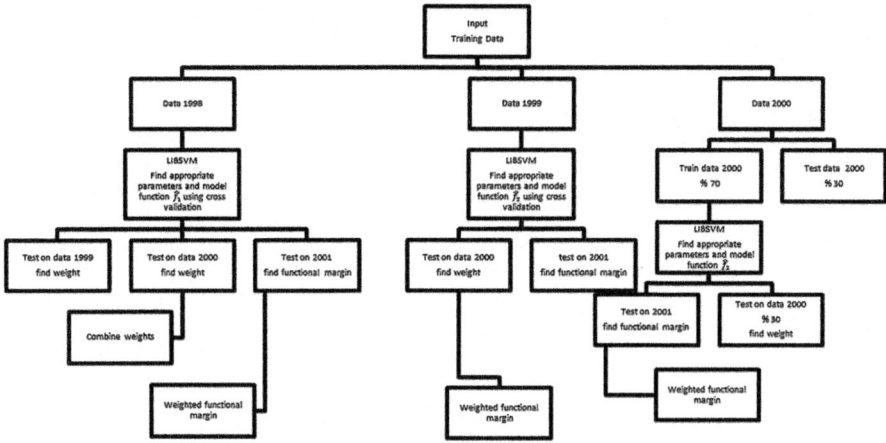

Fig. 1 Training and test algorithm flowchart

The overall testing for the year 2001 is determined by using the Eq. (12). The flow chart of the algorithm is given in Fig. 1.

We compared the performance of the proposed ensemble learning in this study with a single classifier constructed by SVMs with Gaussian and linear kernels separately in which a pooled 3-year information (1998–2000) is used for training purposes and tested in the year of 2001 data.

4.5   Results

Before evaluating the proposed ensemble method, we have implemented standard SVM procedures on the pooled data set to compare the results with the proposed method putting some emphasis on the effect of different standardization procedures. The performance measures are presented in Table 2. In Table 2, it is obvious that just by aggregation of the 3 years data in the training set, neither Gaussian nor linear kernel based SVMs give a better predictive results than the method proposed in this study. Correct classification rate (0.51) is not different from a random guess and sensitivity is zero.

Table 2 The prediction performance using simply pooled data without second round normalization (1998–2000 pooled training set, 2001 test set)

Kernel type	CCR	SPE	SEN
Gaussian	0.51	1.00	0.00
Linear	0.51	1.00	0.00

Table 3 The prediction performance using two step normalization in the training set (1998–2000 pooled training set, 2001 test set)

Kernel type	CCR	SPE	SEN
Gaussian	0.73	0.67	0.80
Linear	0.80	1.00	0.60

Table 4 The prediction performance using two step normalization in the training set (1998–2000 training set, 2001 test set) for the proposed weighted ensemble model

Kernel type	CCR	SPE	SEN
Gaussian	0.78	0.95	0.60
Linear	0.80	1.00	0.60
Gaussian and Linear	0.56	0.33	0.80

When we implemented a second round normalization, i.e. within pooled data, after POMS is used for bank blocks individually, better performance is achieved for both Gaussian and linear kernel based SVMs which are presented in Table 3. It should be noted that in Table 3, linear kernel based SVMs classify all the solid banks as "successful" while specification ended up with 1 and the sensitivity measures are more meaningful now for both kernel types.

Finally, we have constructed our proposed ensemble classifier using the method explained in Sect. 4.4 and we achieved the results given in Table 4.

It is interesting to see from Table 4 that linear kernel based weighted ensemble and from Table 3 linear kernel based pooled data models have given the same performance values. The models are good at finding successful banks but not very good at finding the failed banks. According to Table 4, the Gaussian-based weighted ensemble model is unsuccessful in the same sense.

When we have used a combination of different kernel functions for our weighted ensemble model we have managed to improve the sensitivity measure up to 0.80, revealing the weights as $\beta_1 = 0.26, \beta_2 = 0.20$, and $\beta_3 = 0.54$ corresponding to the years 1998, 1999 and 2000. This improvement means that we have managed to foresee several more banks which are going to go bankrupt before the failure time. Besides, in accordance with our skeptical approach, i.e., seeking high sensitivity and low specificity, the specificity was found as 0.33. Since it is more important for us to identify a bank that is likely to fail than to identify a successful bank, incorrectly identifying some banks as unsuccessful is not a mistake, but rather a chance to assess the future performance of the banks.

In the light of the findings, we have examined the banking activities of the banks which were labeled as unsuccessful for the following years. We have seen that among those who were incorrectly identified as "failed"; Imar Bank was taken over by SDIF in 2003, Pamukbank, Sekerbank were transferred to other state banks. Kocbank merged with Yapi Kredi Bank. Turk Dis Ticaret Bank, Turk Ekonomi Bank, and Garanti Bank were at least partly sold (BAT 2017).

(https://www.tbb.org.tr/Content/Upload/Dokuman/1362/Faaliyeti_Sona_Eren_
Bankalar.xls, 11.10.2017)

5 Conclusions

In this study, we have proposed a weighted ensemble approach which generates an ensemble of decision functions from annual data to analyze longitudinal bank financial ratios. A combination of Gaussian and linear kernels are used for our proposed weighted ensemble of SVMs. The results show that using a weighted panel structure of 3 years results in better performance on longitudinal data set to extract useful information when comparing the simply pooled data.

As a classification measure, we have focused on sensitivity which reflects the ability of the model of predicting failed banks, and hence robustness of the model. The results show that using a double normalization for the longitudinal data structure gives more reasonable sensitivity measure.

In this study, we have chosen SVMs as a base classifier and as a future work, including other classifiers would provide valuable information by extending our ensemble approach in new comparative ways.

Acknowledgements The research for this article was supported as a D-Type project (FEN-D-100616-0292) by Marmara University Scientific Research Projects Committee (BAPKO).

References

Allwein, E.L., Schapire, R.E., Singer, Y.: Reducing multiclass to binary: a unifying approach for margin classifiers. J. Mach. Learn. Res. **1**, 113–141 (2000)

Auria, L., Moro, R.A.: Support Vector Machines (SVM) as a Technique for Solvency Analysis, 1 Aug 2008. DIW Berlin Discussion Paper No. 811. Available at SSRN: https://ssrn.com/abstract=1424949 or http://dx.doi.org/10.2139/ssrn.1424949. Accessed 11 Oct 2017

Balcaen, S., Ooghe, H.: 35 years of studies on business failure: an overview of the classic statistical methodologies and their related problems. Br. Account. Rev. **38**(1), 63–93 (2006). https://doi.org/10.1016/j.bar.2005.09.001

Banks Association of Turkey: Banks in Turkey. https://www.tbb.org.tr/en/banks-and-banking-sector-information/statistical-reports/20. Accessed 11 Oct 2017

Boyacioglu, M.A., Kara, Y., Baykan, O.: Predicting bank financial failures using neural networks, support vector machines and multivariate statistical methods: a comparative analysis in the sample of savings deposit insurance fund (SDIF) transferred banks in Turkey. Expert Syst. Appl. **36**, 3355–3366 (2009)

Breiman, L.: Bagging predictors. Mach. Learn. **26**(2), 123–140 (1996), Arcing classifiers. Ann. Stat. **26**, 801–849

Breiman, L.: Arcing classifier (with discussion and a rejoinder by the author). Ann. Stat. **26**(3), 801–849 (2008)

Canbas, S., Cabuk, A., Kilic, S.B.: Prediction of commercial bank failure via multivariate statistical analysis of financial structures. Eur. J. Oper. Res. **166**, 528–546 (2005). https://doi.org/10.1016/j.ejor.2004.03.023

Celik, A.E., Karatepe, Y.: Evaluating and forecasting banking crises through neural network models: an application for Turkish banking sector. Expert Syst. Appl. **33**, 809–815 (2007)

Chang, C.-C., Lin, C.-J.: LIBSVM: a library for support vector machines. ACM Trans. Intell. Syst. Technol. **2**, 27:1–27:27 (2011). Software available at http://www.csie.ntu.edu.tw/~cjlin/libsvm

Cristianini, N., Shave-Taylor, J.: An Introduction to Support Vector Machines. Cambridge University Press, Cambridge (2000)

Demyanyk, Y., Hasan, I.: Financial crises and bank failures: a review of prediction methods. Omega-Int. J. Manag. Sci. **38**(5), 315–324 (2010). https://doi.org/10.1016/j.omega.2009.09.007

Erdogan, B.E.: Prediction of bankruptcy using support vector machines: an application to bank bankruptcy. J. Stat. Comput. Simul. **83**(8), (2013). https://doi.org/10.1080/00949655.2012.666550

Erdogan, B.E.: Long-term Examination of Bank Crashes Using Panel Logistic Regression: Turkish Banks Failure Case. Int. J. Stat. Prob. **5**(3), 42 (2016)

Fethi, M.D., Pasiouras, F.: Assessing bank efficiency and performance with operational research and artificial intelligence techniques: A survey. Eur. J. Oper. Res. **204**, 189–198 (2010)

Freund, Y., Schapire, R.E.: Experiments with a new boosting algorithm. In: Proceedings of the International Conference on Machine Learning (ICML), pp. 148–156 (1996). http://citeseerx.ist.psu.edu/viewdoc/download;jsessionid=6776E1594247795E2E9B1C4E4F5A534E?doi=10.1.1.29.3868&rep=rep1&type=pdf. Accessed 11 Oct 2017

Freund, Y., Schapire, R.: A decision theoretic generalization of on-line learning and an application to boosting. J. Comput. Syst. Sci. **55**, 119–139. MR1473055 (1997)

Hamel, L.: Knowledge Discovery with Support Vector Machines. Wiley, NJ (2009). ISBN 978-0-470-37192-3

Hansen, L.K., Salomon, P.: Neural network ensembles. IEEE Trans. Pattern Anal. Mach. Intell. **12**(10), 993–1001 (1990)

Hossari, G.: A Ratio-Based Multi-Level Modeling Approach for signaling Corporate Collapse: A Study of Australian Corporations, A Theses submitted to the fulfillment of the requirements for the degree of Doctor of Philosophy, Australian Graduate School of Entrepreneurship, Swinburne University of Technology (2006)

Hastie, T., Tibshirani, R., Friedman, J.: The Elements of Statistical Learning, Springer (2009)

Härdle, W.K., Moro, R.A., Schäfer, D.: Rating Companies with Support Vector Machines, DIW Discussion Paper No. 416, Berlin (2004)

Keerthi, S.S., Lin, C.J.: Asymptotic behaviors of support vector machines with Gaussian kernel. Neural Comput. **15**(7), 1667–1689 (2003)

Kleinberg, E.M.: An overtraining-resistant stochastic modeling method for pattern recognition. Ann. Stat. **4**, 2319–2349 (1996)

Kleinberg, E.M.: A mathematically rigorous foundation for supervised learning. In: Kittler, J., Roli, F. (eds.) Multiple Classifier Systems. First International Workshop, MCS 2000, Cagliari, Italy, vol. 1857 of Lecture Notes in Computer Science, pp. 67–76. Springer (2000)

Kong, E., Dietterich, T.: Error—correcting output coding correct bias and variance. In: The XII International Conference on Machine Learning, pp. 313–321, San Francisco, CA. Morgan Kauffman (1995)

Kumar, P.R., Ravi, V.: Bankruptcy prediction in banks and firms via statistical and intelligent techniques - A review. Eur. J. Oper. Res.—EJOR **180**(1), 1–28 (2007). https://doi.org/10.1016/j.ejor.2006.08.043

Kuncheva, L.I.: Combining Pattern Classifiers: Methods and Algorithms. Wiley-Interscience, New York (2004)

Mason, L., Bartlett, P., Baxter, J.: Improved generalization through explicit optimization of margins. Mach. Learn. **38**, 243–255 (2000)

Moeller, J.: A word on standardization in longitudinal studies: don't. Front. Psychol. **6**, 1389 (2015). https://doi.org/10.3389/fpsyg.2015.01389. Accessed 15 Sept 2015

Moguerza, J.M., Munoz, A.: Support vector machines with applications. Stat. Sci. **21**, 322–336 (2006)

Motai, Y.: Data Variant Kernel Analysis. Wiley, NJ, USA (2015)

Özöğür-Akyüz, S., Windeatt, T., Smith, R.: Pruning of error correcting output code by optimization of accuracy-diversity trade off. Mach. Learn. **101**(1), 253–269 (2015)

Schapire, R.E.: A brief introduction to boosting. In: Dean, T. (ed.) 16th International Joint Conference on Artificial Intelligence, pp. 1401–1406 (1999)

Utami, I.T., Sartono, B., Sadik, K.: Comparison of single and ensemble classifiers of support vector machine and classification tree. J. Math. Sci. Appl. **2**(2), 17–20 (2014)

Vapnik, V.N.: Statistical Learning Theory. Wiley, NJ (1998)

Zhang, Y., Burer, S., Street, W.N.: Ensemble pruning via semi-definite programming. J. Mach. Learn. Res. **7**, 1315–1338 (2006)

The Complementary Exponential Phase Type Distribution

Serkan Eryilmaz

Abstract In this chapter, the distribution which is called the complementary exponential phase type distribution is studied. This distribution appears as the distribution of the random maxima defined by $\max(X_1, X_2, ..., X_N)$, where $X_1, X_2, ...$ is a sequence of independent and identically distributed random variables having exponential distribution, and independently N has a discrete phase-type distribution. Bivariate extension of the distribution is also presented.

Keywords Bivariate distribution · Exponential distribution · Maximum likelihood estimation · Phase-type distributions

1 Introduction

Let $X_1, X_2, ...$ be a sequence of independent and identically distributed (iid) random variables having exponential distribution. Independently, let N have a geometric distribution. The distribution of the random variable $T = \max(X_1, X_2, ..., X_N)$ has been called a complementary exponential geometric distribution by Louzada et al. (2011). Such a random variable is useful in several fields including actuarial science, and reliability. As stated by Louzada et al. (2011), in the latent complementary risk scenario, the number of causes N and the lifetime X_i corresponding to a particular cause are not observable, and only the maximum lifetime value T among all causes is observed. If X_i denotes the lifetime of a component in a system, then T corresponds to the lifetime of a parallel system having random number of components (Eryilmaz 2017).

Louzada et al. (2013) studied the random variable T when the common distribution of $X_1, X_2, ...$ is exponentiated exponential and N has geometric distribution. In this chapter, we study the distribution of T when X_is have common exponential

S. Eryilmaz (✉)
Department of Industrial Engineering, Atilim University, Incek, 06836 Ankara, Turkey
e-mail: serkan.eryilmaz@atilim.edu.tr

© Springer International Publishing AG 2018   105
M. Tez and D. von Rosen (eds.), *Trends and Perspectives in Linear Statistical Inference*, Contributions to Statistics,
https://doi.org/10.1007/978-3-319-73241-1_7

distribution, and independently N has a discrete phase-type distribution. Assuming a phase-type distribution for N enables us to obtain a more general distribution than the one obtained by Louzada et al. (2011) since geometric distribution is the simplest phase-type distribution. Eryilmaz (2016) introduced a new class of lifetime distributions by considering the distribution of random minima $\min(X_1, ..., X_N)$ when N has a phase-type distribution. Therefore the present work can be seen as complementary to that of Eryilmaz (2016).

A discrete phase type distribution can be seen as the distribution of the time to absorption in an absorbing Markov chain with d transient states and one absorbing state. If a random variable N has a discrete phase-type distribution of order d, then its probability mass function (pmf) is given by

$$P\{N = n\} = \mathbf{a}\mathbf{Q}^{n-1}\mathbf{t}', \tag{1}$$

for $n \in \mathbb{N}$, where $\mathbf{Q} = \left(q_{ij}\right)_{d \times d}$ is a matrix which includes the transition probabilities among the d transient states, and $\mathbf{t}' = (\mathbf{I} - \mathbf{Q})\mathbf{e}'$ is a vector which includes the transition probabilities from transient states to the absorbing state, $\mathbf{a} = (a_1, ..., a_d)$ is the initial probability vector with the entry corresponding to the absorption state removed, $\sum_{i=1}^{d} a_i = 1$, $\mathbf{I}$ is the identity matrix, and $\mathbf{e} = (1, ..., 1)_{1 \times d}$ (see, e.g. He (2014)). We will use the notation $PH_d(\mathbf{a}, \mathbf{Q})$ to represent a discrete phase-type distribution of order d which has pmf given by (1).

The simplest discrete phase-type distribution is geometric distribution which can be defined by $PH_1(1, 1 - p)$, where p is the parameter of the geometric distribution.

If $N \sim PH_d(\mathbf{a}, \mathbf{Q})$, then the probability generating function of N is given by

$$\phi_N(z) = 1 - \mathbf{a}\mathbf{e}' + \mathbf{a}z\left(\mathbf{I} - z\mathbf{Q}\right)^{-1}\mathbf{t}', \tag{2}$$

$0 < z < 1$ (see, e.g. He (2014)).

The chapter is organized as follows. In Sect. 2, the new distribution is introduced and some of its properties are presented. Section 3 contains extension of the new distribution to the bivariate case.

2   The Distribution and Its Properties

For a sequence $X_1, X_2, ...$ of independent and identically distributed (iid) random variables having common cumulative distribution function (cdf) $F(x) = 1 - e^{-\lambda x}, x \geq 0$, define

$$T = \max(X_1, X_2, ..., X_N),$$

where N is independent of X_is and $N \sim PH_d(\mathbf{a}, \mathbf{Q})$.

The cdf of the random variable T can be written as

$$
\begin{aligned}
G(x) &= P\left\{\max(X_1, X_2, ..., X_N) \le x\right\} \\
&= \sum_n F^n(x)P\{N = n\} \\
&= E(F^N(x)) \\
&= \phi_N(F(x))
\end{aligned}
$$

which is, in fact, the probability generating function of N at $F(x)$. Thus, from (2), the cdf of the random variable T is obtained as

$$
\begin{aligned}
G(x) &= 1 - \mathbf{a}e' + \mathbf{a}F(x)\left(\mathbf{I} - F(x)\mathbf{Q}\right)^{-1}\mathbf{t}' \\
&= 1 - \mathbf{a}e' + \mathbf{a}(1 - e^{-\lambda x})\left(\mathbf{I} - (1 - e^{-\lambda x})\mathbf{Q}\right)^{-1}\mathbf{t}'
\end{aligned}
\tag{3}
$$

The random variable whose cdf is given by (3) will be said to have a complementary exponential phase-type distribution. From (3), the probability density function of T is found to be

$$
\begin{aligned}
g(x) &= f(x)\mathbf{a}\left(\mathbf{I} - F(x)\mathbf{Q}\right)^{-2}\mathbf{t}' \\
&= \lambda e^{-\lambda x}\mathbf{a}\left(\mathbf{I} - (1 - e^{-\lambda x})\mathbf{Q}\right)^{-2}\mathbf{t}'.
\end{aligned}
\tag{4}
$$

From (3), various distributions can be obtained for a phase random variable having representation $PH_d(\mathbf{a}, \mathbf{Q})$. For an illustration, let N have a geometric distribution of order k which has phase representation $PH_k(\mathbf{a}, \mathbf{Q})$ with $\mathbf{a} = (1, 0, ..., 0)$, $\mathbf{t} = (0, 0, ..., p)$, and the $k \times k$ matrix

$$
\mathbf{Q} = \begin{bmatrix}
1 - p & p & 0 & \cdots & 0 \\
1 - p & 0 & p & \cdots & 0 \\
\vdots & \vdots & \vdots & \ddots & \vdots \\
1 - p & 0 & 0 & \cdots & 0
\end{bmatrix}
$$

(see, e.g. Tank and Eryilmaz (2015)). It should be noted that the random variable N has a geometric distribution of order k if it denotes the number of trials to get k consecutive successes in a sequence of binary trials having two possible outcomes as either success or failure with respective probabilities p and $1 - p$. For $k = 1$, the random variable N has a geometric distribution. In this case from (4), the pdf of the distribution is obtained as

$$
g(x) = \frac{p\lambda e^{-\lambda x}}{\left[1 - (1 - p)(1 - e^{-\lambda x})\right]^2}
\tag{5}
$$

which has been introduced and studied by Louzada et al. (2011). Let $k = 2$, then from (4) we obtain the following new distribution

Fig. 1 Plots of the pdf of
the distribution given by (6)

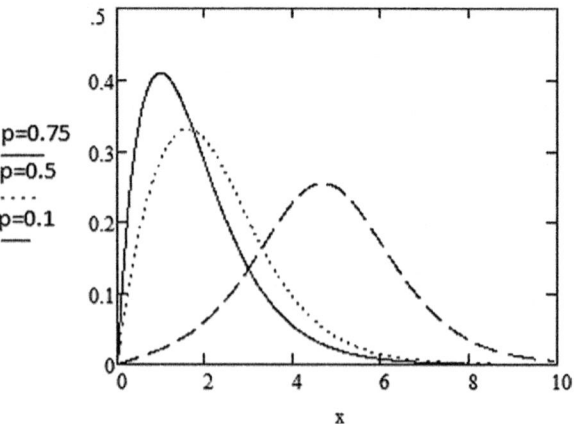

$$g(x) = \frac{\lambda e^{-\lambda x}(1 - e^{-\lambda x})p^2 \left[1 + e^{-\lambda x} + p(1 - e^{-\lambda x})\right]}{\left[1 - (1 - e^{-\lambda x}) + p(1 - e^{-\lambda x}) - p(1 - e^{-\lambda x})^2 + p^2(1 - e^{-\lambda x})^2\right]^2}, \quad (6)$$

for $0 < p < 1$, and $x \geq 0$.

In Fig. 1, we plot the pdf given by (6) for selected values of the parameter p when $\lambda = 1$. The larger p the more right skewed distribution. The distribution becomes symmetric when p tends to zero. Such a flexible distribution might be useful to model various real life data sets.

2.1 Inference

Suppose $x_1, x_2, ..., x_n$ is a random sample from the distribution with pdf (4). Assume that the phase-type random variable N has only one unknown parameter, say p. Then the log-likelihood function of the two parameters is given by

$$l(\lambda, p) = n \log(\lambda) - \lambda \sum_{i=1}^{n} x_i + \sum_{i=1}^{n} \log(\mathbf{a} \left(\mathbf{I} - (1 - e^{-\lambda x_i})\mathbf{Q}\right)^{-2} \mathbf{t}'). \quad (7)$$

The maximum likelihood estimators of the parameters p and λ can be obtained by solving

$$\frac{\partial l(\lambda, p)}{\partial \lambda} = 0 \,, \quad \frac{\partial l(\lambda, p)}{\partial p} = 0.$$

The maximum likelihood estimators of p and λ must be derived numerically. The EM algorithm is suitable for this purpose. EM algorithm based estimators of p and λ can be obtained following the similar steps in Eryilmaz (2016).

3 Extension to the Bivariate Case

For a sequence $(X_1, Y_1), (X_2, Y_2), \ldots$ of iid random vectors having common bivariate cdf $F(x, y) = P\{X_i \le x, Y_i \le y\}, i \ge 1$, define

$$T_1 = \max(X_1, X_2, \ldots, X_N),$$
$$T_2 = \max(Y_1, Y_2, \ldots, Y_N),$$

where N is independent of (X_i, Y_i)s and $N \sim PH_d(\mathbf{a}, \mathbf{Q})$. Manifestly, the random variables T_1 and T_2 are dependent.

The joint cdf of the random variables T_1 and T_2 is obtained as

$$
\begin{aligned}
H(x, y) &= P\{\max(X_1, X_2, \ldots, X_N) \le x, \max(Y_1, Y_2, \ldots, Y_N) \le y\} \\
&= \sum_n F^n(x, y) P\{N = n\} \\
&= \phi_N(F(x, y)) \\
&= 1 - \mathbf{a}\mathbf{e}' + \mathbf{a}F(x, y)(\mathbf{I} - F(x, y)\mathbf{Q})^{-1}\mathbf{t}'.
\end{aligned}
\tag{8}
$$

For particular choices of $F(x, y)$, various bivariate distributions can be generated from (8). Let

$$F(x, y) = (1 - e^{-\lambda x})(1 - e^{-\theta y})\left[1 + \alpha e^{-\lambda x} e^{-\theta y}\right], \tag{9}$$

for $x, y \ge 0$ and $-1 \le \alpha \le 1$. That is, the joint distribution of X_i and Y_i is bivariate exponential of FGM type.

For an illustration, let N have a geometric distribution, i.e. $N \sim PH_1(1, 1 - p)$. Then from (8) and (9),

$$
\begin{aligned}
H(x, y) &= \frac{pF(x, y)}{1 - (1 - p)F(x, y)} \\
&= \frac{p(1 - e^{-\lambda x})(1 - e^{-\theta y})\left[1 + \alpha e^{-\lambda x} e^{-\theta y}\right]}{1 - (1 - p)(1 - e^{-\lambda x})(1 - e^{-\theta y})\left[1 + \alpha e^{-\lambda x} e^{-\theta y}\right]},
\end{aligned}
\tag{10}
$$

for $0 < p < 1; -1 \le \alpha \le 1; x, y \ge 0$. The marginals of $H(x, y)$ are both have complementary exponential geometric distributions.

Clearly, inferential issues become more complicated for the bivariate case. However, it is worthy of investigation. Estimation of unknown parameters p, α, λ, and θ which are included in the model (10) will be among our future research problems.

References

Eryilmaz, S.: A new class of lifetime distributions. Stat. Probab. Lett. **112**, 63–71 (2016)

Eryilmaz, S.: A note on optimization problems of a parallel system with a random number of units. Int. J. Reliab. Qual. Saf. Eng. **24**(1750022–1), 1750022–9 (2017)

He, Q.-M.: Fundamentals of Matrix-Analytic Methods. Springer, NY (2014)

Louzada, F., Roman, M., Cancho, V.G.: The complementary exponential geometric distribution: model, properties, and a comparison with its counterpart. Comput. Stat. Data Anal. **55**, 2516–2524 (2011)

Louzada, F., Marchi, V., Carpenter, J.: The complementary exponentiated exponential geometric lifetime distribution. J. Probab. Stat. 2013, Article ID: 502159 (2013)

Tank, F., Eryilmaz, S.: The distributions of sum, minima and maxima of generalized geometric random variables. Stat. Pap. **56**, 1191–1203 (2015)

Some Properties of Linear Prediction Sufficiency in the Linear Model

Jarkko Isotalo, Augustyn Markiewicz and Simo Puntanen

Abstract A linear statistic $\mathbf{Fy}$ is called linearly prediction sufficient, or shortly BLUP-sufficient, for the new observation $\mathbf{y}_*$, say, if there exists a matrix $\mathbf{A}$ such that $\mathbf{AFy}$ is the best linear unbiased predictor, BLUP, for $\mathbf{y}_*$. We review some properties of linear prediction sufficiency that have not been received much attention in the literature and provide some clarifying comments. In particular, we consider the best linear unbiased prediction of the error term related to $\mathbf{y}_*$. We also explore some interesting properties of mixed linear models including the connection between a particular extended linear model and its transformed version.

Keywords Best linear unbiased estimator · BLUE · Best linear unbiased predictor · BLUP · Linear sufficiency · Orthogonal projector · Transformed linear model

MSC: 62J05 · 62J10

1 Introduction

To make the article more self-readable, we go through some basic concepts related to linear sufficiency. So, let us get started with the general linear model $\mathbf{y} = \mathbf{X}\boldsymbol{\beta} + \boldsymbol{\varepsilon}$, shortly denoted as a triplet

$$\mathcal{M} = \{\mathbf{y}, \mathbf{X}\boldsymbol{\beta}, \mathbf{V}\},$$

J. Isotalo
Department of Forest Sciences, University of Helsinki, P.O. Box 27, 00014 Helsinki, Finland
e-mail: jarkko.isotalo@helsinki.fi

A. Markiewicz
Department of Mathematical and Statistical Methods, Poznań University of Life Sciences,
Wojska Polskiego 28, 60637 Poznań, Poland
e-mail: amark@up.poznan.pl

S. Puntanen (✉)
Faculty of Natural Sciences, University of Tampere, 33014, Tampere, Finland
e-mail: simo.puntanen@uta.fi

© Springer International Publishing AG 2018
M. Tez and D. von Rosen (eds.), *Trends and Perspectives in Linear Statistical Inference*, Contributions to Statistics,
https://doi.org/10.1007/978-3-319-73241-1_8

111

where $\mathbf{X}_{n\times p}$ is a known model matrix, the vector $\mathbf{y}$ is an observable n-dimensional random vector, $\boldsymbol{\beta}$ is a $p \times 1$ vector of unknown parameters, and $\boldsymbol{\varepsilon}$ is an unobservable vector of random errors with expectation $E(\boldsymbol{\varepsilon}) = \mathbf{0}$, and covariance matrix $cov(\boldsymbol{\varepsilon}) = \mathbf{V}$, where the nonnegative definite matrix $\mathbf{V}$ is known and can be singular. Premultiplying the model $\mathcal{M}$ by $\mathbf{F}_{f\times n}$ yields the transformed model

$$\mathcal{M}_t = \{\mathbf{Fy}, \mathbf{FX}\boldsymbol{\beta}, \mathbf{FVF'}\},$$

which will have a crucial role in our considerations.

Let $\mathbf{y}_*$ denote a $q \times 1$ unobservable random vector containing new future observations. The new observations are assumed to follow the linear model

$$\mathbf{y}_* = \mathbf{X}_*\boldsymbol{\beta} + \boldsymbol{\varepsilon}_*,$$

where $\mathbf{X}_*$ is a known $q \times p$ matrix, $\boldsymbol{\beta}$ is the same vector of unknown parameters as in $\mathcal{M}$, and $\boldsymbol{\varepsilon}_*$ is a q-dimensional random error vector. The expectation and the covariance matrix are

$$E\begin{pmatrix}\mathbf{y}\\\mathbf{y}_*\end{pmatrix} = \begin{pmatrix}\mathbf{X}\boldsymbol{\beta}\\\mathbf{X}_*\boldsymbol{\beta}\end{pmatrix} = \begin{pmatrix}\mathbf{X}\\\mathbf{X}_*\end{pmatrix}\boldsymbol{\beta}, \quad cov\begin{pmatrix}\mathbf{y}\\\mathbf{y}_*\end{pmatrix} = \begin{pmatrix}\mathbf{V} & \mathbf{V}_{12}\\\mathbf{V}_{21} & \mathbf{V}_{22}\end{pmatrix} = \boldsymbol{\Gamma},$$

where the covariance matrix matrix $\boldsymbol{\Gamma}$ is assumed to be known. For brevity, we denote the linear model with new observations as

$$\mathcal{M}_* = \left\{\begin{pmatrix}\mathbf{y}\\\mathbf{y}_*\end{pmatrix}, \begin{pmatrix}\mathbf{X}\\\mathbf{X}_*\end{pmatrix}\boldsymbol{\beta}, \begin{pmatrix}\mathbf{V} & \mathbf{V}_{12}\\\mathbf{V}_{21} & \mathbf{V}_{22}\end{pmatrix}\right\}.$$

Our main interest in $\mathcal{M}_*$ lies in predicting $\mathbf{y}_*$ on the basis of observable $\mathbf{y}$.

Suppose we transform $\mathcal{M}$ into $\mathcal{M}_t$ and do the prediction in this situation. Corresponding to $\mathcal{M}_*$ we have now the following setup:

$$\mathcal{M}_{t*} = \left\{\begin{pmatrix}\mathbf{Fy}\\\mathbf{y}_*\end{pmatrix}, \begin{pmatrix}\mathbf{FX}\\\mathbf{X}_*\end{pmatrix}\boldsymbol{\beta}, \begin{pmatrix}\mathbf{FVF'} & \mathbf{FV}_{12}\\\mathbf{V}_{21}\mathbf{F'} & \mathbf{V}_{22}\end{pmatrix}\right\}.$$

As for notation, let $\mathbb{R}^{m\times n}$ denote the set of $m \times n$ real matrices. The symbols $\mathbf{A'}$, $\mathbf{A}^-$, $\mathbf{A}^+$, $\mathscr{C}(\mathbf{A})$, and $\mathscr{C}(\mathbf{A})^\perp$, denote, respectively, the transpose, a generalized inverse, the Moore–Penrose inverse, the column space, and the orthogonal complement of the column space of the matrix $\mathbf{A}$. By $(\mathbf{A} : \mathbf{B})$ we denote the partitioned matrix with $\mathbf{A}_{m\times n}$ and $\mathbf{B}_{m\times k}$ as submatrices. By $\mathbf{A}^\perp$ we denote any matrix satisfying $\mathscr{C}(\mathbf{A}^\perp) = \mathscr{C}(\mathbf{A})^\perp$. Furthermore, we will write $\mathbf{P}_\mathbf{A} = \mathbf{AA}^+ = \mathbf{A}(\mathbf{A'A})^-\mathbf{A'}$ to denote the orthogonal projector (with respect to the standard inner product) onto $\mathscr{C}(\mathbf{A})$, and $\mathbf{Q}_\mathbf{A} = \mathbf{I} - \mathbf{P}_\mathbf{A}$. In particular, we denote $\mathbf{M} = \mathbf{I}_n - \mathbf{P}_\mathbf{X}$. One choice for $\mathbf{X}^\perp$ is of course $\mathbf{M}$.

The linear estimator $\mathbf{Gy}$ is the best linear unbiased estimator, BLUE, of $\mathbf{X}\boldsymbol{\beta}$ whenever $\mathbf{Gy}$ is unbiased and it has the smallest covariance matrix (in the Löwner sense) among all linear unbiased estimators of $\mathbf{X}\beta$. The following lemma characterizes the BLUE; see, e.g., Drygas (1970, p. 55), Rao (1973, p. 282) and more recently Baksalary and Trenkler (2009).

Lemma 1.1 *Consider the general linear model* $\mathcal{M} = \{\mathbf{y}, \mathbf{X}\beta, \mathbf{V}\}$*. Then, the estimator* $\mathbf{Gy}$ *is the* BLUE *for* $\mathbf{X}\beta$ *if and only if* $\mathbf{G}$ *satisfies the equation*

$$\mathbf{G}(\mathbf{X} : \mathbf{VX}^{\perp}) = (\mathbf{X} : \mathbf{0}). \tag{1.1}$$

The corresponding condition for $\mathbf{Ay}$ *to be the* BLUE *of an estimable parametric function* $\mathbf{K}\beta$*, i.e.,* $\mathscr{C}(\mathbf{K}') \subset \mathscr{C}(\mathbf{X}')$*, is*

$$\mathbf{A}(\mathbf{X} : \mathbf{VX}^{\perp}) = (\mathbf{K} : \mathbf{0}).$$

We assume the model $\mathcal{M}$ to be consistent in the sense that the observed value of $\mathbf{y}$ lies in $\mathscr{C}(\mathbf{X} : \mathbf{V})$ with probability 1. Hence, we assume that under $\mathcal{M}$

$$\mathbf{y} \in \mathscr{C}(\mathbf{X} : \mathbf{V}) = \mathscr{C}(\mathbf{X} : \mathbf{VX}^{\perp}) = \mathscr{C}(\mathbf{X} : \mathbf{VM}).$$

The corresponding consistency is assumed in all models that we will consider. Moreover, in the consistent linear model $\mathcal{M}$, the estimators $\mathbf{G}_1\mathbf{y}$ and $\mathbf{G}_2\mathbf{y}$ are said to be equal with probability 1 if

$$\mathbf{G}_1\mathbf{y} = \mathbf{G}_2\mathbf{y} \quad \text{for all } \mathbf{y} \in \mathscr{C}(\mathbf{X} : \mathbf{V}).$$

The linear predictor $\mathbf{By}$ is said to be unbiased for $\mathbf{y}_*$ if $\mathrm{E}(\mathbf{y}_* - \mathbf{By}) = \mathbf{0}$ for all $\beta \in \mathbb{R}^p$. This is equivalent to $\mathbf{X}'_* = \mathbf{X}'\mathbf{B}'$. The inclusion $\mathscr{C}(\mathbf{X}'_*) \subset \mathscr{C}(\mathbf{X}')$ is the well-known condition for the estimability of $\mathbf{X}_*\beta$ under $\mathcal{M}$. When $\mathscr{C}(\mathbf{X}'_*) \subset \mathscr{C}(\mathbf{X}')$ holds, we say that $\mathbf{y}_*$ is predictable under $\mathcal{M}_*$. Now a linear unbiased predictor $\mathbf{By}$ is the best linear unbiased predictor, BLUP, for $\mathbf{y}_*$, if the Löwner ordering

$$\mathrm{cov}(\mathbf{y}_* - \mathbf{By}) \leq_{\mathrm{L}} \mathrm{cov}(\mathbf{y}_* - \mathbf{Cy})$$

holds for all $\mathbf{C}$ such that $\mathbf{Cy}$ is an unbiased linear predictor for $\mathbf{y}_*$.

The following lemma characterizes the BLUP; for the proof, see, e.g., Christensen (2011, p. 294) and Isotalo and Puntanen (2006, p. 1015).

Lemma 1.2 *Consider the linear model* $\mathcal{M}_*$*, where* $\mathscr{C}(\mathbf{X}'_*) \subset \mathscr{C}(\mathbf{X}')$*, i.e.,* $\mathbf{y}_*$ *is predictable. The linear predictor* $\mathbf{By}$ *is the best linear unbiased predictor* (BLUP) *for* $\mathbf{y}_*$ *if and only if* $\mathbf{B}$ *satisfies the equation*

$$\mathbf{B}(\mathbf{X} : \mathbf{VX}^{\perp}) = (\mathbf{X}_* : \mathbf{V}_{21}\mathbf{X}^{\perp}) = (\mathbf{X}_* : \mathrm{cov}(\mathbf{y}_*, \mathbf{y})\mathbf{X}^{\perp}).$$

We will frequently utilize Lemma 2.2.4 of Rao and Mitra (1971), which says that for nonnull matrices $\mathbf{A}$ and $\mathbf{C}$ the following holds:

$$\mathbf{A}\mathbf{B}^-\mathbf{C} = \mathbf{A}\mathbf{B}^+\mathbf{C} \iff \mathscr{C}(\mathbf{C}) \subset \mathscr{C}(\mathbf{B}) \ \& \ \mathscr{C}(\mathbf{A}') \subset \mathscr{C}(\mathbf{B}'). \qquad (1.2)$$

One well-known solution for $\mathbf{G}$ in (1.1) (which is always solvable) is

$$\mathbf{P}_{\mathbf{X};\mathbf{W}^-} := \mathbf{X}(\mathbf{X}'\mathbf{W}^-\mathbf{X})^-\mathbf{X}'\mathbf{W}^-,$$

where $\mathbf{W}$ is a matrix belonging to the set of nonnegative definite matrices defined as

$$\mathscr{W} = \left\{ \mathbf{W} \in \mathbb{R}^{n \times n} : \mathbf{W} = \mathbf{V} + \mathbf{X}\mathbf{U}\mathbf{U}'\mathbf{X}', \ \mathscr{C}(\mathbf{W}) = \mathscr{C}(\mathbf{X} : \mathbf{V}) \right\}.$$

Denoting

$$\mathbf{P}_{\mathbf{X};\mathbf{W}^+} := \mathbf{X}(\mathbf{X}'\mathbf{W}^-\mathbf{X})^-\mathbf{X}'\mathbf{W}^+,$$

we observe, in view of (1.2), that $\mathbf{P}_{\mathbf{X};\mathbf{W}^-}\mathbf{y} = \mathbf{P}_{\mathbf{X};\mathbf{W}^+}\mathbf{y}$ for all $\mathbf{y} \in \mathscr{C}(\mathbf{W})$.

The structure of this contribution is as follows. In Sect. 2 we recall some well-known conditions for the BLUE- and BLUP-sufficiency and in particular clarify and extend some concepts related to BLUP-sufficiency. In Sect. 3 we introduce some representations for the BLUPs and explore the corresponding sufficiency relations. Section 4 provides some representations for the BLUPs and BLUEs and in Sect. 5 we apply our results to the linear mixed models. While writing this contribution, our attempt has been to call well-known (or pretty well-known) results Lemmas, while Theorems refer to our own contributions or clarifications.

2 Conditions for Linear Sufficiency and Linear Prediction Sufficiency

A linear statistic $\mathbf{F}\mathbf{y}$, where $\mathbf{F} \in \mathbb{R}^{f \times n}$, is called linearly sufficient for $\mathbf{X}\beta$ under the model $\mathscr{M} = \{\mathbf{y}, \mathbf{X}\beta, \mathbf{V}\}$, if there exists a matrix $\mathbf{A} \in \mathbb{R}^{n \times f}$ such that $\mathbf{A}\mathbf{F}\mathbf{y}$ is the BLUE for $\mathbf{X}\beta$. Correspondingly, $\mathbf{F}\mathbf{y}$ is linearly sufficient for estimable $\mathbf{K}\beta$, where $\mathbf{K} \in \mathbb{R}^{k \times p}$, if there exists a matrix $\mathbf{A} \in \mathbb{R}^{k \times f}$ such that $\mathbf{A}\mathbf{F}\mathbf{y}$ is the BLUE for $\mathbf{K}\beta$. To have a slightly shorter terminology, we often will use the phrase "BLUE-sufficient" and the notation $\mathbf{F}\mathbf{y} \in \mathscr{S}(\mathbf{K}\beta)$.

For the following Lemma 2.1 and Lemma 2.2, see, e.g., Baksalary and Kala (1981, 1986), Drygas (1983), Tian and Puntanen (2009, Th. 2.8), and Kala et al. (2017, Th. 2).

Lemma 2.1 *The statistic* $\mathbf{F}\mathbf{y}$ *is BLUE-sufficient for* $\mathbf{X}\beta$ *under the model* $\mathscr{M} = \{\mathbf{y}, \mathbf{X}\beta, \mathbf{V}\}$ *if and only if any of the following equivalent statements holds:*

(a) $\mathscr{C}\begin{pmatrix} \mathbf{X}' \\ \mathbf{0} \end{pmatrix} \subset \mathscr{C}\begin{pmatrix} \mathbf{X}'\mathbf{F}' \\ \mathbf{M}\mathbf{V}\mathbf{F}' \end{pmatrix}$,

(b) $\mathscr{C}(\mathbf{X}) \subset \mathscr{C}(\mathbf{W}\mathbf{F}')$, *where* $\mathbf{W} \in \mathscr{W}$,

(c) $\mathscr{C}(\mathbf{X}'\mathbf{F}') = \mathscr{C}(\mathbf{X}')$ *and* $\mathscr{C}(\mathbf{F}\mathbf{X}) \cap \mathscr{C}(\mathbf{F}\mathbf{V}\mathbf{X}^\perp) = \{\mathbf{0}\}$.

Let $\mathbf{K}\beta$ *be estimable under* $\mathcal{M}$. *Then,* $\mathbf{Fy}$ *is* BLUE-*sufficient for* $\mathbf{K}\beta$ *if and only if*

(d) $\mathscr{C}\begin{pmatrix} \mathbf{K}' \\ \mathbf{0} \end{pmatrix} \subset \mathscr{C}\begin{pmatrix} \mathbf{X}'\mathbf{F}' \\ \mathbf{MVF}' \end{pmatrix}$.

Let $\mathbf{F}_0$ be a matrix with property $\mathscr{C}(\mathbf{F}_0') = \mathscr{C}(\mathbf{F}')$. Then, Lemma 2.1 immediately implies the following:

$$\mathbf{Fy} \in \mathscr{S}(\mathbf{K}\beta) \iff \mathbf{F}_0\mathbf{y} \in \mathscr{S}(\mathbf{K}\beta). \tag{2.1}$$

If $\mathscr{C}(\mathbf{F}_0') \subset \mathscr{C}(\mathbf{F}')$, then the implication "$\Longleftarrow$" is holding in (2.1).

Lemma 2.2 *Consider the model* $\mathcal{M} = \{\mathbf{y}, \mathbf{X}\beta, \mathbf{V}\}$ *and its transformed version* $\mathcal{M}_t = \{\mathbf{Fy}, \mathbf{FX}\beta, \mathbf{FVF}'\}$, *and let* $\mathbf{K}\beta$ *be estimable under* $\mathcal{M}$ *and* $\mathcal{M}_t$. *Then, the following statements are equivalent:*

(a) $\mathbf{Fy}$ *is* BLUE-*sufficient for* $\mathbf{K}\beta$.
(b) BLUE($\mathbf{K}\beta \mid \mathcal{M}$) = BLUE($\mathbf{K}\beta \mid \mathcal{M}_t$) *with probability* 1.
(c) *There exists at least one representation of* BLUE *of* $\mathbf{K}\beta$ *under* $\mathcal{M}$ *which is the* BLUE *also under the transformed model* $\mathcal{M}_t$.

Notice that the parametric function $\mathbf{K}\beta$ is estimable under $\mathcal{M}$ as well as under $\mathcal{M}_t$ if and only if

$$\mathscr{C}(\mathbf{K}') \subset \mathscr{C}(\mathbf{X}') \cap \mathscr{C}(\mathbf{X}'\mathbf{F}') = \mathscr{C}(\mathbf{X}'\mathbf{F}'), \tag{2.2}$$

while $\mathbf{X}\beta$ is estimable under $\mathcal{M}_t$ whenever

$$\mathscr{C}(\mathbf{X}') = \mathscr{C}(\mathbf{X}'\mathbf{F}'), \quad \text{i.e.,} \quad \text{rank}(\mathbf{X}) = \text{rank}(\mathbf{FX}).$$

The concept of linear prediction sufficiency is defined analogically as follows: Let $\mathbf{y}_*$ be predictable under the model $\mathcal{M}_*$, i.e., $\mathscr{C}(\mathbf{X}_*') \subset \mathscr{C}(\mathbf{X}')$. Then, $\mathbf{Fy}$ is called linearly prediction sufficient for $\mathbf{y}_*$ if there exists a matrix $\mathbf{A}$ such that $\mathbf{AFy}$ is the BLUP for $\mathbf{y}_*$; that is, there exists a matrix $\mathbf{A}$ such that

$$\mathbf{AF}(\mathbf{X} : \mathbf{VM}) = (\mathbf{X}_* : \mathbf{V}_{21}\mathbf{M}). \tag{2.3}$$

Corresponding to the phrase "BLUE-sufficient", we may use the term "BLUP-sufficient" and the notation $\mathbf{Fy} \in \mathscr{S}(\mathbf{y}_*)$.

The following theorem collects together some important properties of the linear prediction sufficiency.

Theorem 2.1 *Suppose that* $\mathbf{y}_*$ *is predictable under* $\mathcal{M}_*$ *and* $\mathcal{M}_{t*}$. *Then:*

(a) *If* $\mathbf{Fy}$ *is* BLUP-*sufficient for* $\mathbf{y}_*$, *then every representation of the* BLUP *for* $\mathbf{y}_*$ *under the transformed model* $\mathcal{M}_{t*}$ *is* BLUP *also under the original model* $\mathcal{M}_*$.

Moreover, the following statements are equivalent:

(b) **Fy** *is BLUP-sufficient for* $\mathbf{y}_*$, *or shortly* $\mathbf{Fy} \in \mathscr{S}(\mathbf{y}_*)$.

(c) $\mathscr{C}\begin{pmatrix} \mathbf{X}'_* \\ \mathbf{MV}_{12} \end{pmatrix} \subset \mathscr{C}\begin{pmatrix} \mathbf{X}'\mathbf{F}' \\ \mathbf{MVF}' \end{pmatrix}.$

(d) $\mathrm{BLUP}(\mathbf{y}_* \mid \mathscr{M}_*) = \mathrm{BLUP}(\mathbf{y}_* \mid \mathscr{M}_{t*})$ *with probability* 1.

(e) *There exists at least one representation of BLUP of* $\mathbf{y}_*$ *under* $\mathscr{M}_*$ *which is BLUP also under the transformed model* $\mathscr{M}_{t*}$.

Proof The claim (a) was proved by Isotalo and Puntanen (2006, Th. 3.2); see also Remark 2.1 below. The equivalence of (b) and (c) is obvious because (b) means that there exists a matrix **A** such that (2.3) holds. Suppose that (2.3) holds for some **A**. Then, the same multiplier **AF** gives the BLUP for $\mathbf{y}_*$ under the transformed model $\mathscr{M}_{t*}$ if and only if

$$\mathbf{A}(\mathbf{FX} : \mathbf{FVF}'\mathbf{Q}_{\mathbf{FX}}) = (\mathbf{X}_* : \mathbf{V}_{21}\mathbf{F}'\mathbf{Q}_{\mathbf{FX}}). \tag{2.4}$$

In view of Markiewicz and Puntanen (2017, Lemma 5) and Rao and Mitra (1971, Compl. 7, p. 118), the following holds:

$$\mathscr{C}(\mathbf{F}'\mathbf{Q}_{\mathbf{FX}}) = \mathscr{C}(\mathbf{F}') \cap \mathscr{C}(\mathbf{M}), \tag{2.5a}$$

$$\mathbf{F}'\mathbf{Q}_{\mathbf{FX}} = \mathbf{MF}'\mathbf{Q}_{\mathbf{FX}}. \tag{2.5b}$$

Substituting (2.5b) into (2.4), we immediately see that (2.3) implies (2.4) i.e., (b) implies (e). The statement (d) means that we have the equality

$$\mathbf{B}(\mathbf{X} : \mathbf{V}) = \mathbf{CF}(\mathbf{X} : \mathbf{V}) \tag{2.6}$$

for some **B** and **C** satisfying

$$\mathbf{B}(\mathbf{X} : \mathbf{VM}) = (\mathbf{X}_* : \mathbf{V}_{21}\mathbf{M}),$$
$$\mathbf{C}(\mathbf{FX} : \mathbf{FVF}'\mathbf{Q}_{\mathbf{FX}}) = (\mathbf{X}_* : \mathbf{V}_{21}\mathbf{F}'\mathbf{Q}_{\mathbf{FX}}).$$

Now, **By** is BLUP for $\mathbf{y}_*$ under $\mathscr{M}_*$ and and hence, in light of (2.6), **CFy** is also BLUP for $\mathbf{y}_*$ under $\mathscr{M}_*$, and thus by definition, **Fy** is BLUP-sufficient for $\mathbf{y}_*$. Hence, we have shown that (d) implies (b). It is obvious that (e) implies (d) and thereby the proof is completed. $\square$

The above proof is parallel to that of Kala et al. (2017, Th. 2) concerning Lemma 2.2.

Remark 2.1 Regarding the claim (a) in Theorem 2.1, (Isotalo and Puntanen 2006, Th. 3.2) state the following: "Every representation of the BLUP for $\mathbf{y}_*$ under the transformed model $\mathscr{M}_{t*}$ is BLUP also under the original model $\mathscr{M}_*$ and vice versa." Stated in this way, the vice versa part is not quite correct and may result in wrong or confusing interpretations. Hence, we will clarify the meaning of the vice versa part below. The corresponding considerations for the BLUE of estimable parametric function are done in Kala et al. (2017, Sec. 4) and here we proceed along their lines.

To do this, we take a look at the multipliers of the response vector $\mathbf{y}$ when obtaining the BLUPs. Let $\mathbf{y}_*$ be predictable under the models $\mathcal{M}_*$ and $\mathcal{M}_{t*}$ and denote

$$
\begin{aligned}
\mathscr{A} &= \left\{ \mathbf{A} : \mathbf{AFy} = \mathrm{BLUP}(\mathbf{y}_* \mid \mathcal{M}_*) \right\} \\
&= \left\{ \mathbf{A} : \mathbf{AF}(\mathbf{X} : \mathbf{VM}) = (\mathbf{X}_* : \mathbf{V}_{21}\mathbf{M}) \right\}, \\
\mathscr{C} &= \left\{ \mathbf{C} : \mathbf{CFy} = \mathrm{BLUP}(\mathbf{y}_* \mid \mathcal{M}_{t*}) \right\} \\
&= \left\{ \mathbf{C} : \mathbf{C}(\mathbf{FX} : \mathbf{FVF'Q_{FX}}) = (\mathbf{X}_* : \mathbf{V}_{21}\mathbf{F'Q_{FX}}) \right\}.
\end{aligned}
$$

Proceeding along the same lines as (Kala et al. 2017, Th. 3) in their BLUE considerations, we can obtain the following result.

Theorem 2.2 *Suppose that $\mathbf{Fy}$ is BLUP-sufficient for the predictable $\mathbf{y}_*$ under the model $\mathcal{M}_*$, and let the sets of matrices $\mathscr{A}$ and $\mathscr{C}$ be defined as above. Then $\mathscr{A} = \mathscr{C}$.*

To describe more statistically the meaning of Theorem 2.2, let $\mathbf{Fy}$ be BLUP-sufficient for $\mathbf{y}_*$ under $\mathcal{M}_*$. Then, for each matrix $\mathbf{C}$ such that $\mathbf{CFy}$ is the BLUP of $\mathbf{y}_*$ in the transformed model $\mathcal{M}_{t*}$, the statistic $\mathbf{CFy}$ is also the BLUP of $\mathbf{y}_*$ in the original model $\mathcal{M}_*$, and vice versa. Notice that in this statement the "vice versa" means that we consider such $\mathbf{C}$ for which $\mathbf{CFy}$ is BLUP under $\mathcal{M}_*$, not the set of matrices $\mathbf{B}$ such that $\mathbf{By}$ is BLUP under $\mathcal{M}_*$. □

3 Some Representations for the BLUPs

Let us start by considering the BLUP for $\boldsymbol{\varepsilon}_*$ which offers interesting views. Theorem 3.1 below could be proved directly using Lemma 1.2 by choosing $\boldsymbol{\varepsilon}_*$ as the "new future observations". However, we find it illustrative to give an alternative proof.

Theorem 3.1 *Under the model $\mathcal{M}_*$, the statistic $\mathbf{Cy}$ is the BLUP for $\boldsymbol{\varepsilon}_*$ if and only if*

$$
\mathbf{C}(\mathbf{X} : \mathbf{VM}) = (\mathbf{0} : \mathbf{V}_{21}\mathbf{M}),
$$

or, equivalently, $\mathbf{C} = \mathbf{AM}$ for some matrix $\mathbf{A}$ such that

$$
\mathbf{AMVM} = \mathbf{V}_{21}\mathbf{M}. \tag{3.1}
$$

Proof The predictor $\mathbf{Cy}$ is unbiased for $\boldsymbol{\varepsilon}_*$ if and only if $\mathrm{E}(\boldsymbol{\varepsilon}_* - \mathbf{Cy}) = \mathbf{0}$ and so $\mathbf{CX} = \mathbf{0}$ and hence necessarily $\mathbf{C} = \mathbf{AM}$ for some matrix $\mathbf{A}$. Now $\mathbf{AMy}$ is the BLUP for $\boldsymbol{\varepsilon}_*$ if $\mathbf{A}$ is such that the covariance matrix of the prediction error $\boldsymbol{\varepsilon}_* - \mathbf{AMy}$ is minimal in the Löwner sense. We recall that for any matrix $\mathbf{A}$, we have the Löwner ordering

$$\mathrm{cov}(\boldsymbol{\varepsilon}_* - \mathbf{AMy}) \geq_{\mathrm{L}} \mathrm{cov}[\boldsymbol{\varepsilon}_* - \mathbf{V}_{21}\mathbf{M}(\mathbf{MVM})^-\mathbf{My}]\,, \tag{3.2}$$

where

$$\mathrm{cov}(\boldsymbol{\varepsilon}_*, \mathbf{My})[\mathrm{cov}(\mathbf{My})]^- = \mathbf{V}_{21}\mathbf{M}(\mathbf{MVM})^-.$$

For the Löwner inequality in (3.2), see Puntanen et al. (2011, Th. 9). We have thus found that BLUP($\boldsymbol{\varepsilon}_*$) has a representation

$$\mathrm{BLUP}(\boldsymbol{\varepsilon}_*) = \mathbf{V}_{21}\mathbf{M}(\mathbf{MVM})^-\mathbf{My}.$$

On the other hand, according to Puntanen et al. (2011, Cor. 9.1), for any matrix $\mathbf{A}$,

$$\mathrm{cov}(\boldsymbol{\varepsilon}_* - \mathbf{AMy}) \geq_{\mathrm{L}} \mathrm{cov}(\boldsymbol{\varepsilon}_* - \mathbf{A}_1\mathbf{My})$$

if and only if $\mathbf{A}_1$ is a solution to (3.1). $\square$

In view of the identity, see Haslett et al. (2014, Sec. 2),

$$\begin{aligned}
\mathbf{P}_{\mathbf{X};\mathbf{W}^+} &= \mathbf{X}(\mathbf{X}'\mathbf{W}^-\mathbf{X})^-\mathbf{X}'\mathbf{W}^+ \\
&= \mathbf{P}_{\mathbf{W}} - \mathbf{VM}(\mathbf{MVM})^-\mathbf{MP}_{\mathbf{W}}\,, \tag{3.3}
\end{aligned}$$

the BLUP($\boldsymbol{\varepsilon}_*$) can be expressed, for example, as follows:

$$\begin{aligned}
\mathrm{BLUP}(\boldsymbol{\varepsilon}_*) &= \mathbf{V}_{21}\mathbf{M}(\mathbf{MVM})^-\mathbf{My} \\
&= \mathbf{V}_{21}\mathbf{W}^-(\mathbf{I}_n - \mathbf{G})\mathbf{y} \\
&= \mathbf{V}_{21}\mathbf{V}^-(\mathbf{I}_n - \mathbf{G})\mathbf{y},
\end{aligned}$$

where $\mathbf{W} \in \mathcal{W}$, $\mathbf{y} \in \mathcal{C}(\mathbf{W})$, and $\mathbf{G} = \mathbf{X}(\mathbf{X}'\mathbf{W}^-\mathbf{X})^-\mathbf{X}'\mathbf{W}^- = \mathbf{P}_{\mathbf{X};\mathbf{W}^-}$.

It is well known that the general solution to $\mathbf{A}(\mathbf{X} : \mathbf{VM}) = (\mathbf{X}_* : \mathbf{0})$ can be written, for example, as

$$\mathbf{A}_0 = (\mathbf{X}_* : \mathbf{0})(\mathbf{X} : \mathbf{VM})^+ + \mathbf{N}_1\mathbf{Q}_{\mathbf{W}} := \mathbf{A}_1 + \mathbf{N}_1\mathbf{Q}_{\mathbf{W}}\,,$$

where $\mathbf{N}_1 \in \mathbb{R}^{q \times n}$ is free to vary and $\mathbf{Q}_{\mathbf{W}} = \mathbf{I}_n - \mathbf{P}_{\mathbf{W}}$, $\mathbf{W} \in \mathcal{W}$. Similarly, the general solution to $\mathbf{B}(\mathbf{X} : \mathbf{VM}) = (\mathbf{X}_* : \mathbf{V}_{21}\mathbf{M})$ can be written as

$$\mathbf{B}_0 = (\mathbf{X}_* : \mathbf{V}_{21}\mathbf{M})(\mathbf{X} : \mathbf{VM})^+ + \mathbf{N}_2\mathbf{Q}_{\mathbf{W}} := \mathbf{B}_1 + \mathbf{N}_2\mathbf{Q}_{\mathbf{W}}\,,$$

where the matrix $\mathbf{N}_2 \in \mathbb{R}^{q \times n}$ is free to vary. Consider then the equation

$$\mathbf{C}(\mathbf{X} : \mathbf{VM}) = (\mathbf{0} : \mathbf{V}_{21}\mathbf{M})\,,$$

for which the general solution is

$$C_0 = (0 : V_{21}M)(X : VM)^+ + N_3 Q_W := C_1 + N_3 Q_W,$$

where the matrix $N_3 \in \mathbb{R}^{q \times n}$ is free to vary. Then $B_1 = A_1 + C_1$ and

$$B_0 = A_0 + C_0 + N_0 Q_W,$$

where N_0 is free to vary. In other words, if

$$\begin{pmatrix} A \\ C \end{pmatrix} (X : VM) = \begin{pmatrix} X_* & 0 \\ 0 & V_{21}M \end{pmatrix},$$

then

$$(A + C)(X : VM) = (X_* : V_{21}M),$$

and so

$$(A + C)y = \mathrm{BLUP}(y_*).$$

Of course,

$$Ay = \mathrm{BLUE}(X_* \beta), \quad Cy = \mathrm{BLUP}(\varepsilon_*),$$

so that we have obtained the following result:

Theorem 3.2 *Under the linear model $\mathcal{M}_*$, where y_* is predictable, the following decomposition holds (with probability 1):*

$$\mathrm{BLUP}(y_*) = \mathrm{BLUE}(X_* \beta) + \mathrm{BLUP}(\varepsilon_*).$$

Next, we consider the BLUP-sufficiency of Fy for ε_*.

Theorem 3.3 *The statistic Fy is BLUP-sufficient for ε_* under $\mathcal{M}_*$ if and only if any of the following equivalent conditions holds:*

(a) $\mathscr{C} \begin{pmatrix} 0 \\ MV_{12} \end{pmatrix} \subset \mathscr{C} \begin{pmatrix} X'F' \\ MVF' \end{pmatrix}.$

(b) $\mathscr{C}(MV_{12}) \subset \mathscr{C}(MVF'Q_{FX}) = \mathscr{C}(MVMF'Q_{FX}).$

(c) $\mathrm{BLUP}(\varepsilon_* \mid \mathcal{M}_*) = \mathrm{BLUP}(\varepsilon_* \mid \mathcal{M}_{t*})$ *with probability 1.*

(d) *There exists at least one representation of BLUP of ε_* under $\mathcal{M}_*$ which is BLUP also under the transformed model $\mathcal{M}_{t*}$.*

In particular, if Fy is BLUE-sufficient for $X\beta$, then (b) becomes

(e) $\mathscr{C}(MV_{12}) \subset \mathscr{C}(MVF').$

Proof The statistic Fy is BLUP-sufficient for ε_* under $\mathcal{M}_*$ if the equation

$$AF(X : VM) = (0 : V_{21}M) \tag{3.4}$$

has a solution for $\mathbf{A}$ which obviously happens if and only if (a) holds. The condition (a) means that there exists a matrix $\mathbf{N}$ such that

$$0 = \mathbf{X}'\mathbf{F}'\mathbf{N}, \quad \mathbf{MV}_{12} = \mathbf{MVF}'\mathbf{N}.$$

Hence, $\mathbf{N} = \mathbf{Q}_{\mathbf{FX}}\mathbf{N}_1$ for some matrix $\mathbf{N}_1$ and

$$\mathbf{MV}_{12} = \mathbf{MVF}'\mathbf{Q}_{\mathbf{FX}}\mathbf{N}_1. \tag{3.5}$$

The equality (3.5) holds for some matrix $\mathbf{N}_1$ if and only if

$$\mathscr{C}(\mathbf{MV}_{12}) \subset \mathscr{C}(\mathbf{MVF}'\mathbf{Q}_{\mathbf{FX}}) = \mathscr{C}(\mathbf{MVMF}'\mathbf{Q}_{\mathbf{FX}}),$$

where we have used (2.5b).

Suppose that (a) holds so that there exists some matrix $\mathbf{A}$ such that (3.4) holds. Then, the same multiplier $\mathbf{AF}$ gives the BLUP for $\boldsymbol{\varepsilon}_*$ under the transformed model $\mathscr{M}_{t*}$ if and only if $\mathbf{A}$ satisfies the equation

$$\mathbf{A}(\mathbf{FX} : \mathbf{FVF}'\mathbf{Q}_{\mathbf{FX}}) = (\mathbf{0} : \mathbf{V}_{21}\mathbf{F}'\mathbf{Q}_{\mathbf{FX}}).$$

Proceeding onwards as in the proof of Theorem 2.1, the equivalence between (a), (c) and (d) can be shown.

To prove (e), let us assume that $\mathbf{Fy}$ is BLUE-sufficient for $\mathbf{X}\boldsymbol{\beta}$. It is clear that

$$\mathscr{C}(\mathbf{MVF}'\mathbf{Q}_{\mathbf{FX}}) \subset \mathscr{C}(\mathbf{MVF}'). \tag{3.6}$$

Using the rank rule of the matrix product, see Marsaglia and Styan (1974, Cor. 6.2),

$$\text{rank}(\mathbf{MVF}'\mathbf{Q}_{\mathbf{FX}}) = \text{rank}(\mathbf{MVF}') - \dim \mathscr{C}(\mathbf{FVM}) \cap \mathscr{C}(\mathbf{FX})$$
$$= \text{rank}(\mathbf{MVF}'), \tag{3.7}$$

because in view of part (c) of Lemma 2.1, we have $\dim \mathscr{C}(\mathbf{FVM}) \cap \mathscr{C}(\mathbf{FX}) = \{\mathbf{0}\}$. This means that we get equality in (3.6) and so the proof of (e) is completed.

It is of course clear that the corresponding property as (a) in Theorem 2.1, holds as well for the BLUP($\boldsymbol{\varepsilon}_*$).

Theorem 3.4 *Consider the following three statements:*

(a) $\mathbf{Fy}$ *is BLUE-sufficient for* $\mathbf{X}_*\boldsymbol{\beta}$.
(b) $\mathbf{Fy}$ *is BLUP-sufficient for* $\boldsymbol{\varepsilon}_*$.
(c) $\mathbf{Fy}$ *is BLUP-sufficient for* $\mathbf{y}_*$.

Then above, any two conditions together imply the third one. Moreover, if

$$\mathscr{C}(\mathbf{X}_*) \cap \mathscr{C}(\mathbf{V}_{21}\mathbf{M}) = \{\mathbf{0}\},$$

then

$$(c) \implies (a) \text{ and } (b).$$

Proof Denote

$$\mathbf{A} = \begin{pmatrix} \mathbf{X}_* \\ \mathbf{0} \end{pmatrix}, \quad \mathbf{B} = \begin{pmatrix} \mathbf{0} \\ \mathbf{MV}_{12} \end{pmatrix} \quad \mathbf{C} = \begin{pmatrix} \mathbf{X}'\mathbf{F}' \\ \mathbf{MVF}' \end{pmatrix}.$$

Now (c) holds if and only if

$$\mathbf{P_C}(\mathbf{A} + \mathbf{B}) = \mathbf{A} + \mathbf{B}, \tag{3.8}$$

which is equivalent to

$$\mathbf{P_C}\mathbf{A} - \mathbf{A} = -(\mathbf{P_C}\mathbf{B} - \mathbf{B}),$$

from which the first part of the theorem follows. To prove the second part, we have to show that if

$$\mathscr{C}(\mathbf{A}') \cap \mathscr{C}(\mathbf{B}') = \{\mathbf{0}\}, \tag{3.9}$$

then

$$\mathscr{C}(\mathbf{A} + \mathbf{B}) \subset \mathscr{C}(\mathbf{C}) \implies \mathscr{C}(\mathbf{A}) \subset \mathscr{C}(\mathbf{C}). \tag{3.10}$$

Postmultiplying (3.8) by $\mathbf{Q_{B'}}$ yields

$$\mathbf{P_C}\mathbf{A}\mathbf{Q_{B'}} = \mathbf{A}\mathbf{Q_{B'}}. \tag{3.11}$$

If $\text{rank}(\mathbf{AQ_{B'}}) = \text{rank}(\mathbf{A})$, which happens if and only if (3.9) holds, we can, in light of the rank cancellation rule of Marsaglia and Styan (1974, Th. 2), cancel the rightmost $\mathbf{Q_{B'}}$ in each side of (3.11) and obtain $\mathbf{P_C}\mathbf{A} = \mathbf{A}$ as claimed in (3.10). □

Remark 3.1 The notion of linear error-sufficiency was introduced by Groß (1998), while considering linear sufficient statistics for the prediction of the random error term ε in the general linear model. This is nothing but the BLUP-sufficiency of ε. Proceeding along the lines of Theorem 3.1, we can conclude that under the model $\mathscr{M}$, the statistic $\mathbf{Cy}$ is the BLUP for ε if and only if

$$\mathbf{C}(\mathbf{X} : \mathbf{VM}) = (\mathbf{0} : \mathbf{VM}),$$

and one explicit solution is

$$\text{BLUP}(\varepsilon \mid \mathscr{M}) = \mathbf{VM}(\mathbf{MVM})^{-}\mathbf{My} = \mathbf{y} - \text{BLUE}(\mathbf{X}\boldsymbol{\beta} \mid \mathscr{M}).$$

Obviously, **Fy** is BLUP-sufficient for ε if and only if

$$\mathscr{N}(\mathbf{FX} : \mathbf{FVM}) \subset \mathscr{N}(\mathbf{0} : \mathbf{VM}).$$

For the BLUP of ε, see also Arendacká and Puntanen (2015, Lemma 1). $\qquad\square$

4 Representations for the BLUP in the Transformed Model

When doing the "BLUP-hunting" in models $\mathscr{M}_*$ and $\mathscr{M}_{t*}$ we assume that the parametric function $\mathbf{X}_*\boldsymbol{\beta}$ is estimable under $\mathscr{M}$ as well as under $\mathscr{M}_t$, which, in light of (2.2), happens if and only if $\mathscr{C}(\mathbf{X}_*') \subset \mathscr{C}(\mathbf{X}'\mathbf{F}')$, so that

$$\mathbf{X}_* = \mathbf{LFX} \quad \text{for some matrix } \mathbf{L}. \tag{4.1}$$

Similarly, $\mathbf{X}\boldsymbol{\beta}$ is required to be estimable under $\mathscr{M}_t$ so that $\mathscr{C}(\mathbf{X}') = \mathscr{C}(\mathbf{X}'\mathbf{F}')$.
Denote

$$\mathbf{G} = \mathbf{X}(\mathbf{X}'\mathbf{W}^-\mathbf{X})^-\mathbf{X}'\mathbf{W}^- = \mathbf{P}_{\mathbf{X};\mathbf{W}^-},$$
$$\mathbf{P}_{\mathbf{FX};(\mathbf{FWF}')^-} = \mathbf{FX}[\mathbf{X}'\mathbf{F}'(\mathbf{FWF}')^-\mathbf{FX}]^-\mathbf{X}'\mathbf{F}'(\mathbf{FWF}')^-,$$
$$\mathbf{G}_t = \mathbf{X}[\mathbf{X}'\mathbf{F}'(\mathbf{FWF}')^-\mathbf{FX}]^-\mathbf{X}'\mathbf{F}'(\mathbf{FWF}')^-\mathbf{F},$$

so that $\mathbf{FG}_t = \mathbf{P}_{\mathbf{FX};(\mathbf{FWF}')^-}\mathbf{F}$.
Estimator $\mathbf{BFy}$ is the BLUE($\mathbf{FX}\boldsymbol{\beta} \mid \mathscr{M}_t$) if and only if $\mathbf{B}$ satisfies

$$\mathbf{B}(\mathbf{FX} : \mathbf{FVF}'\mathbf{Q}_{\mathbf{FX}}) = (\mathbf{FX} : \mathbf{0}),$$

so that one expression for $\mathbf{B}$ is $\mathbf{B} = \mathbf{P}_{\mathbf{FX};(\mathbf{FWF}')^-} := \mathbf{FXA}$ and then

$$\mathbf{FXA}(\mathbf{FX} : \mathbf{FVF}'\mathbf{Q}_{\mathbf{FX}}) = (\mathbf{FX} : \mathbf{0}). \tag{4.2}$$

Because rank($\mathbf{FX}$) = rank($\mathbf{X}$), we can cancel the left-most $\mathbf{F}$ from both sides of (4.2) resulting

$$\mathbf{X}[\mathbf{X}'\mathbf{F}'(\mathbf{FWF}')^-\mathbf{FX}]^-\mathbf{X}'\mathbf{F}'(\mathbf{FWF}')^-(\mathbf{FX} : \mathbf{FVF}'\mathbf{Q}_{\mathbf{FX}}) = (\mathbf{X} : \mathbf{0}).$$

Thus, $\mathbf{G}_t\mathbf{y}$ is the BLUE for $\mathbf{X}\boldsymbol{\beta}$ under $\mathscr{M}_t$ and

$$\mathbf{G}_t(\mathbf{X} : \mathbf{VF}'\mathbf{Q}_{\mathbf{FX}}) = (\mathbf{X} : \mathbf{0}). \tag{4.3}$$

An alternative expression for BLUE($\mathbf{FX}\boldsymbol{\beta} \mid \mathscr{M}_t$) can be obtained using the corresponding identity as in (3.3):

$$\mathbf{P}_{\mathbf{FX};(\mathbf{FWF}')^+} = \mathbf{FX}[\mathbf{X}'\mathbf{F}'(\mathbf{FWF}')^-\mathbf{FX}]^-\mathbf{X}'\mathbf{F}'(\mathbf{FWF}')^+$$
$$= \mathbf{P}_{\mathbf{FW}} - \mathbf{FVF}'\mathbf{Q}_{\mathbf{FX}}(\mathbf{Q}_{\mathbf{FX}}\mathbf{FVF}'\mathbf{Q}_{\mathbf{FX}})^-\mathbf{Q}_{\mathbf{FX}}\mathbf{P}_{\mathbf{FW}}\,.$$

Namely, for $\mathbf{y} \in \mathscr{C}(\mathbf{W})$ and, noting that $\mathbf{P}_{\mathbf{FW}}\mathbf{Fy} = \mathbf{Fy}$, we get

$$\text{BLUE}(\mathbf{FX}\beta \mid \mathscr{M}_t) = \mathbf{FG}_t\mathbf{y}$$
$$= \mathbf{P}_{\mathbf{FX};(\mathbf{FWF}')^+}\mathbf{Fy}$$
$$= \mathbf{Fy} - \mathbf{FVF}'\mathbf{Q}_{\mathbf{FX}}(\mathbf{Q}_{\mathbf{FX}}\mathbf{FVF}'\mathbf{Q}_{\mathbf{FX}})^-\mathbf{Q}_{\mathbf{FX}}\mathbf{Fy}.$$

It is interesting to observe that the matrix

$$\mathbf{G}_\# = \mathbf{I}_n - \mathbf{VF}'\mathbf{Q}_{\mathbf{FX}}(\mathbf{Q}_{\mathbf{FX}}\mathbf{FVF}'\mathbf{Q}_{\mathbf{FX}})^-\mathbf{Q}_{\mathbf{FX}}\mathbf{F}$$

satisfies (4.3), i.e.,

$$\mathbf{G}_\#(\mathbf{X} : \mathbf{VF}'\mathbf{Q}_{\mathbf{FX}}) = (\mathbf{X} : \mathbf{0})\,.$$

However, $\mathbf{G}_\#$ and $\mathbf{G}_t$ are not necessarily equal; their difference is

$$\mathbf{G}_t - \mathbf{G}_\# = \mathbf{N}\mathbf{Q}_{(\mathbf{X}:\mathbf{VF}'\mathbf{Q}_{\mathbf{FX}})}$$

for some matrix $\mathbf{N}$.

Consider then the expressions for the BLUP of ε_* under the transformed model $\mathscr{M}_{t*}$. One way to do this is to use Theorem 3.1, which says that $\mathbf{DFy}$ is the BLUP($\varepsilon_* \mid \mathscr{M}_{t*}$) if $\mathbf{D}$ is a solution to

$$\mathbf{D}(\mathbf{FX} : \mathbf{FVF}'\mathbf{Q}_{\mathbf{FX}}) = (\mathbf{0} : \mathbf{V}_{21}\mathbf{F}'\mathbf{Q}_{\mathbf{FX}})\,.$$

Thus, the BLUP of ε_* under $\mathscr{M}_{t*}$ can be expressed as

$$\text{BLUP}(\varepsilon_* \mid \mathscr{M}_{t*}) = \mathbf{V}_{21}\mathbf{F}'\mathbf{Q}_{\mathbf{FX}}(\mathbf{Q}_{\mathbf{FX}}\mathbf{FVF}'\mathbf{Q}_{\mathbf{FX}})^-\mathbf{Q}_{\mathbf{FX}}\mathbf{Fy}. \qquad (4.4)$$

Recall that in (4.4), $\mathbf{F}'\mathbf{Q}_{\mathbf{FX}}$ can be replaced with $\mathbf{MF}'\mathbf{Q}_{\mathbf{FX}}$. One alternative expression is

$$\text{BLUP}(\varepsilon_* \mid \mathscr{M}_{t*}) = \mathbf{V}_{21}\mathbf{F}'(\mathbf{FVF}')^-\mathbf{F}(\mathbf{I}_n - \mathbf{G}_t)\mathbf{y}.$$

We complete this section by giving some alternative expressions for the BLUP of $\mathbf{y}_*$. Using (4.1), let us denote

$$\mu_* = \mathbf{X}_*\beta = \mathbf{LFX}\beta, \quad \mu = \mathbf{X}\beta\,.$$

The BLUP($\mathbf{y}_*$) under $\mathscr{M}_*$ can be written as

$$
\begin{aligned}
\text{BLUP}(\mathbf{y}_* \mid \mathscr{M}_*) &= \text{BLUE}(\boldsymbol{\mu}_* \mid \mathscr{M}) + \mathbf{V}_{21}\mathbf{V}^-[\mathbf{y} - \text{BLUE}(\boldsymbol{\mu} \mid \mathscr{M})] \\
&= \mathbf{LFGy} + \mathbf{V}_{21}\mathbf{V}^-(\mathbf{I}_n - \mathbf{G})\mathbf{y} \\
&= \mathbf{LFGy} + \mathbf{V}_{21}\mathbf{M}(\mathbf{MVM})^-\mathbf{My} \\
&= \text{BLUE}(\boldsymbol{\mu}_* \mid \mathscr{M}) + \text{BLUP}(\boldsymbol{\varepsilon}_* \mid \mathscr{M}_*),
\end{aligned}
\tag{4.5}
$$

or shortly,

$$
\tilde{\mathbf{y}}_* = \tilde{\boldsymbol{\mu}}_* + \tilde{\boldsymbol{\varepsilon}}_* .
$$

Under the transformed model, we have

$$
\begin{aligned}
\text{BLUP}(\mathbf{y}_* \mid \mathscr{M}_{t*}) &= \text{BLUE}(\boldsymbol{\mu}_* \mid \mathscr{M}_t) + \mathbf{V}_{21}\mathbf{F}'(\mathbf{FVF}')^-\mathbf{F}[\mathbf{y} - \text{BLUE}(\boldsymbol{\mu} \mid \mathscr{M}_t)] \\
&= \mathbf{LFG}_t\mathbf{y} + \mathbf{V}_{21}\mathbf{F}'(\mathbf{FVF}')^-\mathbf{F}(\mathbf{I}_n - \mathbf{G}_t)\mathbf{y} \\
&= \mathbf{LFG}_t\mathbf{y} + \mathbf{V}_{21}\mathbf{F}'\mathbf{Q}_{\mathbf{FX}}(\mathbf{Q}_{\mathbf{FX}}\mathbf{FVF}'\mathbf{Q}_{\mathbf{FX}})^-\mathbf{Q}_{\mathbf{FX}}\mathbf{Fy} \\
&= \text{BLUE}(\boldsymbol{\mu}_* \mid \mathscr{M}_t) + \text{BLUP}(\boldsymbol{\varepsilon}_* \mid \mathscr{M}_{t*}),
\end{aligned}
\tag{4.6}
$$

or shortly,

$$
\tilde{\mathbf{y}}_{t*} = \tilde{\boldsymbol{\mu}}_{t*} + \tilde{\boldsymbol{\varepsilon}}_{t*} .
\tag{4.7}
$$

In (4.5) and (4.6), the matrix $\mathbf{V}$ can be replaced with $\mathbf{W} \in \mathscr{W}$. In passing we may notice that under $\mathscr{M}_*$, $\tilde{\boldsymbol{\mu}}_*$ and $\tilde{\boldsymbol{\varepsilon}}_*$ are uncorrelated, and hence $\text{cov}(\tilde{\mathbf{y}}_*) = \text{cov}(\tilde{\boldsymbol{\mu}}_*) + \text{cov}(\tilde{\boldsymbol{\varepsilon}}_*)$. The corresponding property holds also for the terms of (4.7). For further representations for the BLUP($\mathbf{y}_* \mid \mathscr{M}_*$), we refer to Haslett et al. (2014).

5 Linear Mixed Model

One application of the model $\mathscr{M}_*$ is the linear mixed model

$$
\mathbf{y} = \mathbf{X}\boldsymbol{\beta} + \mathbf{Zu} + \boldsymbol{\varepsilon}, \quad \text{or shortly,} \quad \mathscr{L} = \{\mathbf{y}, \mathbf{X}\boldsymbol{\beta} + \mathbf{Zu}, \mathbf{D}, \mathbf{R}, \mathbf{S}\},
$$

where $\mathbf{X}_{n\times p}$ and $\mathbf{Z}_{n\times q}$ are known matrices, $\boldsymbol{\beta} \in \mathbb{R}^p$ is a vector of unknown fixed effects, $\mathbf{u}$ is an unobservable vector (q elements) of random effects with $\text{E}(\mathbf{u}) = \mathbf{0}$, $\text{cov}(\mathbf{u}) = \mathbf{D}_{q\times q}$, $\text{cov}(\boldsymbol{\varepsilon}, \mathbf{u}) = \mathbf{S}_{n\times q}$, and $\text{E}(\boldsymbol{\varepsilon}) = \mathbf{0}$, $\text{cov}(\boldsymbol{\varepsilon}) = \mathbf{R}_{n\times n}$. In this situation

$$
\text{cov}\begin{pmatrix} \boldsymbol{\varepsilon} \\ \mathbf{u} \end{pmatrix} = \begin{pmatrix} \mathbf{R} & \mathbf{S} \\ \mathbf{S}' & \mathbf{D} \end{pmatrix},
$$

and

$$
\text{cov}(\mathbf{y}) = \mathbf{ZDZ}' + \mathbf{R} + \mathbf{ZS}' + \mathbf{SZ}' := \boldsymbol{\Sigma}.
$$

The mixed model can be expressed as a version of the model with "new observations", the new observations being now in $\mathbf{u}$:

$$\left\{ \begin{pmatrix} \mathbf{y} \\ \mathbf{u} \end{pmatrix}, \begin{pmatrix} \mathbf{X} \\ \mathbf{0} \end{pmatrix} \boldsymbol{\beta}, \begin{pmatrix} \boldsymbol{\Sigma} & \mathbf{ZD}+\mathbf{S} \\ \mathbf{DZ}'+\mathbf{S}' & \mathbf{D} \end{pmatrix} \right\}.$$

Moreover, choosing the "new observations" as $\mathbf{g} = \mathbf{X}\boldsymbol{\beta} + \mathbf{Zu}$, we get

$$\left\{ \begin{pmatrix} \mathbf{y} \\ \mathbf{g} \end{pmatrix}, \begin{pmatrix} \mathbf{X} \\ \mathbf{X} \end{pmatrix} \boldsymbol{\beta}, \begin{pmatrix} \boldsymbol{\Sigma} & (\mathbf{ZD}+\mathbf{S})\mathbf{Z}' \\ \mathbf{Z}(\mathbf{DZ}'+\mathbf{S}') & \mathbf{ZDZ}' \end{pmatrix} \right\}.$$

Thus, see, e.g., Haslett et al. (2015), under the mixed model $\mathscr{L}$ the following statements hold:

(a) $\mathbf{Ay}$ is the BLUE for $\mathbf{X}\boldsymbol{\beta}$ if and only if

$$\mathbf{A}(\mathbf{X} : \boldsymbol{\Sigma}\mathbf{M}) = (\mathbf{X} : \mathbf{0}). \tag{5.1}$$

(b) $\mathbf{By}$ is the BLUP for $\mathbf{u}$ if and only if

$$\mathbf{B}(\mathbf{X} : \boldsymbol{\Sigma}\mathbf{M}) = \begin{bmatrix} \mathbf{0} : (\mathbf{DZ}'+\mathbf{S}')\mathbf{M} \end{bmatrix} = \begin{bmatrix} \mathbf{0} : \text{cov}(\mathbf{u}, \mathbf{y})\mathbf{M} \end{bmatrix}.$$

(c) $\mathbf{Cy}$ is the BLUP for $\mathbf{g} = \mathbf{X}\boldsymbol{\beta} + \mathbf{Zu}$ if and only if

$$\mathbf{C}(\mathbf{X} : \boldsymbol{\Sigma}\mathbf{M}) = \begin{bmatrix} \mathbf{X} : \mathbf{Z}(\mathbf{DZ}'+\mathbf{S}')\mathbf{M} \end{bmatrix} = \begin{bmatrix} \mathbf{X} : \text{cov}(\mathbf{g}, \mathbf{y})\mathbf{M} \end{bmatrix}. \tag{5.2}$$

Thus, we have, corresponding to Theorem 3.2,

$$\text{BLUP}(\mathbf{X}\boldsymbol{\beta} + \mathbf{Zu} \mid \mathscr{L}) = \text{BLUE}(\mathbf{X}\boldsymbol{\beta} \mid \mathscr{L}) + \text{BLUP}(\mathbf{Zu} \mid \mathscr{L}),$$

so that one representation for the BLUP of $\mathbf{g}$ under $\mathscr{L}$ is

$$\begin{aligned} \text{BLUP}(\mathbf{g}) &= \mathbf{Ty} + \mathbf{Z}(\mathbf{DZ}'+\mathbf{S}')\mathbf{W}_{\boldsymbol{\Sigma}}^{-}(\mathbf{y} - \mathbf{Ty}) \\ &= \mathbf{Ty} + \mathbf{Z}(\mathbf{DZ}'+\mathbf{S}')\mathbf{M}(\mathbf{M}\boldsymbol{\Sigma}\mathbf{M})^{-}\mathbf{My}, \end{aligned}$$

where $\mathbf{T} = \mathbf{X}(\mathbf{X}'\mathbf{W}_{\boldsymbol{\Sigma}}^{-}\mathbf{X})^{-}\mathbf{X}'\mathbf{W}_{\boldsymbol{\Sigma}}^{-}$ and

$$\mathbf{W}_{\boldsymbol{\Sigma}} = \boldsymbol{\Sigma} + \mathbf{XUU}'\mathbf{X}', \quad \mathscr{C}(\mathbf{W}_{\boldsymbol{\Sigma}}) = \mathscr{C}(\mathbf{X} : \boldsymbol{\Sigma}).$$

Conditions for $\mathbf{Fy}$ being linearly sufficient or linearly prediction sufficient for $\mathbf{X}\boldsymbol{\beta}$, $\mathbf{u}$, and $\mathbf{g} = \mathbf{X}\boldsymbol{\beta} + \mathbf{Zu}$, respectively, can be straightforwardly derived from (5.1)–(5.2). For example, $\mathbf{Fy}$ is BLUP-sufficient for $\mathbf{g}$ if and only if

$$\mathscr{C} \begin{pmatrix} \mathbf{X}' \\ \mathbf{M}(\mathbf{ZD}+\mathbf{S})\mathbf{Z}' \end{pmatrix} \subset \mathscr{C} \begin{pmatrix} \mathbf{X}'\mathbf{F}' \\ \mathbf{M}\boldsymbol{\Sigma}\mathbf{F}' \end{pmatrix}. \tag{5.3}$$

Corresponding properties as under $\mathcal{M}_*$ in Theorem 3.4 for $\mathbf{X}_*\boldsymbol{\beta}$, $\boldsymbol{\varepsilon}_*$, and $\mathbf{y}_*$ hold also under $\mathcal{L}$ for $\mathbf{X}\boldsymbol{\beta}$, $\mathbf{Zu}$, and $\mathbf{g}$.

For the linear sufficiency in the mixed model, see also Liu et al. (2008, Sec. 3). They defined the BLUP-sufficiency in a slightly different manner which we will not handle here. Inspired by their Theorem 3.1, we will now show that $\mathbf{Fy}$ is BLUP-sufficient for $\mathbf{g} = \mathbf{X}\boldsymbol{\beta} + \mathbf{Zu}$ if

$$\mathscr{C}(\mathbf{X} : \mathbf{ZD} + \mathbf{S}) \subset \mathscr{C}(\mathbf{W}_{\Sigma}\mathbf{F}') . \tag{5.4}$$

In view of part (b) of Lemma 2.1, the "first part" of (5.4), $\mathscr{C}(\mathbf{X}) \subset \mathscr{C}(\mathbf{W}_{\Sigma}\mathbf{F}')$, is equivalent to

$$\mathscr{C}\begin{pmatrix} \mathbf{X}' \\ \mathbf{0} \end{pmatrix} \subset \mathscr{C}\begin{pmatrix} \mathbf{X}'\mathbf{F}' \\ \mathbf{M}\boldsymbol{\Sigma}\mathbf{F}' \end{pmatrix} , \tag{5.5}$$

which means that $\mathbf{Fy} \in \mathscr{S}(\mathbf{X}\boldsymbol{\beta})$. If $\mathbf{Fy}$ would be also BLUP-sufficient for $\mathbf{Zu}$, that is,

$$\mathscr{C}\begin{pmatrix} \mathbf{0} \\ \mathbf{M}(\mathbf{ZD} + \mathbf{S})\mathbf{Z}' \end{pmatrix} \subset \mathscr{C}\begin{pmatrix} \mathbf{X}'\mathbf{F}' \\ \mathbf{M}\boldsymbol{\Sigma}\mathbf{F}' \end{pmatrix} , \tag{5.6}$$

then (5.3) would hold. Now (5.6) can be equivalently expressed as

$$\mathscr{C}[\mathbf{M}(\mathbf{ZD} + \mathbf{S})\mathbf{Z}'] \subset \mathscr{C}(\mathbf{M}\boldsymbol{\Sigma}\mathbf{F}'\mathbf{Q}_{\mathbf{FX}}) = \mathscr{C}(\mathbf{M}\boldsymbol{\Sigma}\mathbf{F}') = \mathscr{C}(\mathbf{M}\mathbf{W}_{\Sigma}\mathbf{F}') , \tag{5.7}$$

where the equality follows from (3.7). Premultiplying (5.4) by $\mathbf{M}$ gives (5.7) at once. Thus, we have proved that (5.4) implies (5.3). Notice that in the light of the second part of Theorem 3.4 the implication (5.3) $\implies$ (5.5) holds in the situation when

$$\mathscr{C}(\mathbf{X}) \cap \mathscr{C}[\mathbf{Z}(\mathbf{DZ}' + \mathbf{S}')\mathbf{M}] = \{\mathbf{0}\} .$$

There is one further interesting link connecting the mixed model and the following extended partitioned model:

$$\mathscr{A} = \{\dot{\mathbf{y}}, \dot{\mathbf{X}}\boldsymbol{\pi}, \dot{\mathbf{V}}\}$$
$$= \left\{ \begin{pmatrix} \mathbf{y} \\ \mathbf{y}_0 \end{pmatrix}, \begin{pmatrix} \mathbf{X} & \mathbf{Z} \\ \mathbf{0} & -\mathbf{I}_q \end{pmatrix} \begin{pmatrix} \boldsymbol{\beta} \\ \boldsymbol{\gamma} \end{pmatrix}, \begin{pmatrix} \mathbf{R} & \mathbf{S} \\ \mathbf{S}' & \mathbf{D} \end{pmatrix} \right\} ,$$

where both $\boldsymbol{\beta}$ and $\boldsymbol{\gamma}$ are *fixed* effects parameters. Expressed in error terms we have

$$\mathbf{y} = \mathbf{X}\boldsymbol{\beta} + \mathbf{Z}\boldsymbol{\gamma} + \boldsymbol{\varepsilon} ,$$
$$\mathbf{y}_0 = \qquad -\boldsymbol{\gamma} + \boldsymbol{\varepsilon}_0 ,$$

where $\operatorname{cov}\begin{pmatrix} \mathbf{y} \\ \mathbf{y}_0 \end{pmatrix} = \operatorname{cov}\begin{pmatrix} \boldsymbol{\varepsilon} \\ \boldsymbol{\varepsilon}_0 \end{pmatrix} = \dot{\mathbf{V}}$. Premultiplying the model $\mathscr{A}$ by the matrix

$$\mathbf{T} = (\mathbf{I}_n : \mathbf{Z}),$$

as in Arendacká and Puntanen (2015, Sec. 2), yields the equation

$$\mathbf{y} + \mathbf{Z}\mathbf{y}_0 = \mathbf{X}\boldsymbol{\beta} + \mathbf{Z}\boldsymbol{\varepsilon}_0 + \boldsymbol{\varepsilon}. \tag{5.8}$$

We see that (5.8) defines a mixed model, say $\mathscr{B}$, where the observable response is $\mathbf{w} = \mathbf{y} + \mathbf{Z}\mathbf{y}_0$ and $\boldsymbol{\varepsilon}_0$ is the unobservable random effect, and

$$\mathrm{cov}(\mathbf{y} + \mathbf{Z}\mathbf{y}_0) = \mathrm{cov}(\mathbf{Z}\boldsymbol{\varepsilon}_0 + \boldsymbol{\varepsilon}) = \mathbf{Z}\mathbf{D}\mathbf{Z}' + \mathbf{R} + \mathbf{Z}\mathbf{S}' + \mathbf{S}\mathbf{Z}' = \boldsymbol{\Sigma}.$$

We can denote the resulting mixed model as

$$\mathscr{B} = \{\mathbf{y} + \mathbf{Z}\mathbf{y}_0, \mathbf{X}\boldsymbol{\beta} + \mathbf{Z}\boldsymbol{\varepsilon}_0, \mathbf{D}, \mathbf{R}, \mathbf{S}\}.$$

We can also interpret $\mathscr{B}$ as a fixed effect model and write it as $\mathscr{B} = \{\mathbf{w}, \mathbf{X}\boldsymbol{\beta}, \boldsymbol{\Sigma}\}$, where the random effect is not written up explicitly.

It is now interesting to know whether the BLUEs of $\mathbf{X}\boldsymbol{\beta}$ under $\mathscr{A}$ and $\mathscr{B}$ are equal. We answer to this question using the linear sufficiency concept, while Haslett et al. (2015) and Arendacká and Puntanen (2015) solved this problem using different approaches. To do this, we write $\mathscr{A}$ as

$$\mathscr{A} = \{\dot{\mathbf{y}}, \dot{\mathbf{X}}\boldsymbol{\pi}, \dot{\mathbf{V}}\} = \{\dot{\mathbf{y}}, \dot{\mathbf{X}}_1\boldsymbol{\beta} + \dot{\mathbf{X}}_2\boldsymbol{\gamma}, \dot{\mathbf{V}}\}.$$

First, we notice that $\dot{\mathbf{X}}_1\boldsymbol{\beta}$ (and thereby $\mathbf{X}\boldsymbol{\beta}$) is estimable because $\mathscr{C}(\dot{\mathbf{X}}_1)$ and $\mathscr{C}(\dot{\mathbf{X}}_2)$ are disjoint. Then, we observe that

$$\mathbf{T}' = \begin{pmatrix} \mathbf{I}_n \\ \mathbf{Z}' \end{pmatrix} \in \left\{ \begin{pmatrix} \mathbf{Z} \\ -\mathbf{I}_q \end{pmatrix}^{\perp} \right\} = \{\dot{\mathbf{X}}_2^{\perp}\},$$

i.e., $\mathbf{T}\dot{\mathbf{X}}_2 = \mathbf{0}$ and $\mathrm{rank}(\mathbf{T}) = \mathrm{rank}(\dot{\mathbf{X}}_2^{\perp})$. It is well known by Frisch–Waugh–Lowell theorem that premultiplying $\mathscr{A}$ by orthogonal projector $\dot{\mathbf{M}}_2 = \mathbf{I}_{n+q} - \mathbf{P}_{\dot{\mathbf{X}}_2}$ yields the reduced model under which the BLUE of $\dot{\mathbf{X}}_1\boldsymbol{\beta}$ is the same as in $\mathscr{A}$, that is, $\dot{\mathbf{M}}_2\dot{\mathbf{y}}$ is linearly sufficient for $\dot{\mathbf{X}}_1\boldsymbol{\beta}$. Now $\mathscr{C}(\mathbf{T}') = \mathscr{C}(\dot{\mathbf{M}}_2)$ and hence, in view of (2.1), $\mathbf{T}\dot{\mathbf{y}}$ is also linearly sufficient for $\dot{\mathbf{X}}_1\boldsymbol{\beta}$ and thereby

$$\mathrm{BLUE}(\mathbf{X}\boldsymbol{\beta} \mid \mathscr{A}) = \mathrm{BLUE}(\mathbf{X}\boldsymbol{\beta} \mid \mathscr{B}).$$

For the linear sufficiency in the partitioned model, see also Kala et al. (2017, Sec. 5).

Haslett et al. (2015) and Arendacká and Puntanen (2015) also showed the following:

$$\mathrm{BLUP}(\boldsymbol{\varepsilon}_0 \mid \mathscr{A}) = \mathrm{BLUP}(\boldsymbol{\varepsilon}_0 \mid \mathscr{B}) = \mathrm{BLUE}(\boldsymbol{\gamma} \mid \mathscr{A}) + \mathbf{y}_0.$$

The connection between the models $\mathscr{A}$ and $\mathscr{B}$ can be used as a tool to calculate the BLUEs and BLUPs in mixed model and it is often referred to as a Henderson's method; see, e.g., Henderson et al. (1959) and McCulloch et al. (2008, Ch. 8).

Acknowledgements Thanks go to Professor Xu-Qing Liu and the anonymous referees for helpful comments. Part of this research was done during the meeting of an International Research Group on Multivariate and Mixed Linear Models in the Mathematical Research and Conference Center, Będlewo, Poland, November 2016, supported by the Stefan Banach International Mathematical Center.

References

Arendacká, B., Puntanen, S.: Further remarks on the connection between fixed linear model and mixed linear model. Stat. Pap. **56**, 1235–1247 (2015). https://doi.org/10.1007/s00362-014-0634-2

Baksalary, J.K., Kala, R.: Linear transformations preserving best linear unbiased estimators in a general Gauss–Markoff model. Ann. Stat. **9**, 913–916 (1981). https://doi.org/10.1214/aos/1176345533

Baksalary, J.K., Kala, R.: Linear sufficiency with respect to a given vector of parametric functions. J. Stat. Plan. Inf. **14**, 331–338 (1986). https://doi.org/10.1016/0378-3758(86)90171-0

Baksalary, O.M., Trenkler, G.: A projector oriented approach to the best linear unbiased estimator. Stat. Pap. **50**, 721–733 (2009). https://doi.org/10.1007/s00362-009-0252-6

Christensen, R.: Plane Answers to Complex Questions: The Theory of Linear Models, 4th edn. Springer, New York (2011)

Drygas, H.: The Coordinate-Free Approach to Gauss-Markov Estimation. Springer, Berlin (1970)

Drygas, H.: Sufficiency and completeness in the general Gauss-Markov model. Sankhyā Ser. A **45**, 88–98 (1983)

Groß, J.: A note on the concepts of linear and quadratic sufficiency. J. Stat. Plan. Inf. **70**, 88–98 (1998)

Haslett, S.J., Isotalo, J., Liu, Y., Puntanen, S.: Equalities between OLSE, BLUE and BLUP in the linear model. Stat. Pap. **55**, 543–561 (2014). https://doi.org/10.1007/s00362-013-0500-7

Haslett, S.J., Puntanen, S., Arendacká, B.: The link between the mixed and fixed linear models revisited. Stat. Pap. **56**, 849–861 (2015). https://doi.org/10.1007/s00362-014-0611-9

Henderson, C.R., Kempthorne, O., Searle, S.R., von Krosigh, C.N.: The estimation of environmental and genetic trends from records subject to culling. Biometrics **15**, 192–218 (1959)

Isotalo, J., Puntanen, S.: Linear prediction sufficiency for new observations in the general Gauss–Markov model. Commun. Stat. Theory Methods **35**, 1011–1023 (2006). https://doi.org/10.1080/03610920600672146

Kala, R., Markiewicz, A., Puntanen, S.: Some further remarks on the linear sufficiency in the linear model. In: Natália Bebiano (ed.) Applied and Computational Matrix Analysis: MatTriad, Coimbra, Portugal, Sept 2015, Selected, Revised Contributions. Springer Proceedings in Mathematics and Statistics, vol. 192, 275–294 (2017). https://doi.org/10.1007/978-3-319-49984-0_19

Kala, R., Puntanen, S., Tian, Y.: Some notes on linear sufficiency. Stat. Pap. **58**, 1–17 (2017). https://doi.org/10.1007/s00362-015-0682-2

Liu, X.-Q., Rong, J.-Y., Liu, J.-Y.: Best linear unbiased prediction for linear combinations in general mixed linear models. J. Multivar. Anal. **99**, 1503–1517 (2008). https://doi.org/10.1016/j.jmva.2008.01.004

Markiewicz, A., Puntanen, S.: Further properties of the linear sufficiency in the partitioned linear model. In: Matrices, Statistics and Big Data: Proceedings of the 25th International Workshop on Matrices and Statistics, IWMS-2016, held in Funchal, Madeira. Portugal, 6–9 June 2016. Springer (in press) (2017)

Marsaglia, G., Styan, G.P.H.: Equalities and inequalities for ranks of matrices. Linear Multilinear Algebra **2**, 269–292 (1974). https://doi.org/10.1080/03081087408817070

McCulloch, C.E., Searle, S.R., Neuhaus, J.M.: Generalized, Linear, and Mixed Models, 2nd edn. Wiley, New York (2008)

Puntanen, S., Styan, G.P.H., Isotalo, J.: Matrix Tricks for Linear Statistical Models: Our Personal Top Twenty. Springer, Heidelberg (2011). https://doi.org/10.1007/978-3-642-10473-2

Rao, C.R.: Representations of best linear estimators in the Gauss–Markoff model with a singular dispersion matrix. J. Multivar. Anal. **3**, 276–292 (1973). https://doi.org/10.1016/0047-259X(73)90042-0

Rao, C.R., Mitra, S.K.: Generalized Inverse of Matrices and Its Applications. Wiley, New York (1971)

Tian, Y., Puntanen, S.: On the equivalence of estimations under a general linear model and its transformed models. Linear Algebra Appl. **430**, 2622–2641 (2009). https://doi.org/10.1016/j.laa.2008.09.016

A Note on Circular
m-consecutive-k-out-of-n:F Systems

Cihangir Kan

Abstract In this chapter, more generalized version of m-consecutive-k-out-of-n:F system is introduced in circular case, that is named as circular m-consecutive-k,l-out-of-n:F system. This system consists of n circularly ordered components such that the system fails if and only if there are at least m l-overlapping runs of k consecutive failed components. The parameter l is a leverage in this system which provides that the reliability of this system is bounded by overlapping ($0 < l < k, k > 1$) and nonoverlapping ($l = 0$) cases. The main aim of this contribution is finding $r_i(n)$ which denotes the number of path sets of circular m-consecutive-k,l-out-of-n:F system including i working components. A combinatorial formula, which calculates the exact reliability of this system, is given. Signature-based analysis are illustrated and numerics are provided.

Keywords System signature · Reliability · Combinatorial method · Survival function

Notation

k_ϕ	minimum number of failed components that may cause system failure
z_ϕ	maximum number of failed components such that system functions
$r_i(n)$	number of path sets including i working components
$T_{k,m:n}^{(l)}$	lifetime of circular m-consecutive-k,l-out-of-n:F system

C. Kan (✉)
Department of Mathematical Sciences, Xi'an Jiaotong-Liverpool University,
Suzhou, China
e-mail: Cihangir.Kan@xjtlu.edu.cn; kancihangir@gmail.com

© Springer International Publishing AG 2018
M. Tez and D. von Rosen (eds.), *Trends and Perspectives
in Linear Statistical Inference*, Contributions to Statistics,
https://doi.org/10.1007/978-3-319-73241-1_9

1 Introduction

If the ordered components are arranged in a line then the system is called linear whereas if they are arranged to form a circle then it is called a circular system. In words, n components can be ordered linearly or circularly (where first component and nth components are adjacent). A circular m-consecutive-k-out-of-n:F system with nonoverlapping runs consists of n components which are circularly ordered such that system fails iff there are at least m nonoverlapping runs of k consecutive failed components. This system has been introduced by Griffith (1986) in 1986. Later, Boland and Papastavridis (1999), Papastavridis (1990), Makri and Philippou (1996), Agarwal et al. (2007), and Eryilmaz et al. (2011b) studied on this system. Similarly, a circular m-consecutive-k-out-of-n:F system with overlapping runs consists of n components, which are circularly ordered, such that the system fails iff there are at least m overlapping runs of k consecutive failed components. This system has been studied by Agarwal and Mohan (2008) and Eryilmaz (2012a). Overlapping means that runs have common components. These linear and circular system models generalize the consecutive-k-out-of-n:F system which is first introduced by Chiang and Niu (1981) in 1981 and then studied by Bollinger and Salvia (1982), Shanthikumar (1982), Derman et al. (1982). Recent discussions on such systems are in Gera (2011), Levitin and Dai (2011), Eryilmaz (2012b), and Cui et al. (2015). The application areas of such system are oil pipeline systems, vacuum system in an electron accelerator, computer ring networks, and microwave stations of a telecom network. For instance, consider a microwave signal transmitting system combined of many stations which are ordered linearly or circularly. The system fails iff at least k adjacent stations fail in the system. This chapter mentions a more generalized version of m-consecutive-k-out-of-n:F system in circular cases, that is named as circular m-consecutive-k, l-out-of-n:F system where the linear case is introduced by Eryilmaz and Mahmoud (2012). For $l = 0$ and $l = k - 1$, this system turns into circular m-consecutive-k-out-of-n:F system with nonoverlapping runs and circular m-consecutive-k-out-of-n:F system with overlapping runs, respectively. For $m = 1$, ordinary consecutive-k-out-of-n:F system is obtained. The circular m-consecutive-k,l-out-of-n:F system combined of n circularly ordered components such that the system fails iff there are at least m l-overlapping runs of k consecutive components ($n \geq m(k - l) + l, l < k$). This system has wider applications in specific areas such as infrared detecting and bank automatic payment systems. For instance, consider a system which is constructed by the circularly arranged identical and independent transmitters. The main principle of the given system is collecting the sufficient amount of information and transferring it. Basic functioning principle for the system can be defined as follows:

The information is gathered by the k consecutive transmitters. New information can be provided with the reuse of the given number of these transmitters which is denoted by "l". When at least m blocks of k consecutive working transmitters collect information then it is transferred. This is an example for G-system. On the other hand,

Fig. 1 Circularly ordered
transmitters

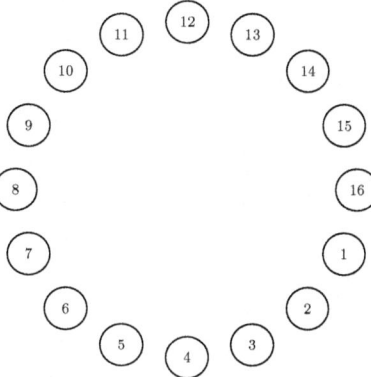

for F systems, the system fails iff there are at least m blocks of length k consecutive disfunctioning transmitters, consisting l reused ones.

In Fig. 1, the system contains 16 transmitters that are circularly ordered (where first component and 16th components are adjacent) and 0s and 1s represent functioning and disfunctioning transmitters respectively. For m = 3 and k = 4 the following line up 1111101111110110 will function for $l = 0, 1$ and will fail for $l = 2, 3$.

In this work, the number of path sets including a certain number of working components in a circular arrangement is obtained by using the binomial distribution of order k for l-overlapping runs of length k, studied by Aki and Hirano (2000) and Makri and Philippou (2005). After derivation of this formula, reliability and system signature of this system has been computed. All proofs can be found in the Appendix. The statistical inference procedure can be done based on a random sample of lifetimes of independent circular *m*-consecutive-*k,l*-out-of-*n*:F systems. The problem of estimating the parameter of the common distribution of components' lifetimes from system's lifetime data can be studied similarly as Eryilmaz (2011a) did.

2 Reliability Evaluation

Assume that a system contains independent and identically distributed components having common reliability p, then the reliability of whole system can be computed as

$$R = \sum_{i=n-z_\phi}^{n} r_i(n)p^i(1-p)^{n-i} \tag{1}$$

where $r_i(n)$ denotes the number of path sets of a system including i working components and z_ϕ is the minimum number of working components such that the system can still work successfully.

Lemma 1 *For a circular m-consecutive-k,l-out-of-n:F system,*

$$k_\phi = k + (m-1)(k-l)$$

and

$$z_\phi = n - 2 - \left\lfloor \frac{n-k-(m-1)(k-l)-1}{k} \right\rfloor$$

where $n \geq m(k-l) + l, l < k$ and $\lfloor x \rfloor$ denotes the integer part of x.

Lemma 2 (Makri et al. (2007)) *Let $C(\beta; \alpha, r - \alpha; m_1 - 1, m_2 - 1)$ be the number of allocations of β indistinguishable balls into r distinguishable cells, α specified of which have capacity $m_1 - 1$ and each of the rest $r - \alpha$ has capacity $m_2 - 1$. Then*

$$C(\beta; \alpha, r - \alpha; m_1 - 1, m_2 - 1)$$

$$= \sum_{j_1=0}^{\left\lfloor \frac{\beta}{m_1} \right\rfloor} \sum_{j_2=0}^{\left\lfloor \frac{\beta - m_1 j_1}{m_2} \right\rfloor} (-1)^{j_1+j_2} \binom{\alpha}{j_1} \binom{r-\alpha}{j_2} \binom{\beta - m_1 j_1 - m_2 j_2 + r - 1}{r-1}$$

Note that $C(n - i; i; k - 1)$ represents the number of ways of $n - i$ success can be placed into i linear cells with no cell receiving more than $k - 1$. Considering cyclic arrangement, each such arrangements gives n arrangements by rotation. But the set of the $nC(n - i; i; k - 1)$ arrangements is partitioned into sets of i like arrangements. So, in circular case

$$C^c(n - i; i; k - 1) = \frac{n}{i} C(n - i; i; k - 1)$$

where $C^c(n - i, i, k - 1)$ denotes the number of cyclic arrangement of $n - i$ successes and i cells such that each cell contains at most $k - 1$ consecutive success. Now, by using Lemma 2, $r_i(n)$ can be calculated for the circular m-consecutive-k,l-out-of-n:F system.

Theorem 3 *The number of path sets of circular m-consecutive-k,l-out-of-n:F system including i working components is*

$$r_i(n) = \frac{n}{i} [C(n - i; i, 0; k - 1, k - 1)$$

$$+ \sum_{s=1}^{m-1} \sum_{a=1}^{\min(i,s)} \binom{i}{a} \binom{s-1}{a-1} C(n - i - al - s(k-l); a, i - a; k - l - 1, k - 1)]$$

for $n - z_\phi \leq i \leq n$.

By using Theorem, one can compute the reliability of the circular m-consecutive-k,l-out-of-n:F system.

In Fig. 2, it can be easily seen that the reliability of the circular 2-consecutive-3,l-out-of-10:F system decreases by l and it is bounded by circular 2-consecutive-3,l-out-of-10:F system with overlapping and nonoverlapping runs for $l = 0, 1, 2$. For

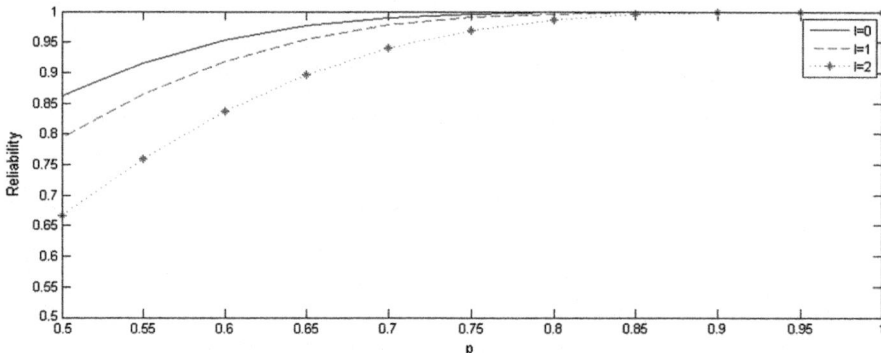

Fig. 2 The reliability of the circular 2-consecutive-3, 1-out-of-10:F system for l = 0, 1 and 2

Table 1 All possible binary sequences for $l = 0$ and $l = 1$

Number of working components

$l = 0$	1	2	3	4	5
	–	00111	00011	00001	00000
		01011	00101	00010	
		10011	01001	00100	
		01101	10001	01000	
		10101	00110	10000	
		11001	01010		
		01110	10010		
		10110	01100		
		11010	10100		
		11100	11000		
$r_i(5)$	0	10	10	5	1
$l = 1$	1	2	3	4	5
	–	01011	00011	00001	00000
		01101	00101	00010	
		10101	01001	00100	
		10110	10001	01000	
		11010	00110	10000	
			01010		
			10010		
			01100		
			10100		
			11000		
$r_i(5)$	0	5	10	5	1

example, one can illustrate the computation of reliability not only by using $r_i(n)$ but also by hand for the values $n = 5$, $m = 2$, and $k = 2$. In Table 1, the possible binary states are listed for both $l = 0$ and $l = 1$.

For $l = 0$, the reliability of circular 2-consecutive-2,0-out-of-5:F system is

$$R = 10p^2(1 - p)^3 + 10p^3(1 - p)^2 + 5p^4(1 - p) + p^5$$

which can also be computed by using Eq. (1) for $r_i(5) = (0, 10, 10, 5, 1)$.

By using the same way, we can calculate the reliability of circular 2-consecutive-2,1-out-of-5:F system. The functioning states are shown in Table 1, therefore the reliability of circular 2-consecutive-2 ,1-out-of-5:F system is

$$R = 5p^2(1 - p)^3 + 10p^3(1 - p)^2 + 5p^4(1 - p) + p^5$$

which can verify that the computations are true with the corresponding $r_i(5) = (0, 5, 10, 5, 1)$ for $l = 1$.

3 System Signature

Let T_i denote the lifetime of the ith component in a coherent system with the structure function and lifetime T. Then

$$T = \phi(T_1, T_2, \ldots, T_n)$$

If T_i's are s-independent and have common absolutely continuous distribution function, then the survival function can be stated as

$$P(T > t) = \sum_{i=1}^{n} p_i P(T_{i:n} > t) \tag{2}$$

where $T_{1:n} \leq T_{2:n} \leq \cdots \leq T_{n:n}$ is the order statistics associated with $T_1, T_2, \ldots, T_n$ and $p_i = P(T = T_{i:n})$, in other words,

$$p_i = \frac{\text{The number of orderings for which the } i\text{th failure causes the system failure}}{n!}$$

for $i = 1, 2, \ldots, n$. which is well known as Samaniego's Signature (Samaniego 2007). The ith element of the signature vector can be easily computed from

$$p_i = \frac{r_{n-i+1}(n)}{\binom{n}{n-i+1}} - \frac{r_{n-i}(n)}{\binom{n}{n-i}}, \quad \text{for } i = 1, 2, \ldots, n.$$

(Boland 2001). Hence, applying the formula (2) for the structure of circular *m*-consecutive-*k,l*-out-of-*n* : F system, we can restate it as follows

$$P(T > t) = \sum_{i=k_\phi}^{z_\phi+1} p_i P(T_{i:n} > t) \tag{3}$$

In Table 2, the signatures of circular *m*-consecutive-*k,l*-out-of-*n* :*F* system for some values are presented

By using the formula (3), the reliability of circular 2-consecutive-3,*l*-out-of-10:*F* system can be written as

$$P(T^0_{3,2:10} > t) = 0.0119P(T_{6:10} > t) + 0.381P(T_{7:10} > t) + 0.5P(T_{8:10} > t)$$

$$P(T^1_{3,2:10} > t) = 0.0397P(T_{5:10} > t) + 0.2222P(T_{6:10} > t) + 0.4881P(T_{7:10} > t)$$
$$+ 0.25P(T_{8:10} > t)$$

$$P(T^2_{3,2:10} > t) = 0.476P(T_{4:10} > t) + 0.1508P(T_{5:10} > t) + 0.3492P(T_{6:10} > t)$$
$$+ 0.3690P(T_{7:10} > t) + 0.0833P(T_{8:10} > t)$$

and mean time to failure (MTTF) of the system are as

Table 2 The signatures of circular *m*-consecutive-*k* ,*l*-out-of-*n*:*F* system for some values

n	*m*	*k*	*l*	**p**
10	2	3	0	$(0,0,0,0,0,0.119,0.381,0.5,0,0)$
			1	$(0,0,0,0,0.0397,0.2222,0.4881,0.25,0,0)$
			2	$(0,0,0,0.476,0.1508,0.3492,0.3690,0.0833,0,0)$
10	3	3	0	$(0,0,0,0,0,0,0,0,1,0)$
			1	$(0,0,0,0,0,0,0.0833,0.5833,0.3333,0)$
			2	$(0,0,0,0,0.0397,0.1508,0.4762,0.3333,0,0)$
10	2	4	0	$(0,0,0,0,0,0,0,0.3333,0.6667,0)$
			1	$(0,0,0,0,0,0,0.0833,0.4722,0.4444,0)$
			2	$(0,0,0,0,0,0.0476,0.2024,0.5278,0.2222,0)$
			3	$(0,0,0,0,0.0397,0.1518,0.3095,0.5,0,0)$
12	2	4	0	$(0,0,0,0,0,0,0,0.0606,0.2667,0.4909,0.1818,0)$
			1	$(0,0,0,0,0,0,0.0152,0.1182,0.3576,0.5091,0,0)$
			2	$(0,0,0,0,0,0.013,0.0628,0.203,0.4303,0.2909,0,0)$
			3	$(0,0,0,0,0.0152,0.0628,0.1494,0.2939,0.3515,0.1273,0,0)$
12	3	3	0	$(0,0,0,0,0,0,0,0,0.1818,0.5455,0.2727,0)$
			1	$(0,0,0,0,0,0,0.0152,0.1545,0.5030,0.3273,0,0)$
			2	$(0,0,0,0,0.0152,0.0628,0.21,0.3788,0.3333,0,0,0)$

$$E(T^0_{3,2:10}) = 0.0119E(T_{6:10}) + 0.381E(T_{7:10}) + 0.5E(T_{8:10})$$

$$E(T^1_{3,2:10}) = 0.0397E(T_{5:10}) + 0.2222E(T_{6:10}) + 0.4881E(T_{7:10}) + 0.25E(T_{8:10})$$

$$E(T^2_{3,2:10}) = 0.476E(T_{4:10}) + 0.1508E(T_{5:10}) + 0.3492E(T_{6:10}) + 0.3690E(T_{7:10})$$
$$+ 0.0833E(T_{8:10})$$

The signature of a system has been found to be useful for comparing systems in terms of various stochastic orderings. Let X and Y be two independent random variables having cumulative distribution F and G. X is said to be smaller than Y in stochastic ordering (denoted by $X \overset{st}{\leq} Y$) if their respective survival functions satisfy the inequality $P(X > x) \leq P(Y > x)$ for all x, or equivalently $E(\psi(X)) \leq E(\psi(Y))$ for all increasing functions ψ for which the expectation exists. Let $\mathbf{p} = (p_1, \ldots, p_n)$ and $\mathbf{q} = (q_1, \ldots, q_n)$ be, respectively, the signatures of coherent systems $T = \phi(T_1, \ldots, T_n)$ and $Z = \psi(T_1, \ldots, T_n)$, both used on n i.i.d. components. In 1999, Kochar et al. (1999) proved that if $\mathbf{p} \leq_{st} \mathbf{q}$, then $T \leq_{st} Z$. By using Table 2, it is easy to see that signature of circular m-consecutive-k,l-out-of-n:F system is stochastically less or equal than the signature of circular m-consecutive-k,$l - 1$-out-of-n:F system for $l = 1, \ldots, k - 1$. Assuming lifetime distributions of system components are same and only changing the l parameter it can be expected that

$$T^{(l)}_{k,m:n} \leq_{st} T^{(l-1)}_{k,m:n} \text{ for } l = 1, \ldots, k - 1.$$

Table 3 MTTF of circular m-consecutive-k,l-out-of-n:F system for some values of n, m, k, and l

n	m	k	l	$MTTF$
10	2	3	0	1.2325
			1	1.1056
			2	0.9389
10	3	3	0	1.9290
			1	1.5679
			2	1.1512
10	2	4	0	1.7623
			1	1.6234
			2	1.4448
			3	1.2067
12	2	4	0	1.5699
			1	1.4032
			2	1.2798
			3	1.1224
12	3	3	0	1.6790
			1	1.3335
			2	1.0305

Assuming T_i's are s-independent and have common exponential distribution function with mean 1, the expected value of the *i*th smallest component is equal to

$$E(T_{i:n}) = \sum_{j=1}^{i} \frac{1}{n-j+1}$$

for $i = 1, 2, \ldots, n$. By using this expectation one can easily compute the MTTF of circular *m*-consecutive-*k*,*l*-out-of-*n*:F system for some values of n, m, k, and l as follows:

In Table 3, we see that the MTTF is increasing in *m* and *k* and decreasing in *n* and *l* which is consistent with Fig. 2.

4 Summary and Conclusions

In this chapter, more generalized version of *m*-consecutive-*k*-out-of-*n*:F system is introduced in circular case, that is named as circular *m*-consecutive-*k*,*l*-out-of-*n*:F system. The parameter *l* is a leverage in this system which provides that the reliability of this system is bounded by overlapping $(0 < l < k, k > 1)$ and nonoverlapping $(l = 0)$ cases. The main aim of this contribution is finding $r_i(n)$ which denotes the number of path sets of circular *m*-consecutive-*k*,*l*-out-of-*n*:F system including *i* working components. A combinatorial formula, which calculates the exact reliability of this system, is given. Signature-based analysis are illustrated and numerics are provided.

Acknowledgements The author is grateful to the referees and associate editors for their helpful comments and suggestions, which have significantly improved the contribution.

Appendix

Proof of Lemma 2. A circular *m*-consecutive-*k*,*l*-out-of-*n*:F system combined of *n* circularly ordered components such that the system fails iff there are at least *m* *l*-overlapping runs of *k* consecutive components $(n \geq m(k - l) + l, l < k)$. So, the minimum number of failed components such that system fails can be found as

$$k_\phi = k + (m - 1)(k - l)$$

For finding the maximum number of failed components such that system still functions, which is denoted by z_ϕ, we can consider a binary sequence of runs which are cyclically arranged as follows

$$\underbrace{011\ldots10}_{x}\underbrace{11\ldots}_{n-x-2}$$

where x denotes the number of failed components such that total number of l-overlapping runs is $m-1$, and the remaining part $n-x-2$ runs can obtain a maximum of

$$n-x-2-\left\lfloor\frac{n-x-2}{k}\right\rfloor$$

failures. The maximum value of x is $k+(m-1)(k-l)-1$ so

$$z_\phi = n-k-(m-1)(k-l)+1-2-\left\lfloor\frac{n-k-(m-1)(k-l)+1-2}{k}\right\rfloor$$
$$= n-2-\left\lfloor\frac{n-1-m(k-l)-l}{k}\right\rfloor$$

∎

Proof of Theorem 3. Consider a binary sequence of runs which are cyclically arranged as follows

$$\underbrace{11\ldots10}_{x_1}\underbrace{1\ldots10}_{x_2}\ldots0\underbrace{1\ldots10}_{x_i}$$

such that for $s = 0, 1, \ldots, m-1$,

$$x_1 + x_2 + \cdots + x_i = n - i \tag{4}$$

where

$$\left\lfloor\frac{x_1-l}{k-l}\right\rfloor + \cdots + \left\lfloor\frac{x_a-l}{k-l}\right\rfloor = s$$

for a of $x_j \times s \geq k$ and $i - a$ of $x_j \times s < k$. and $\left\lfloor\frac{x_j-l}{k-l}\right\rfloor$ denotes the number of l-overlapping runs of length k in the ith failure run.

By using Theorem 4.1 of Makri et al. (2007), the number of path sets of a circular m-consecutive-k, l-out-of-n system can be calculated as follows

$$r_i(n) = \frac{n}{i}\sum_{s=0}^{m-1}\sum_{a}\binom{i}{a}N(i,a,k,l,s,n)$$

where $N(i,a,k,l,s,n)$ denotes the number of integer solution to (4). Let $y_j = x_j - l$ for $j = 1, 2, \ldots, a$ and $y_j = x_j$ for $j = a+1, a+2, \ldots, i$.
Then (4) is equivalent to

$$y_1 + y_2 + \cdots + y_i = n - i - al \tag{5}$$

st

$$\left\lfloor \frac{y_1}{k-l} \right\rfloor + \cdots + \left\lfloor \frac{y_a}{k-l} \right\rfloor = s$$

$y_1, y_2, \ldots, y_a \geq k - l$ and $0 \leq y_{a+1}, y_{a+2}, \ldots, y_i < k$.
Let $\left\lfloor \frac{y_j}{k-l} \right\rfloor = z_j$ for $j = 1, 2, \cdots, a$. Then (5) is equivalent to

$$y_1 + y_2 + \cdots + y_i = n - i - al \tag{6}$$

st

$$z_j(k - l) \leq y_j < z_j(k - l) + k - l \text{ for } j = 1, 2, \ldots, a$$
$$0 \leq y_j < k \text{ for } j = a + 1, a + 2, \ldots, i$$

and

$$z_1 + z_2 + \cdots + z_a = s \tag{7}$$

st

$$z_i > 0 \text{ for } i = 1, 2, \ldots, a.$$

The number of integer solutions to (7) is $\binom{s-1}{a-1}$ and by using Lemma 2.1 of Makri et al. (2007) and taking $u_j = y_j - (k - l)z_j$ for $j = 1, 2, \ldots, a$ we can obtain the number of integer solutions to (6) as $N(i, a, k, l, s, n) = \binom{s-1}{a-1} C(n - i - al - s(k - l); a, i - a; k - l - 1, k - 1)$ for $s = 1, 2, \ldots, m - 1$. For $s = 0$, the number of integer solutions can be obtained as $N(i, a, k, l, s, n) = C(n - i; i, 0; k - 1, k - 1)$. ∎

References

Agarwal, M., Mohan, P., Sen, K.: GERT analysis of *m*-consecutive-*k*-out-of-*n*:F systems with dependence. Econ. Quality Cont. **22**, 141–157 (2007). https://doi.org/10.1515/EQC.2007.141

Agarwal, M., Mohan, P.: GERT analysis of *m*-consecutive-*k*-out-of-*n*:F system with overlapping runs and $(k - 1)$-step markov dependence. Int. J. Oper. Res. **3**, 36–51 (2008). https://doi.org/10.1504/IJOR.2008.016153

Aki, S., Hirano, K.: Numbers of success runs of specified length until certain stopping time rules and generalized binomial distributions of order *k*. Annals Inst. Stat. Math. **52**, 767–777 (2000). https://doi.org/10.1023/A:1017585512412

Boland, P.J., Papastavridis, S.: Consecutive *k*-out-of-*n* systems with cycle *k*. Stat. Prob. Lett. **44**, 155–160 (1999). https://doi.org/10.1016/S0167-7152(99)00003-6

Boland, P.J.: Signatures of indirect majority systems. J. Appl. Prob. **38**, 597–603 (2001). https://doi.org/10.1017/S0021900200020064

Bollinger, R.C., Salvia, A.A.: Consecutive-*k*-out-of-*n* networks. IEEE Trans. Reliab. **31**, 53–56 (1982). https://doi.org/10.1109/TR.1982.5221227

Cui L., Lin, C., Du, S.: m-Consecutive-k,l-Out-of-n Systems. IEEE Trans. Rel. **64**, 386–393 (2015). https://doi.org/10.1109/TR.2014.2337091

Chiang, D.T., Niu, S.C.: Reliability of consecutive-k-out-of-n:F system. IEEE Trans. Reliab. **30**, 87–89 (1981). https://doi.org/10.1109/TR.1981.5220981

Derman, C., Lieberman, G.J., Ross, S.M.: On the consecutive-k-out-of-n:F system. IEEE Trans. Reliab. **31**, 57–63 (1982)

Eryilmaz, S.: Estimation in coherent reliability systems through copulas. Reliab. Eng. Syst. Safety **96**, 564–568 (2011a). https://doi.org/10.1016/j.ress.2010.12.024

Eryilmaz, S., Koutras, M.V., Triantafyllou, I.S.: Signature based analysis of m-consecutive-k-out-of-n:F systems with exchangeable components. Naval Res. Log. **58**, 344–354 (2011b). https://doi.org/10.1002/nav.20449

Eryilmaz, S.: m-consecutive-k-out-of-n:F system with overlapping runs: signature-based reliability analysis. Int. J. Oper. Res. **15**, 64–73 (2012a). https://doi.org/10.1504/IJOR.2012.048292

Eryilmaz, S.: Reliability of Combined m-consecutive-k-out-of-n:F and consecutive k_c-out-of-n:F systems. IEEE Trans. Reliab. **61**, 215–219 (2012b). https://doi.org/10.1109/TR.2011.2182401

Eryilmaz, S., Mahmoud, B.: Linear m-consecutive-k,l-out-of-n :F system. IEEE Trans. Reliab. **61**, 787–791 (2012). https://doi.org/10.1109/TR.2012.2207573

Gera, A.E.: Combined m_1-consecutive-k_{c_1}-out-of-n and m_2-consecutive-k_{c_2}-out-of-n systems. IEEE Trans. Reliab. **60**, 493–497 (2011). https://doi.org/10.1109/TR.2011.2136550

Griffith, W.S.: On consecutive-k-out-of-n: failure systems and their generalizations. In: Basu, A.P. (ed.), Reliability and Quality Cont., pp. 157–165. Elsevier, Amsterdam, The Netherlands (1986)

Kochar, S., Mukerjee H., Samaniego, F.: The "signature" of a coherent system and its application to comparison among systems. Naval Res. Log. **46**, 507–523 (1999). https://doi.org/10.1002/(SICI)1520-6750(199908)46:5<507::AID-NAV4>3.0.CO;2-D

Levitin, G., Dai, Y.: Linear m-consecutive-k-out-of-r-from-n:F systems. IEEE Trans. Reliab. **60**, 640–646 (2011). https://doi.org/10.1016/j.ress.2010.09.009

Makri, F.S., Philippou, A.N.: Exact reliability formulas for linear and circular m-consecutive-k-out-of-n:F systems. Microelectr. Reliab. **36**, 657–660 (1996). https://doi.org/10.1016/0026-2714(95)00153-0

Makri, F.S., Philippou, A.N.: On binomial and circular binomial distributions of order k for l-overlapping success runs of length k. Stat. Papers **46**, 411–432 (2005)

Makri, F.S., Philippou, A.N., Psillakis, Z.M.: Polya, inverse polya, and circular polya distributions of order k for l-overlapping success runs. Comm. Stat. Theory Methods **36**, 657–668 (2007)

Papastavridis, S.: m-consecutive-k-out-of-n systems. IEEE Trans. Reliab. **39**, 386–387 (1990)

Samaniego, F.J.: System Signatures and their Applications in Engineering Reliability. Springer, New York (2007)

Shanthikumar, J.G.: Recursive algorithm to evaluate the reliability of a consecutive-k-out-of-n:F system. IEEE Trans. Reliab. **31**, 442–443 (1982). https://doi.org/10.1109/TR.1982.5221423

A Categorical Principal Component Regression on Computer-Assisted Instruction in Probability Domain

Tuğba Kapucu, Ozlem Ilk and İnci Batmaz

Abstract The purpose of this study is to examine the effects of computer-assisted instructional material (CAIM) prepared in R program on eighth grade students' permutation–combination and probability achievement and their attitudes toward computer-assisted learning. In the study, we collect survey data from 74 conveniently selected students and their schools; data consists of 45 highly correlated explanatory variables with different measurement levels. To deal with the multicollinearity problem among mixed type of explanatory variables, first, we apply categorical principal components analysis (CATPCA), and hence, reduce the dimension of data. In the following, we use uncorrelated components instead of the original correlated variables to fit the multiple linear regression (MLR) model in order to question whether CAIM has been effective in teaching probability domain. Results show that the general success of the students and basic socioeconomic and technological factors affecting this situation and interaction of those factors with the secondary social situation of the student's family have statistically significant effects on the probability achievement of the students. Instruction method is also found as a statistically significant factor in explaining students' achievement in permutation–combination subjects. However, none of the explanatory variables considered in the study are found statistically significant in explaining the attitudes of students toward computer-assisted instruction (CAI).

İnci Batmaz—On sabatical leave from Middle East Techical University, Department of Statistics, 06800 Ankara, Turkey.

T. Kapucu (✉) · O. Ilk
Department of Statistics, Middle East Technical University, Ankara, Turkey
e-mail: tugbakapucuu@gmail.com

O. Ilk
e-mail: oilk@metu.edu.tr

İ. Batmaz
Department of Mathematics, Imperial College London, Statistics Section, London, UK
e-mail: i.batmaz@imperial.ac.uk

© Springer International Publishing AG 2018
M. Tez and D. von Rosen (eds.), *Trends and Perspectives in Linear Statistical Inference*, Contributions to Statistics,
https://doi.org/10.1007/978-3-319-73241-1_10

Keywords Categorical principal components analysis · Computer-assisted instruction · Linear models · Multicollinearity · R software Teaching probability

1 Introduction

Probability is an old branch of mathematics dealing with calculating probability of a variety of events (HodnikČadež and Škrbec 2011). With the change of world conditions, the importance of probability knowledge has also gained importance. As Lopes and Moura (2002) indicated, knowledge of facts of probability is significant not only to make decisions and predictions but also to provide an earlier accession of population to social and economical arguments in which tables and graphics synthesize studies and analysis, and comparison of indexes is done to support ideas. Due to the fact that teaching of probability and statistics advances students' ability of collecting, organizing, interpreting, and comparing data to acquire and support conclusions, it is appropriate to be placed in mathematics curriculum in the Elementary Education (Lopes and Moura 2002).

Even though lack of people's ability in learning the concepts encompassing judgement under uncertainness is mentioned so much over the last three decades, the research on solutions is insufficient (Keeler and Steinhorst 2001). Another research done by Batanero et al. (2005) showed that teachers have lack of experience in probability and transport their probabilistic misconceptions to their students. Most teachers find it difficult to teach probability. They do not rely on their ability of teaching probability, and they are unconfident in the activity of probability including uncertainty (Stohl 2005).

Garfield and Ahlgren (1988) stated that students dislike probability due to exposure of a highly abstract study in a formal way. So, they suggested introducing probability through activities and simulations for overcoming difficulties in teaching probability. Simulation is a new and fundamental tool in all aspects of education today. Chance and Rossman (2006), as reported in Zieffler and Garfield (2007), said that the use of simulation in probability and statistics course as a supportive of instructional efforts is highly recommended by statisticians and researchers. Simulation is a powerful instrument giving the opportunity of repetition of a study many times (Burrill 2002). Using simulation provides students to perceive active processes, rather than stable configurations and representations (Zieffler and Garfield 2007). Using probabilistic simulations and games by teachers in their lectures can be very beneficial for students, thus, students can learn how to make a statement based on evidence and enlightened decisions on data which may cause a smaller chance of error (Souza 2015).

Probability and statistics have been involved in curriculum of school mathematics for less than 40 years and they supplement the traditional topics of arithmetic, algebra, and geometry (Borovcnik and Kapadia 2010). Although almost all countries accept statistics as an inseparable part of mathematics curriculum, the concept of probability is just introduced to older students (Borovcnik and Kapadia 2010). According to Ferreira et al. (2014), students are required to show more advanced skills while learning probability concepts compared to traditional school tasks, because as Borovcnik and Kapadia (2010) indicated that conceptual errors in probability affect all aspects of people's decisions such as assessment, medical tests, etc. Therefore, it is important that probability should be a vital part of elementary school mathematics and effective teaching strategies and materials should be utilized while teaching probability concepts. In the current elementary school mathematics curriculum in Turkey, however, probability is included only in the eighth grade. Besides, students do not show success in questions related to probability both in national and international exams. For instance, Turkey ranked 30th among 38 countries in terms of probability and data analysis subject's score (TIMSS 1999). Furthermore, students have low scores in mathematics in the examination of transition to secondary education from primary education (TEOG).

There are a number of difficulties that confronted in teaching and learning the concept of probability (Gürbüz 2008). According to him, the lack of suitable teaching materials is among the reasons of these difficulties, and computer technology can provide a solution to these problems by developing materials that influence students' learning, positively by simulation and animation (Gürbüz 2008). However, the number of research studies conducted to examine the effectiveness of simulation on mathematics achievement is not sufficient both in Turkey and abroad. Furthermore, there are only a few research studies conducted to examine the effects of simulation in probability instruction (e.g., Garfield and Zieffler 2007; Braun et al. 2013). One of the purposes of the present study is to examine the effect of CAIM prepared in R program on elementary school students' probability achievement. Another concern of the present study is to investigate how students' attitudes toward CAI change while animation and simulation are used as a material in the mathematics education.

In order to accomplish our aims, we set up experiments in four district schools. Two of them are the state schools located in the villages of Mardin, a city in southeastern Turkey. The other two are private schools located in the city centers of Mardin and Ankara (the capital of Turkey). These schools are particularly selected to consider the effects of socioeconomic and cultural differences in the analysis. Data obtained from the experiments conducted on 74 students contain several exploratory variables with different measurement levels. To inquire about our research questions mentioned above, we intend to develop a multiple linear regression (MLR) model between variables in our study. MLR assumes that exploratory variables in the model are independent of each other. However, in high-dimensional data, such as the one that we work on here, multicollinearity among the variables is unavoidable. Hence, the first step is to reduce the data dimension. We achieve this by applying the categorical principal components analysis (CATPCA).

The chapter is organized as follows. In Sect. 2, literature review on CAI is presented. The CATPCA is described in Sect. 3. In Sect. 4, the application and its findings are given. Concluding remarks are stated in the last section.

2   Literature Review

Educational technology has the meaning of the media intention which concentrates on materials and equipment, i.e., delivery systems (Ely and Plomp 1986). Tickton (1970), as reported in Ely and Plomp (1986), defined the educational technology as "the media born of the communications revolution which can be used for instructional purposes alongside the teacher, textbook and blackboard." The pieces that make up instructional technology are television, films, overhead projectors, computers, and the other items of "hardware" and "software". Instructional technology is a component of educational technology. While educational technology is a broader concept, instructional technology is a more specific concept that contains mechanism of learning and instruction. Computers and computer software are among the most known teaching materials but instructional technology is not limited to these tools.

Computer-assisted instruction (CAI) is a teaching method in which computer is used as a supportive tool for teacher in lectures. CAI strengthens teaching process and students' motivations by making possible for students to learn with respect to their learning speed (Şahin and Yildirim 1999). It is vital to improve and use instructional materials and activities that evoke students' visual and intellectual frame. By extension of technological advances, technological devices, especially computers are started to be used in educational environments to develop audiovisual materials such as animation and simulation. The use of simulations and animations associated with abstract concepts enables students to participate in learning process and to structure the concepts easily in their minds (Karamustafaoğlu et al. 2005).

In the literature, there are some research studies conducted to examine the effectiveness of CAI methods on students' probability achievement. In one of these studies, Gürbüz (2008) presented CAIM for teaching probability topic at primary school. In the light of this study, CAIM including simulations and animations was developed through the instrument of Dreamweaver MX 2004 and Flash MX 2004. In another study, Ferreira et al. (2014) discussed aspects of high school students' learning of probability in a context where they were supported by the statistical software R. In a study carried out by Braun et al. (2013), how to introduce young students to the ideas of randomness and uncertainty by using statistical program R in elementary schools was investigated.

In the light of these researches, it was found that CAI was more effective than traditional instruction for teaching probability concepts, and CAIMs would be effective in teaching and learning of probability topics. In the present study, a CAI material including a sequence of animation and simulation activities to illustrate basic probabilistic concepts is improved by using statistical program R. It aims to investigate the effect of CAI method compared to traditional instructional method on students' probability achievement. In addition, unlike the other studies, this study also examines how the socioeconomic and cultural factors affect the achievement.

3 The Method

3.1 Categorical Principal Components Analysis (CATPCA)

In the study, there are a large number of correlated categorical variables, which we wish to reduce to a small number of dimensions with as little loss of information as possible. Principal component analysis (PCA) or factor analysis is considered to be appropriate ways to be executed in such a data reduction process. The aim of PCA or factor analysis is to reduce the number of m variables to smaller number of p uncorrelated linear combinations of these variables, called principal components, while maximizing the amount of variance in the data as much as possible (Everitt and Hothorn 2011). However, traditional PCA is not a suitable method of data reduction for categorical variables, since variables in PCA are assumed to be scaled at numeric level (interval or ratio level of measurement) and also linear relationship among variables is required. Alternative to traditional PCA, categorical (also known as nonlinear) principal components analysis (CATPCA/NLPCA), which overcomes limitations of PCA, is used in this study.

CATPCA is an alternative data reduction technique to PCA concerned with identifying fundamental components of a set of variables while maximizing the amount of variance accounted for by the principal components, for the variables which are categorical (e.g., nominal, ordinal, and even numeric). The purpose of CATPCA is equivalent to that of PCA, namely to reduce a dataset consisting of many correlated variables to a smaller number of uncorrelated summary variables (principal components) that represent the observed data as closely as possible. The main difference between the methods is that, while PCA detects a linear relationship between variables, CATPCA can also detect nonlinear relationships by quantifying categorical or nonlinearly related variables in an optimal way to achieve the PCA goal (Linting and Van der Kooji 2012). It is important to underline the fact that when all variables in the dataset are continuous and the linearity assumption between the variables is satisfied, CATPCA and PCA give exactly the same solutions.

Gifi (1990), as cited in Mair and Leeuw (2010), proposed a wide collection of nonlinear multivariate methods based on "optimal quantification". NLPCA/ CATPCA uses optimal quantification (also known as optimal scaling, or optimal scoring) approach to assign numeric values to categories of variables (Linting and Van der Kooji 2012). Optimal quantification is a process which converts the category labels into category quantifications by maximizing the variance accounted for among the quantified variables (Linting and Van der Kooji 2012). A 0–1 dummy matrix based on the data consisting of categorical variables is the starting point of the analysis. Afterward, a loss function, L, comprising the (unknown) object and the category scores are constituted. These variables are stretched during the iterations and category scores are computed in such a way that they are optimal in terms of a minimal loss function. In the Gifi model, the transformations compared to that of regression such as exponential, logarithmic, or square root transformations are unknown: The categories are quantified according to a certain criterion. In the given approach, an objective function is selected and then examined how the target function alters the overall possible transformations of the variables, and eventually, quantifies the categories accordingly (Mair and Leeuw 2010).

Suppose we have measurements of n objects or individuals on m variables collected in an $n \times m$ observed score matrix $\mathbf{X}$, where each variable is denoted by $X_j, j = 1, 2, \ldots, m$, that is the jth column of $\mathbf{X}$. If the variables X_j are nominal or ordinal, then a nonlinear transformation, namely optimal quantification, is required to transform observed scores into category quantification, q_j, given by

$$q_j = \varphi_j(X_j). \tag{1}$$

Thus, $Q = (q_1, q_2, \ldots, q_m)$ is the matrix of category quantification with q_j as the vector of computed category scores.

As mentioned above, we should determine the optimal quantification of categorical variables. For this purpose, an "aspect", ϕ, a function of the correlational matrix $\mathbf{R}$, is defined as the criterion to be optimized. Broadly, the optimization problem can be formulated as

$$Max \, \phi(R(X)). \tag{2}$$

In other words, the observed variables (categories) are scaled in such a way that the correlation matrix based on these scores is maximized (Mair and Leeuw 2010). At this point, we specify "eigenvalue aspects", which aim to maximize the largest eigenvalue λ of $\mathbf{R}$. Here, the aim of the optimal quantification is to optimize the first p eigenvalues of the correlation matrix of the quantified variables, where p represents the number of components defined by the user in the analysis.

Let $\mathbf{S}$ be the $n \times p$ matrix of object scores, which are the scores of the individuals on the principal components obtained by the CATPCA. The object scores are multiplied by a set of optimal weights, named as "component loadings". Also, let $\mathbf{A}$ be an $m \times p$ matrix of the component loadings, where the jth column is denoted by a_j. Then the loss function, L, for minimization of difference between the original data and principal components is obtained as

$$L(Q, A, S) = n^{-1} \sum_{j=1}^{m} tr\left(q_j a_j^T - S\right)^T \left(q_j a_j^T - S\right), \tag{3}$$

where tr is the trace function, that is, for any matrix $\mathbf{A}$, $tr\left(A^T A\right) = \sum_i \sum_j a_{ij}^2$.

As a result, the CATPCA is carried out by minimizing the least squares (LS) loss function L given in (3), in which the matrix $\mathbf{X}$ is replaced by the matrix $\mathbf{Q}$. Here, the loss function L is exposed to some limitations. First, the transformed variables are standardized to overcome the indeterminacy between q_j and a_j, that is $q_j^T q_j = n$. This standardization implies that q_j contains z-scores and as a result, the component loadings in a_j are correlations among transformed variables and principal components. Also, the object scores are restricted to avoid the trivial solution by $S^T S = nI$, where I is the identity matrix. However, the object scores are centered, i.e., $1^T S = 0$, where 1 is a vector of ones. Linting et al. (2007a) stated that these restrictions imply the columns of S to be orthogonal z-scores (as cited in Kemalbay and Korkmazoğlu 2014). According to Gifi (1990), as cited in Kemalbay and Korkmazoğlu (2014), the minimization of restricted loss function given in (3) is obtained by means of an alternating LS (ALS) algorithm.

Consequently, CATPCA converts categories into numeric values as mentioned above. The most important issues to be considered in this regard are that the correlation matrix in CATPCA is not fixed as opposed to the correlation matrix in the PCA; rather, "analysis level" that is chosen for each of the variables in the active CATPCA process determines the type of quantification. The specified analysis level also decides the amounting freedom permitted in converting the category values into category quantifications (Linting and Kooji, 2012).

3.2 Sample Size Consideration in Data Reduction

Sample size is an important consideration in data reduction methods such as principal component analysis (PCA) or factor analysis (FA), because small sample may lead to erroneous conclusions, and thus, unreliable results (Osborne and Costello 2004). Some guidelines to determine an adequate sample size suggested in literature depend on either absolute minimum size or number of observations to variables ratio (Arrindell and Van der Ende 1985; Guadagnoli and Velicer 1988; Osborne and Costello 2004). It is also noted that more observations are needed to analyze binary data than those needed for continuous data (Pearson and Mundfrom 2008). In some other studies, however, it is stated that setting a minimum sample size is not "valid or useful", and according to some others, "no absolute rules" can exist on this matter but simply, "more is better (Osborne and Costello 2004)". However, in another studies, it is indicated that how "strong" the data is more important than the size of the sample studied (Costello and Osborne 2005). It is even said that "If factors stay the same, more variables could lead to better results, with small number of observations (Preacher and MacCallum 2002)." Depending on these statements, we may say that there has not been an agreement on this matter yet among researchers. Moreover, according to same papers reviewed (Lingard and Rowlinson 2006), almost half of studies considered use much less observations or smaller number of observations to variables ratio than suggested in literature.

To the best of our knowledge, there are no suggestions regarding minimum size requirements, particularly, for CATPCA, in literature. However, there are few studies on how to determine if the sample size is large enough for this analysis. One suggestion is to use KMO statistic, which measures if we can factorize original variables efficiently (Mendes and Ganga 2013). The other one is measuring "instability" resulted from not having enough sample size by using bootstrapping technique (Linting et al. 2007b).

4 Application and Results

4.1 Experimental Setup

In this study, quasi-experimental research design is conducted in order to investigate the impacts of the simulation–animation activities prepared in R program on the eighth grade students' achievement and attitudes on CAI. During the study, in the experimental group, CAI based on animation–simulation prepared in R program is used as a supplementary teaching tool. On the other hand, the control group is instructed through traditional teaching method during the study.

The subjects of this study are ($N = 74$) eighth grade students from four distinct schools in two different regions of Turkey in the spring semester of the academic year 2015–2016. Two of the schools, "Dereyani Ortaokulu" and "Şehit Öğretmen Fasih Söğüt", are state village elementary schools in Mardin; the third one, "Bahçeşehir College", is a private elementary school in Mardin, and the last one, "İhsan Doğramaci Foundation Bilkent College", is a private elementary school in Ankara. In order to investigate the effectiveness of the CAIM including a sequence of simulation–animation activities to illustrate basic probabilistic concepts, improved by the authors, "Dereyani Elementary State School" is conveniently assigned to experimental group while "Şehit Öğretmen Fasih Söğüt Elementary State School" is assigned to the control group. However, it is difficult to determine the effectiveness of CAIM, because these two schools do not have enough technological equipment, and the students are not familiar with technology. Despite the possible deviations that may occur, two private schools, "Bahçeşehir" and "Bilkent" Colleges, are included in the study to examine the cultural impact. Only CAI is given to both of these private elementary schools to identify potential cultural impact.

4.2 Description of Variables

There are five groups of variables in this study. Three of them are independent variables and two groups are dependent variables. Independent variables consist of 1. Treatment (Traditional method versus CAI with R program), 2. Students' demographic characteristics such as gender, income, parents' education level, number of siblings, city they live in, type of the computer equipment used at home, computer/Internet usage time per day, and purpose of computer/technology usage, and 3. Technological equipment such as type of technological equipment in class, school environment in terms of technology, and attitude of school administrator toward CAI. The first group is collected according to an experimental plan. The second and the third groups are collected through demographic surveys of students and technology equipment survey of schools. Dependent variables, on the other hand, contain 1. Achievement in permutation–combination and probability subjects and 2. Computer-assisted learning attitude. Achievements of students are measured by Objective Comprehension Tests: posttests (OCT1 and OCT2) whereas their attitudes are measured by Computer-Assisted Learning Attitude Scale (CALAS).

As a result, data consist of 45 explanatory variables with different measurement levels (nominal, ordinal and numeric), which we wish to reduce to a small number of dimension with as little loss of information as possible.

4.3 Categorical Principal Components Analysis

Clearly, the number of explanatory variables is quite many to build a MLR model. Therefore, we reduce the dimension of the mixed type dataset by using CATPCA to construct a new set of variables, called principal components. Then, a MLR model is developed to investigate the effects of CAI to mathematical achievement and attitude of students.

"Gender" and "City" variables are not included in CATPCA, because they are included in regression model directly to see their effects on the response of interest. Therefore, a total of 43 variables obtained from surveys are used for CATPCA. Description of variables that are included in CATPCA is given in Table 1. The variables "Projection School", "Desktop Computer in Class", "Smartphone in Class", and "Projection in Class" have zero variance since they are never selected in the survey. Hence, they are excluded from the analysis leaving 39 variables in total. In the following, we give a step-by-step description of the data reduction method by using the program CATPCA available in SPSS 20 (IBM Corp 2011).

CATPCA Step 1: Determination of Appropriateness of CATPCA

Bartlett's test of sphericity (BTS) and Kaiser–Meyer–Olkin (KMO) criteria are checked to determine the appropriateness of traditional PCA. Bartlett's test of sphericity is used to check whether the correlation matrix between variables is an identity matrix, which means the variables are uncorrelated. Rejecting the null hypothesis means that there are dependent variables for PCA to group, and hence PCA is appropriate. KMO is the indicator of sample adequacy and measures the degree of correlation between variables. Small values of KMO indicate that correlations between variables cannot be explained accurately by other variables and therefore PCA may not be suitable. However, high values (between 0.5 and 1.0) indicate that

Table 1 Description of the 43 variables used in the analysis

Variables: Name (Description)	
1. Settlement	25. Math Average Grade (over 100)
2. Mother Education Level	26. Overall GPA (over 100)
3. Father Education Level	27. Location of School
4. Sibling Number	28. School Type
5. Sibling Training	29. Computer Lab in School (Yes/No)
6. Income	30. Net in Computer Lab (Yes/No)
7. Computer Use (Yes/No)	31. *Projection in School (Yes/No)*
8. Laptop Use (Yes/No)	32. Smart board in School (Yes/No)
9. Desktop Use (Yes/No)	33. Tablet in School (Yes/No)
10. Tablet Use (Yes/No)	34. Computer Teacher in School
11. Smartphone (Yes/No)	35. Computer Lesson (Yes/No)
12. PS3 (Yes/No)	36. *Desktop Computer in Class (Yes/No)*
13. Internet Home (Yes/No)	37. Laptop in Class (Yes/No)
14. Internet Work (Yes/No)	38. Tablet in Class (Yes/No)
15. Internet School (Yes/No)	39. *Smartphone in Class (Yes/No)*
16. Internet Dormitory (Yes/No)	40. *Projection in Class (Yes/No)*
17. Internet Cafe (Yes/No)	41. School Environment
18. Internet Everywhere (Yes/No)	42. School Attitude (toward CAI)
19. Net Use Time	43. Schools (4 category)
20. Net for Fun (Yes/No)	
21. Net for Academic Affairs (Yes/No)	
22. Net for Routine Works (Yes/No)	
23. Net for Office Program (Yes/No)	
24. Net for PS3 (Yes/No)	

Note Variables shown in italics are not selected for the analyses

PCA is useful as a means of data reduction. The same criteria to develop the traditional PCA can be used in the CATPCA (Mendes and Ganga 2013). In the light of the information, we apply those methods for the implementation of CATPCA. We apply Bartlett's test of sphericity and KMO to the original data. However, SPSS 20 (IBM Corp 2011) do not provide the results of those methods, rather give the error of "This matrix is not positive definite." It is likely the case that correlation matrix of the data is nonpositive definite, i.e., some of the eigenvalues of correlation matrix are nonpositive numbers. Eigenvalues might take negative values due to linear dependencies among variables. As a result of this analysis, we check correlation between variables. Two things are significant with respect to correlation matrix: the variables have to be intercorrelated, but they cannot correlate too highly since this causes multicollinearity (Field 2009). Therefore, we use Spearman rank correlation coefficients to determine which variables are highly correlated, and then use only one in the analysis. A cutoff for highly correlated variables is determined as 0.90 according to literature (Abbott 2004). Some variables are found to be highly

Table 2 KMO and Bartlett's test results

Kaiser–Meyer–Olkin measure of sampling adequacy			0.786
Bartlett's test of sphericity	Approx. Chi-Square		1612.907
	df		406
	p-value		<0.0001

correlated, and in fact correlation coefficients between some of them are 1 or −1. Therefore, it is useless to include those variables to the analysis at the same time. As a result, "Location School", "School Type", "Computer Lab", "Net Computer Lab", "Computer Teacher", "Computer Lesson", "School Environment", "Tablet School", "Schools", and "Smart board School" variables are eliminated from the analysis due to linear dependency. After these 10 variables are excluded, remaining number of variables for CATPCA analysis is 29. We apply the Bartlett's test of sphericity and KMO with the remaining variables. Considering Table 2, KMO index is 0.786 (that is, between 0.5 and 1) and Bartlett's test of sphericity rejects the existence of an identity matrix. Based on those results, the data sample with 29 variables is considered adequate for the implementation of the CATPCA.

CATPCA Step 2: Specifying Analysis Levels

In the present dataset, we do not wish to assume linearity, and the number of the categories is small compared to the number of individuals. Therefore, we treat all variables ordinally. Since some of the variables such as "Mother/Father Education Level" have numerous categories compared to the others, it might have been useful to apply a monotonic spline analysis level to those variables; however, a spline analysis level is more restrictive than ordinal analysis level and thus gives lower Variance Accounted For (VAF). In summary, we specify an ordinal analysis level for the 29 variables.

CATPCA Step 3: Missing Values

The number of missing values in the data is considerably small: Three students have missing values only in "Overall GPA" scores. Therefore, the missing values are treated passively, which is the default option for the treatment of missing data in CATPCA. That means, students with missing values are deleted only for those variables on which they have missing values.

CATPCA Step 4: Discretizing

The theory of nonlinear PCA is based on categorical variables with integer values (Linting et al. 2007a). Therefore, (positive) integer-valued data is required in CATPCA. This is not the feature of CATPCA, but just a technical requirement in SPSS 20 (IBM Corp 2011) for the analysis. In the data, we have two variables that are continuous; "Math Average Grade" and "Overall GPA". We have to discretize these variables for the analysis by CATPCA. Since sample is not very large, we want small number of categories of those variables to increase stability of results, so grouping option is used as a discretizing option. When the number of categories is chosen as seven for both "Math Average Grade" and "Overall GPA" variables, ties occur for Overall GPA, so the number of categories for that variable is changed to six.

CATPCA Step 5: Evaluating the Number of Components

We have to determine the sufficient number of components/dimensions to retain in the analysis. The survey instruments measure socioeconomic and cultural factors of students, technology ownership of the students, and the purpose of using technology, in addition to technological equipment and environment of the school. Therefore, it seems reasonable to assume that six components/dimensions are called for. Since CATPCA solutions are not nested, it is required to look at scree plots in different dimensions to compare different solutions. Therefore, we generate scree plots of the eigenvalues in four-, five-, six-, and seven-dimensional solutions. Figure 1 shows scree plots for these four solutions. Due to the lack of interpretability of the seventh dimension, and due to the small differences among the graphs of 7-6-5 dimensions, we try the six-dimensional and five-dimensional solutions.

CATPCA Step 6: Preliminary Analysis

After the selection of variables and the decision on the number of dimensions, we apply CATPCA. All six dimensions account for a substantial percentage of

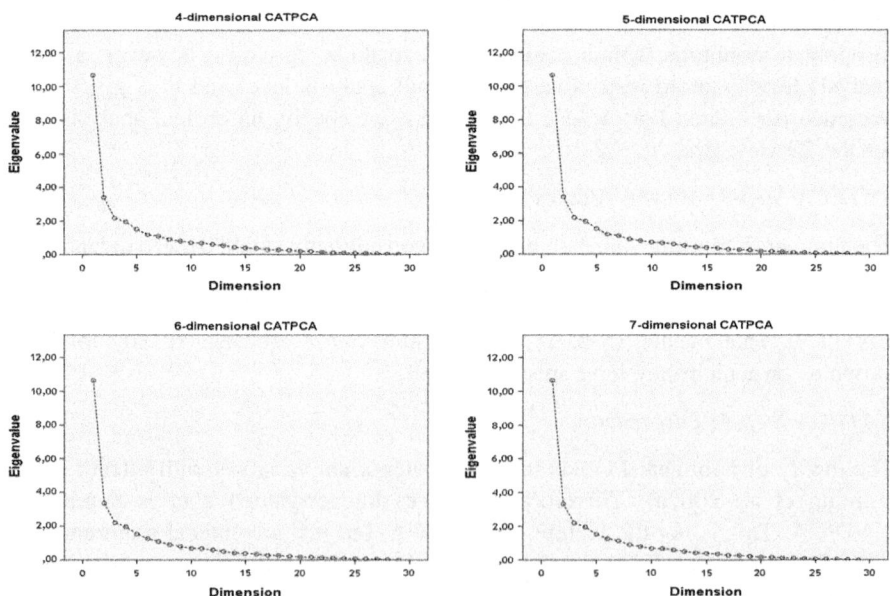

Fig. 1 Scree plots with lines denoting the eigenvalues for a four-, five-, six-, and seven-dimensional CATPCA solutions on 29 variables analyzed at an ordinal analysis level

72.431% of the total variance in transformed variables. According to six-dimensional CATPCA solution, "No Computer", "Laptop Use", "Desktop Use", "Smartphone Use", "PS3", "Internet Work", "Internet School", "Internet Everywhere", and "Routine Works" variables have very small mean coordinate, very close to or below 0.1. Those variables are considered to be excluded from the analysis. Since we could not assign meaningful interpretations to dimensions with the result of this analysis, those variables are excluded from the analysis. CATPCA with six-dimensional solution is repeated with the remaining 20 variables.

We can see in the model summary table, Table 3, that the internal consistency coefficient increases from 0.986 with all 29 variables to 0.991 with only 20 variables. Moreover, 85.52% of the variance is accounted for by these 20 variables. In other words, there are fewer variables, but we are accounting for more of the variance with those 20 items than the amount of variance accounted for by the 29 variables. However, the eigenvalue for the sixth dimension decreases below 1. Therefore, we also apply CATPCA with five dimensions and later decide which one to use in data reduction.

After deciding on the number of dimensions and the number of variables in six-dimensional solution, we check if we could simplify the structure of the solution by rotating the results. Rotation options are not available within CATPCA in SPSS 20 (IBM Corp 2011). However, rotation can be performed by saving the transformed variables and applying them in a linear PCA (Linting and Kooji 2012). As a result, we use VARIMAX rotation as an orthogonal rotation within traditional PCA in SPSS 20 (IBM Corp 2011) to rotate the transformed variables (Table 4).

Since six-dimensional solution with 20 variables involve an eigenvalue smaller than 1, five-dimensional CATPCA is also applied to determine the final analysis. The same procedure is applied. Variance accounted for by the five-dimensional CATPCA with all 29 variables is founded as 68.237%. The variables "No Computer", "Laptop Use", "Desktop Use", "Tablet Use", "Smartphone Use", "PS3", "Internet Work", and "Routine Works", whose mean coordinates are around or below 0.1 are excluded from the analysis, and then CATPCA is repeated with the remaining variables. As a result, the model with five dimensions accounts for 79.80% of the total variance in the optimally scaled variables.

Table 3 Model summary of CATPCA for 6-dimensional solution with 20 variables

Dimension	Variance Accounted For (VAF)		
	Cronbach's Alpha	Total (Eigenvalue)	Percentage of variance
1	0.929	8.539	42.695
2	0.709	3.066	15.330
3	0.512	1.948	9.740
4	0.436	1.707	8.535
5	0.012	1.011	5.055
6	−0.211	0.833	4.165
Total	0.991[a]	17.104	85.520

Note [a]Total Cronbach's Alpha is based on the total eigenvalue

Table 4 Rotated component loadings from a six-dimensional CATPCA on 20 variables, with all variables analyzed ordinally

	Components					
	1	2	3	4	5	6
Overall GPA	0.886					
Tablet use in class	−0.873					
Settlement	0.868					
Income	0.865					
School attitude	−0.865					
Mathematical average grade	0.858					
Father education level	0.706					
Sibling number		0.858				
Use of office program		0.842				
Sibling training		0.796				
Mother education level	0.559	−0.662				
Tablet use			0.819			
Net use time			−0.671			
Internet cafe				0.880		
Laptop class				−0.711		
Internet home			0.528	−0.615		
Net for Ps3					0.898	
Internet dormitory					0.855	
Academic						0.852
Fun						−0.730

Notes 1 Extraction Method: Principal component analysis; *2* Rotation Method: Varimax with Kaiser Normalization; *3* Rotation converged in six iterations

CATPCA Step 7: Final Analysis and Interpretation

In order to decide on which dimensional solution to use in the categorical data reduction, we compare six-dimensional and five-dimensional solutions with respect to VAF. Six-dimensional solution accounts for 85.52% of the total variance, while five-dimensional solution accounts for 79.80% of the total variance. Although the eigenvalue of the sixth dimension in the first solution is below 1, the total variance explained in that model is higher compared to five-dimensional solution. Extracting too many factors may cause undesired error variance, so selecting the most suitable criterion for study when deciding on the number of factors to extract is very important (Young and Pearce 2013). Joliffe's criterion which suggests retaining factors with eigenvalues above 0.70 (Jolliffe 2002) and scree plots are used to decide on how many factors to retain in the study. In conclusion, we decide to use six-dimensional CATPCA solution on 20 variables, with all variables analyzed ordinally (See Table 3). The total VAF across the six dimensions is 85.52%, with clearly dominant first dimension (VAF: Dimension 1 = 42.69%, Dimension 2 = 15.33%, Dimension 3 = 9.74%, Dimension 4 = 8.53%, Dimension 5 = 5.05%,

Dimension 6 = 4.16%). That implies that the six selected dimensions/components explain about 86% of the variance in the 20 ordinally quantified variables, which indicates a good fit.

Dimension-1 (PC1) includes the variables: Overall GPA, Math Average Grade, Father Education Level, Income, Settlement, School Attitude, and Tablet Use in class. We name this factor as "The general success of the students and basic socio-economic and technological factors affecting this situation." Dimension-2 (PC2) includes the variables: Sibling Number, Sibling Training, Mother Education Level, and Use of Office Program. This dimension is named as "the secondary social situation of the family." Net Use Time and Tablet Use variables are involved in Dimension-3 (PC3), while Internet Cafe, Internet Home, and Laptop Class constitute Dimension-4 (PC4). Dimension-5 (PC5) includes the variables: Net for PS3 and Internet Dormitory. Lastly, Dimension-6 (PC6) includes the variables of use of technology for Academic and Fun. As it is seen, Dimension-3 through Dimension-6 are related with Internet ownership and use, so name like that.

4.4 Multiple Linear Regression Analysis (MLR)

The method of this chapter is focused on narrowing down the data containing many mixed type of variables compared to the number of observations in order to see relevant patterns and relationships between variables. Therefore, we initially reduce the dimension of the covariates in the MLR model to avoid multicollinearity. Six principal components are achieved as the explanatory variables after performing the CATPCA. The data for regression model consist of 74 observations of the following variables: gender, city, instruction method, and six principal components—PC1, PC2, PC3, PC4, PC5, and PC6.

First, an MLR model is fitted on the scores of Objective Comprehension Test 1 (OCT1) with the variables: gender, instruction method, and six principal components (Montgomery et al. 2015). OCT1 scores are obtained just for the students having education in Mardin. Therefore, the city variable is essentially constant here, and therefore it is not included into the model. Forward stepwise selection method is used whereby the independent variables with largest absolute correlation with OCT1 are chosen as explanatory variables. As a result, only "instructional method" is found statistically significant. Therefore, instructional method is included in the model as the only major factor explaining success in OCT1, and the final model is fit as given below.

$$OCT1 = 26.32 + 12.75 \ Instruction \ Method. \tag{4}$$

Residual plots in Fig. 2 indicate the validity of the LS regression assumptions. Regression output is displayed in Table 5. The p-value of the ANOVA F-test is approximately zero, which is smaller than the predetermined significance level,

(a)   **(b)**

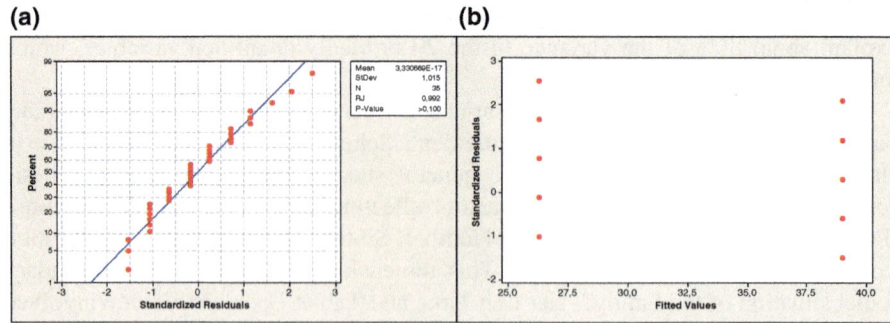

Fig. 2 **a** Normal probability plot of standardized residuals for OCT1. **b** Plot of standardized residuals versus fits for OCT1

Table 5 Regression output of the model given in (4)

Source	SS	df	MS	F	p-value
Regression	1411	1	1411.13	15.25	0.000
Residual error	3053	33	92.51		
Total	4464	34			
s = 9.61798	R-sq = 31.61%	R-sq(adj) = 29.54%			
Coefficients					
Term	*Coef*	*SE Coef*	*t-value*	*p-value*	*VIF*
Constant	26.32	2.21	11.93	0.000	
Ins. Mth.	12.75	3.26	3.91	0.000	1.00

$\alpha = 0.05$; so we can say that there is a statistically significant relationship between the OCT1 score and the instruction method.

Secondly, an MLR model is fitted on OCT2 score with the following variables: gender, city, instruction method, and six principal components (Montgomery et al. 2015). Before making any comments on the model fit, assumptions for the regression are examined. Interaction term is added to overcome the lack-of-fit problem in the model, and the final regression equation is fitted as follows:

$$OCT2 = 52.37 + 17.97\ PC1 + 3.32\ PC2 + 6.41\ PC1*PC2. \tag{5}$$

Residual analysis indicates no violation of assumptions (See Fig. 3). The results of the regression are displayed in Table 6. According to the ANOVA F-test, whose p-value is approximately zero, fitted model in Eq. (5) is a statistically significant model.

Lastly, an MLR model is fitted on the Computer-Assisted Learning Attitude Scale (CALAS) score with the variables: gender, city, instruction method, and six

(a)   **(b)**

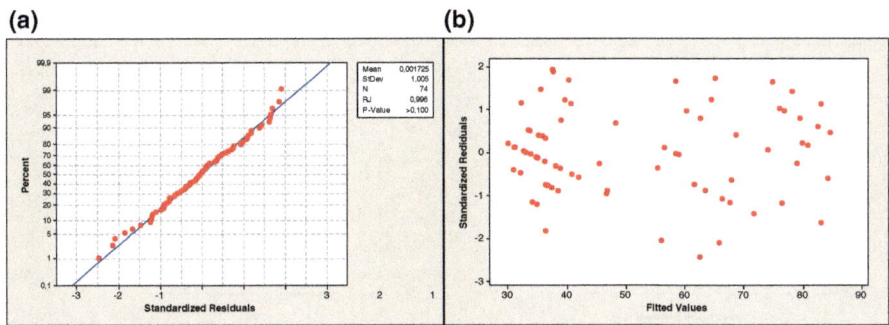

Fig. 3 **a** Normal probability plot of standardized residuals for OCT2. **b** Plot of standardized residuals versus fits for OCT2

Table 6 Regression output of the model given in (5)

Source	SS	df	MS	F	p-value
Regression	24070.6	3	8023.5	32.55	0.000
Residual error	17253.8	70	246.5		
Total	41324.4	73			
s = 15.6998	R-sq = 58.25%	R-sq(adj) = 56.46%			

Coefficients					
Term	*Coef*	*SE Coef*	*t-value*	*p-value*	*VIF*
Constant	52.37	1.83	28.69	0.000	
PC1	17.97	2.10	8.57	0.000	1.33
PC2	3.32	2.71	1.22	0.226	2.23
PC1 * PC2	6.41	2.67	2.40	0.019	2.56

principal components to understand which factors contributed to students' attitude toward CAI (Montgomery et al. 2015). Since normality assumption is not satisfied (See Fig. 4a), we perform Box-Cox transformation on the response variable CALAS as a remedial measure. The optimal value for lambda is found as −3.31 (See Fig. 4b), and hence $1/CALAS^{3.31}$ transformation is applied. Then, the regression procedure is repeated by taking the transformed variable as the response. The LS regression assumption checks on the final fitted model show no violation (See Fig. 5). However, F-test does not indicate a statistically significant relationship between the variables (p-value = 0.537). According to the regression outputs given in Table 7, none of the explanatory variables involved in the analysis are significant. As a result of the analysis, we can say that those independent variables would not explain the variation in the CALAS variable at all.

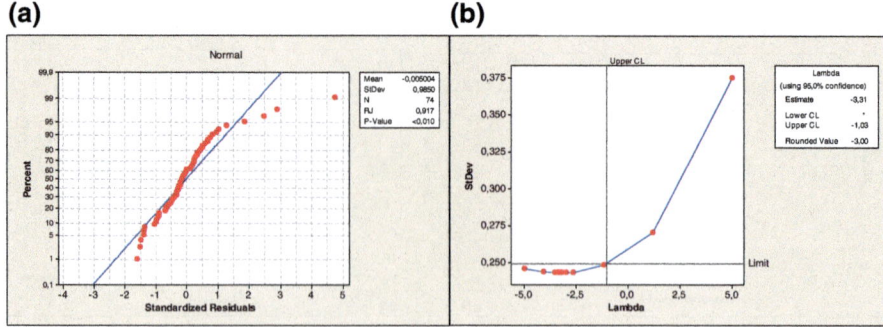

Fig. 4 **a** Normal probability plot of standardized residuals for CALAS. **b** Box-Cox transformation on the CALAS variable

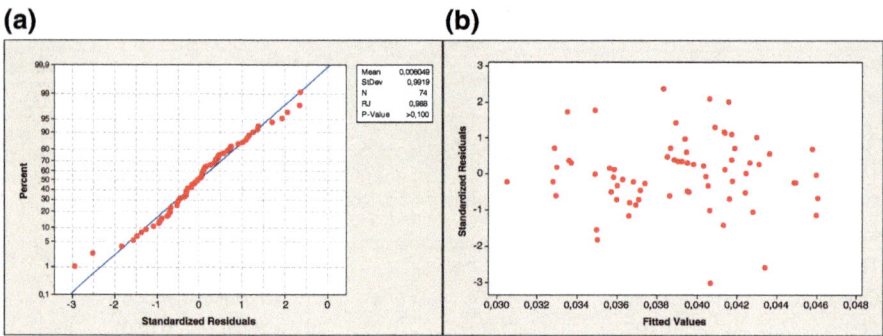

Fig. 5 **a** Normal probability plot of standardized residuals for CALAS. **b** Plot of standardized residuals versus fits for CALAS

4.5 Validation of Sample Adequacy

In Sect. 3.2, we emphasize the importance of sample size consideration in dimension reduction procedures such as PCA. In this study, with the available resources, we could only afford data having 74 observations which is relatively small. So, to determine the adequacy of the sample, we utilized the KMO statistic as stated in Sect. 3.2. In addition, in order to validate the results obtained from CATPCA regression presented in Sects. 4.3 and 4.4, we adopt an alternative approach for reducing data dimension. First, we conduct pairwise comparisons between the schools with respect to all characteristics that we measure. Depending on the results, we treat the two village elementary schools as if they are one "state school". Following, in developing MLR models, we apply stepwise regression method for variable selection. Both approaches result in the same conclusions (See Kapucu (2016) for details of this comparison study).

Table 7 Regression output of the model for CALAS

Source	SS	df	MS	F	p-value
Regression	0.000917	9	0.000102	0.89	0.537
Residual error	0.007307	64	0.000114		
Total	0.008224	73			
s = 0.0106853	R-sq = 11.15%	R-sq(adj) = 0.00%			

Coefficients					
Term	*Coef*	*SE Coef*	*t-value*	*p-value*	*VIF*
Constant	0.03658	0.00947	3.86	0.000	
Gender	−0.00186	0.00304	−0.61	0.542	1.47
City	0.00008	0.00019	0.43	0.668	6.29
Ins. Mth.	0.00053	0.00792	0.07	0.946	7.77
PC1	0.00165	0.00337	0.49	0.627	7.41
PC2	0.00085	0.00246	0.34	0.732	3.94
PC3	0.00178	0.00253	0.70	0.484	4.15
PC4	0.00068	0.00131	0.52	0.607	1.12
PC5	−0.00201	0.00134	−1.50	0.138	1.16
PC6	0.00152	0.00158	0.96	0.342	1.63

5 Conclusion

The aim of this study is to examine how a CAIM, designed for teaching "probability" topics for eighth grade level students, the socioeconomic and cultural factors affect the achievement of the students. For this purpose, we have developed CAIMs consisting of animations and simulations by using R language. In order to achieve our aims, we set up a quasi-experiment involving four elementary schools in two districts of Turkey. Two of the state schools and a private school are in Mardin, a city located in southern Turkey. The other one is a private school in the capital city, Ankara. In the study, both demographic and socioeconomic peculiarities of the students and the competence of the schools with respect to technological equipment are evaluated by surveys. The data collected are analyzed by CATPCA regression, in which, mixed type of data is first reduced in dimension by using CATPCA, and then, MLR models are developed between the response and exploratory variables of interests. The advantage of CATPCA method used in dimension reduction is that the correlated variables are contained in principal components, thereby all the relevant factors are included in the analysis.

In the light of the findings obtained from the CATPCA regression analysis, the following results could be stated:

- "Instruction method" is a significant factor in explaining students' achievement in OCT1 of probability and statistics learning area of eighth grade mathematics curriculum. This test includes the sub-learning area, namely identifying possible cases, which is one of the focused topics in the developed CAIM.

- "Instruction method" is not a significant factor in explaining students' achievement in OCT2 of probability and statistics learning area of eighth grade mathematics curriculum. This test includes the sub-learning area, namely probability and event types, focused by our CAIM. Nevertheless, the first-term mathematics grade as well as socioeconomic and technological factors significantly influence a student's ability to succeed.
- None of the variables included in the analysis contribute to the students' attitude toward CAI. Therefore, it can be concluded that we do not have enough evidence to state a relationship between the CAIM prepared in R program and students' attitude toward the CAI.

This study reveals the importance of using CAI in teaching probability domain. However, there are some limitations of the study such as sample size, content covered, time considerations, and the capability of R for developing animations, which prevent us to generalize our results. As a further study, we are planning to improve the CAIM developed using R program, and repeat a similar study using more content coverage with a larger sample.

References

Abbott, D.: Find correlated variables prior to modeling. In: Data Mining and Predictive Analytics. http://abbottanalytics.blogspot.com.tr/2004/12/find-correlated-variables-prior-to.html (2004). Accessed 12 Oct 2016

Arrindell, W.A., Van der Ende, J.: An empirical test of the utility of the observations-to-variables ratio in factor and components analysis. Appl. Psychol. Meas. **9**, 165–178 (1985)

Batanero, C., Biehler, R., Maxara, C., Engel, J., and Vogel, M.: Using simulation to bridge teachers' content and pedagogical knowledge in probability. In: Proceedings of the 15th ICMI Study Conference: The Professional Education and Development of Teachers of Mathematics, 1–6 Jan 2005, Águas de Lindóia, Brazil

Borovcnik, M.G., Kapadia, R.: Research and developments in probability education internationally. In: Proceedings of the British Congress for Mathematics Education, pp. 41–48 (2010)

Burrill, G.: Simulation as a tool to develop statistical understanding. Paper presented at the Sixth International Conference on Teaching Statistics. Cape Town, South Africa (2002)

Braun, W.J., White, B.J., Craig, G.: R tricks for kids. Teach. Stat. Trust **36**(1), 7–12 (2013)

Costello, A.B., Osborne, J.W.: Best practices in exploratory factor analysis: Four recommendations for getting the most from your analysis. Pract. Assess. Res. Eval. **10**(7), (2005). http://pareonline.net/pdf/v10n7a.pdf. Accessed 20 Mar 2017

Ely, D.P., Plomp, T.: The promises of educational technology: a reassessment. Int. Rev. Educ. **32**(3), 231–249 (1986)

Everitt, B., Hothorn, T.: An Introduction to Applied Multivariate Analysis with R. Springer, New York (2011)

Ferreira, R.D.S., Kataoka, V.Y., Karrer, M.: Teaching probability with the support of the R statistical software. Stat. Educ. Res. J. **13**(2), 132–147 (2014)

Field, A.: Discovering Statistics Using SPSS for Windows. Sage Publications, London, New Delhi (2009)

Garfield, J., Ahlgren, A.: Difficulties in learning basic concepts in probability and statistics: implications for research. J. Res. Math. Educ. **19**(1), 44–63 (1988)

Guadagnoli, E., Velicer, W.F.: Relation of sample size to the stability of component patterns. Psychol. Bull. **103**, 265–275 (1988)

Gürbüz, R.: Olasilik konusunun öğretiminde kullanilabilecek bilgisayar destekli bir materyal. Educ. J. Mehmet Akif Ersoy Üniv. **8**(15), 41–52 (2008)

HodnikČadež, T., Škrbec, M.: Understanding the concepts in probability of pre-school and early school children. Eurasia J. Math. Sci. Technol. Educ. **7**(4), 263–279 (2011)

Jolliffe, I.T.: Principal Components Analysis. Springer, New York (2002)

Kapucu, T.: The effect of computer assisted instruction on eight grade students' permutation-combination-probability achievement and attitudes towards computer assisted instruction. M.Sc. thesis, Middle East Technical University, Ankara, Turkey (2016)

Karamustafaoğlu, O., Aydin, M., Özmen, H.: Bilgisayar destekli fizik etkinliklerinin öğrenci kazanimlarina etkisi: basit harmonik hareket örneği. Turk. Online J. Educ. Technol. **4**(4), 67–81 (2005)

Keeler, C., Steinhorst, K.: A new approach to learning probability in the first statistics course. Online J. Stat. Educ. **9**(3) (2001)

Kemalbay, G., Korkmazoğlu, Ö.B.: Categorical principal component logistic regression: a case study for housing loan approval. Proc.-Soc. Behav. Sci. **19**, 730–736 (2014)

Lingard, H.C., Rowlinson, S.: Sample Size in Factor Analysis: Why Size Matters. University of Hong Kong, Hong Kong (2006)

Linting, M., Meulman, J.J., Groenen, P.J.F., Van der Kooij, A.J.: Nonlinear principal components analysis: Introduction and application. Psychol. Methods **12**, 336–358 (2007a)

Linting, M., Meulman, J.J., Groenen, P.J., Van der Kooij, A.J.: Stability of nonlinear principal components analysis: an empirical study using the balanced bootstrap. Psychol. Methods **12**(3), 359 (2007b)

Linting, M., Van der Kooij, A.: Nonlinear principal components analysis with CATPCA: a tutorial. J. Pers. Assess. **94**(1), 12–25 (2012)

Lopes, C.A.E., Moura, A.R.L.: Probability and statistics in elementary school: a research of teachers' training1. Paper presented at International Association for Statistical Education: ICOTS6, Cape Town, South Afrika (2002)

Mair, P., Leeuw, J.: A general framework for multivariate analysis with optimal scaling: the R package aspect. J. Stat. Softw. **32**(9), 1–23 (2010)

Mendes, G.H.S., Ganga, G.M.D.: Predicting success in product development: the application of principle component analysis to categorical data and binomial logistic regression. J. Technol. Manag. Innov. **8**(3), 83–97 (2013)

Montgomery, D.C., Elizabeth, A.P., Vining, G.G.: Introduction to Linear Regression Analysis. Wiley, New York (2015)

Osborne, J.W., Costello, A.B.: Sample size and subject to item ratio in principal components analysis. Pract. Assess. Res. Eval. **9**(11), (2004). http://PAREonline.net/getvn.asp?v=9&n=11. Accessed 20 Mar 2017

Pearson, R.H., Mundfrom, D.J.: Recommended sample size for conducting exploratory factor analysis on dichotomous data. J. Modern Appl. Stat. Methods **9**(2), 359–368 (2008)

Preacher, K.J., MacCallum, R.C.: Exploratory factor analysis in behavior genetics research: factor recovery with small sample sizes. Behav. Genet. **32**, 153–161 (2002)

Souza, L.O.: Collaborative training for teaching probability and statistics: empirical approaches and simulation with elementary school students. In: Proceedings of the Satellite Conference of the International Association for Statistical Education (IASE), 2–6 July 2015, Rio de Janeiro, Brazil

SPSS: IBM Corp. Released 2011. IBM SPSS Statistics for Windows, Version 20.0. IBM Corp, Armonk, NY

Stohl, H.: Probability in teacher education and development. In: Jones, G.A. (ed.) Exploring Probability in School: Challenges for Teaching and Learning, pp. 297–324. Springer, New York (2005)

Şahin, T., Yıldırım, S.: Öğretim teknolojileri ve materyal geliştirme. Ani, Ankara (1999)

TIMSS: International Mathematics Report. http://timss.bc.edu/timss1999i/pdf/T99i_Math_All.pdf (1999). Accessed 20 May 2016

Young, A.G., Pearce, S.: A beginner's guide to factor analysis: focusing on exploratory factor analysis. Tutor. Quant. Methods Psychol. **9**(2), 79–94 (2013)

Zieffler, A., Garfield, J.B.: Studying the role of simulation in developing students' statistical reasoning. In: Proceedings of the 56th Session of the International Statistical Institute, Lisboa, 22–29 Aug 2007. International Statistical Institute, Voorburg, The Netherlands

Contemporary Robust Optimal Design Strategies

Timothy E. O'Brien

Abstract Researchers often find that nonlinear regression models are more applicable for modeling various biological, physical, and chemical processes than the linear ones since they tend to fit the data well and since these models (and model parameters) are more scientifically meaningful. These researchers are thus often in a position of requiring optimal or near-optimal designs for a given nonlinear model. A common shortcoming of most optimal designs for nonlinear models used in practical settings, however, is that these designs typically focus only on (first-order) parameter variance or predicted variance, and thus ignore the inherent nonlinear of the assumed model function. Another shortcoming of optimal designs is that they often have only p support points, where p is the number of model parameters. Measures of marginal curvature, first introduced in Clarke (1987) and further developed in Haines et al. (2004), provide a useful means of assessing this nonlinearity. Other relevant developments are the second-order volume design criterion introduced in Hamilton and Watts (1985) and extended in O'Brien (1992) and O'Brien et al. (2010), and the second-order MSE criterion developed and illustrated in Clarke and Haines (1995). This chapter underscores and highlights various robust design criteria and those based on second-order (curvature) considerations. These techniques, easily coded in the R and SAS/IML software packages, are illustrated here with several key examples.

Keywords Binary logistic model · Experimental design · Generalized nonlinear modeling · Goodness-of-fit · Lack-of-fit · Robustness

AMS Mathematics Subject Classification Numbers 62J02 · 62J12 62K05

T. E. O'Brien (✉)
Department of Mathematics & Statistics and Institute
of Environmental Sustainability, Loyola University Chicago, Chicago, IL, USA
e-mail: tobrie1@luc.edu

© Springer International Publishing AG 2018   165
M. Tez and D. von Rosen (eds.), *Trends and Perspectives
in Linear Statistical Inference*, Contributions to Statistics,
https://doi.org/10.1007/978-3-319-73241-1_11

1   Introduction

In the context of univariate regression, we let Y denote the dependent random variable, y denote its realization, and x denote the independent variable (or $x_1, x_2 \ldots x_k$ in the case of several independent variables). Following convention, we use the term "model" to include (a) the assumed distribution for the response variable—often chosen from the exponential family, (b) the link function connecting $\mu = E(Y)$ with the explanatory variable(s) and with the model parameters in the p-dimensional vector $\boldsymbol{\theta}$, (c) the (mean) model function $\eta(x, \boldsymbol{\theta})$, which combines the explanatory variable(s) and the model parameters, (d) the variance (denoted σ^2) or variance function (perhaps depending on $\boldsymbol{\theta}$ and/or additional parameters such as σ^2), and (e) the nature of the observations, such as independent or correlated (e.g., nested) measurements.

Linear models are ubiquitous in the statistical literature. In matrix form, simple and multiple linear models are written as

$$y = \eta(x, \boldsymbol{\theta}) + \boldsymbol{\varepsilon} = X\boldsymbol{\theta} + \boldsymbol{\varepsilon} \tag{1}$$

In this expression, y, $\eta(x, \boldsymbol{\theta})$, and $\boldsymbol{\varepsilon}$ are of dimension $n \times 1$, $\boldsymbol{\theta}$ is $p \times 1$ for $p = k + 1$, and X is $n \times p$ and is the design matrix comprised of the first derivatives of the model function with respect to the model parameters. In situations where the parameter p-vector $\boldsymbol{\theta}$ is meaningfully partitioned as $\boldsymbol{\theta} = \begin{pmatrix} \boldsymbol{\theta}_1 \\ \boldsymbol{\theta}_2 \end{pmatrix}$ where $\boldsymbol{\theta}_1$ is $p_1 \times 1$ and $\boldsymbol{\theta}_2$ is $p_2 \times 1$ with $p_1 + p_2 = p$, in matrix form this model becomes

$$y = X\boldsymbol{\theta} + \boldsymbol{\varepsilon} = [X_1 | X_2] \begin{pmatrix} \boldsymbol{\theta}_1 \\ \boldsymbol{\theta}_2 \end{pmatrix} + \boldsymbol{\varepsilon} = X_1 \boldsymbol{\theta}_1 + X_2 \boldsymbol{\theta}_2 + \boldsymbol{\varepsilon} \tag{2}$$

In the case of (normal) nonlinear models, the Jacobian matrix (X) in expressions (1) and (2), now denoted V, is again of dimension $n \times p$ and is such that the ith column is the derivative of $\eta(x, \boldsymbol{\theta})$ with respect to the ith model function parameter, θ_i. In the applied literature, researchers often find nonlinear models to more meaningfully model their processes.

Parameter estimation is typically achieved by maximizing the corresponding likelihood (and obtaining maximum likelihood estimates, denoted MLEs), and in standard-independent normal nonlinear situations, this corresponds to least-squares estimation (and LSEs). To wit, in the case of linear models, the $p \times p$ covariance matrix for $\hat{\boldsymbol{\theta}}$ is proportional to $\left(X^T X \right)^{-1}$; for nonlinear models (omitting constants) the covariance matrix, $\left(V^T V \right)^{-1}$, is approximated by $\left(\hat{V}^T \hat{V} \right)^{-1}$ with $\hat{V} = V(\hat{\boldsymbol{\theta}})$.

Optimal design theory for linear and nonlinear models is discussed in Silvey (1980), Pukelsheim (1993), and Atkinson et al. (2007), as well as the references contained therein. It is not common, however, that model misspecification is incorporated into the design criteria, and so the resulting designs often provide no ability to test for model adequacy, thereby limiting their usefulness. Although

Studden (1982) and some related works do address model robustness, this is only achieved in the case of polynomial models. In the current work, we address the larger class of nonlinear models and in a more direct manner.

2   Confidence Regions and Intervals and Optimal Design

For normal nonlinear models, $(1-\alpha)100\%$ Wald confidence regions for $\boldsymbol{\theta}$ are of the form, $\left\{\boldsymbol{\theta} \in \Theta\colon (\boldsymbol{\theta}-\hat{\boldsymbol{\theta}})^T \hat{V}^T \hat{V} (\boldsymbol{\theta}-\hat{\boldsymbol{\theta}}) \le ps^2 F_\alpha \right\}$. Here, as noted above $\hat{\boldsymbol{\theta}}$ is the least-squares (and maximum likelihood) estimate of $\boldsymbol{\theta}$, $\hat{V}$ is the $n \times p$ Jacobian matrix of first derivatives evaluated at $\hat{\boldsymbol{\theta}}$, s^2 is the mean square error (estimator of σ^2), and F_α is a tabled F percentile with p and $n-p$ degrees of freedom with tail probability of α. In contrast with the above, the $(1-\alpha)100\%$ likelihood-based confidence region here is $\left\{\boldsymbol{\theta} \in \Theta\colon S(\boldsymbol{\theta}) - S(\hat{\boldsymbol{\theta}}) \le ps^2 F_\alpha \right\}$, with $S(\boldsymbol{\theta}) = (\boldsymbol{y}-\boldsymbol{\eta}(x,\boldsymbol{\theta}))^T$ $(\boldsymbol{y}-\boldsymbol{\eta}(x,\boldsymbol{\theta})) = \boldsymbol{\varepsilon}^T \boldsymbol{\varepsilon}$—i.e., the sum of squares. These two regions are nearly equivalent depending upon the degree to which the model function, $\boldsymbol{\eta}(x,\boldsymbol{\theta})$, is well-approximated by the affine representation, $\boldsymbol{\eta}(x,\hat{\boldsymbol{\theta}}) + \hat{V}(\boldsymbol{\theta}-\hat{\boldsymbol{\theta}})$. In normal linear models, this result is exactly met, and only approximately so for normal nonlinear, generalized linear, and generalized nonlinear models. Wald and likelihood confidence intervals can be obtained from these regions by conditioning or profiling; further details are given in Seber and Wild (1989) and Pawitan (2013). Often practitioners wish to choose an experimental design to reduce the length of the resulting confidence interval or the volume of the resulting confidence region.

An n-point design measure, denoted here by ξ, is written as

$$\xi = \left\{ \begin{array}{cccc} x_1 & x_2 & \cdots & x_n \\ \omega_1 & \omega_2 & \cdots & \omega_n \end{array} \right\} \qquad (3)$$

In this expression, the ω_i are nonnegative design weights which sum to one, and the x_i are design points (or vectors in the multivariate case) that belong to the design space; these points are not necessarily distinct. Writing $\Omega = diag\{\omega_1, \omega_2, \ldots, \omega_n\}$, the $p \times p$ Fisher information matrix is therefore given by

$$M(\xi, \boldsymbol{\theta}) = V^T \Omega V \qquad (4)$$

As noted at the end of the previous section, the (asymptotic) variance of $\hat{\boldsymbol{\theta}}$ is proportional to $M^{-1}(\xi, \boldsymbol{\theta})$; thus, in many regression settings, designs are often chosen to minimize some convex function of $M^{-1}(\xi, \boldsymbol{\theta})$. Designs which minimize its determinant are called D-optimal, and these designs minimize the volume of the Wald confidence region given in the previous section. Designs which minimize the trace of M^{-1} are called A-optimal, and important connections are given in Kiefer (1974) and Dette and O'Brien (1999). Since for nonlinear/logistic models, M depends upon $\boldsymbol{\theta}$, so-called local (or Bayesian) designs are typically obtained.

The (first-order approximation) variance of the predicted response at the value x is

$$d(x, \xi, \boldsymbol{\theta}) = \frac{\partial \eta(x, \boldsymbol{\theta})}{\partial \boldsymbol{\theta}^T} \boldsymbol{M}^{-1}(\xi) \frac{\partial \eta(x, \boldsymbol{\theta})}{\partial \boldsymbol{\theta}} \tag{5}$$

Designs which minimize (over the space of designs ξ) the maximum (over x) of $d(x, \xi, \boldsymbol{\theta})$ in (5) are called G-optimal; note that these designs are chosen to minimize the worse-case (largest) predicted variance. As in the case described above, since this predicted variance depends upon $\boldsymbol{\theta}$ for nonlinear and non-normal models, researchers often seek optimal designs either using a "best guess" for $\boldsymbol{\theta}$ (called a local optimal design) or by assuming a plausible prior distribution on $\boldsymbol{\theta}$ and by finding a Bayesian optimal design.

The celebrated General Equivalence Theorem (GET) of Kiefer and Wolfowitz (1960) establishes that D- and G-optimal designs are equivalent in the case of linear models; these results were extended to nonlinear models in White (1973). This theorem also demonstrates that the variance function (5) evaluated using the D-/G-optimal design does not exceed the line $y = p$ (where p is the number of model function parameters)—but that it will exceed this line for all other designs. A corollary of the GET establishes that the maximum of the variance function is achieved for the D-/G-optimal design at the support points of this design. This result is quite useful in demonstrating optimality of a given design by substituting it into (5) and plotting the resulting variance function.

3   Key Illustrations

The following examples demonstrate the estimation methods discussed above and serve to exemplify design strategies given in Sect. 4.

3.1   The Ratio of Two Normal Mean

Cook and Witmer (1985) discusses the Fieller-Creasy problem and model function, written a $\eta(x, \boldsymbol{\theta}) = \theta_1 x + \theta_1 \theta_2 (1 - x)$. With constant variance, n_1 experimental units receive one treatment with mean μ_1, n_2 cases receive a second treatment with mean μ_2, $\theta_1 = \mu_1$ and $\theta_2 = \frac{\mu_2}{\mu_1}$, and x is an indicator variable associated with the first treatment group. Hence, for the first treatment group (i.e., for $i = 1, 2, \ldots n_1$), $\eta(x, \boldsymbol{\theta}) = \theta_1$, and $\eta(x, \boldsymbol{\theta}) = \theta_1 \theta_2$ for subjects in the second treatment group (i.e., for $i = n_1 + 1, n_1 + 2, \ldots n_1 + n_2 = n$). The key parameter here is θ_2 which corresponds to the ratio of the treatment means.

Regarding the partial derivatives in the Jacobian matrix, for subjects receiving treatment one, $\frac{\partial \eta(x_i, \boldsymbol{\theta})}{\partial \theta_1} = 1$ and $\frac{\partial \eta(x_i, \boldsymbol{\theta})}{\partial \theta_2} = 0$, and for subjects receiving treatment two, $\frac{\partial \eta(x_i, \boldsymbol{\theta})}{\partial \theta_1} = \theta_2$ and $\frac{\partial \eta(x_i, \boldsymbol{\theta})}{\partial \theta_2} = \theta_1$. Thus, letting $\mathbf{1}_{n_1}$ and $\mathbf{1}_{n_2}$ denote vectors of one's of lengths n_1 and n_2, respectively, and $\mathbf{0}_{n_1}$ a vector of zeros of length n_1, the corresponding Jacobian matrix is

$$V = \begin{bmatrix} \mathbf{1}_{n_1} & \mathbf{0}_{n_1} \\ \theta_2 \mathbf{1}_{n_2} & \theta_1 \mathbf{1}_{n_2} \end{bmatrix} \tag{6}$$

Nonlinear is demonstrated by noting the dependence in (6) upon $\boldsymbol{\theta}$. Further,

$$V^T V = \begin{bmatrix} n_1 + n_2 \theta_2^2 & n_2 \theta_1 \theta_2 \\ n_2 \theta_1 \theta_2 & n_2 \theta_1^2 \end{bmatrix}, (V^T V)^{-1} = \frac{1}{n_1 n_2 \theta_1^2} \begin{bmatrix} n_2 \theta_1^2 & -n_2 \theta_1 \theta_2 \\ -n_2 \theta_1 \theta_2 & n_1 + n_2 \theta_2^2 \end{bmatrix} \tag{7}$$

In terms of design, the only adjustable quantities for this model are the sample sizes (n_1 and n_2). As such, a more transparent manner of framing the Fieller-Creasy problem is in terms of the design (probability) measure

$$\xi = \left\{ \begin{matrix} 1 & 0 \\ \omega & 1 - \omega \end{matrix} \right\} \tag{8}$$

Since $x = 1$ corresponds here to the first treatment and $\omega \in [0, 1]$ is the (perhaps irrational) proportion of the total experimental units devoted to this first treatment, ω takes the place of $\frac{n_1}{n_1 + n_2}$ in the above notation. The 2×2 Jacobian is $V = \begin{bmatrix} 1 & 0 \\ \theta_2 & \theta_1 \end{bmatrix}$ and the information matrix is

$$M(\xi, \boldsymbol{\theta}) = V^T \Omega V = \begin{bmatrix} 1 & \theta_2 \\ 0 & \theta_1 \end{bmatrix} \begin{bmatrix} \omega & 0 \\ 0 & 1 - \omega \end{bmatrix} \begin{bmatrix} 1 & 0 \\ \theta_2 & \theta_1 \end{bmatrix} = \begin{bmatrix} \omega + \theta_2^2 (1 - \omega) & \theta_1 \theta_2 (1 - \omega) \\ \theta_1 \theta_2 (1 - \omega) & \theta_1^2 (1 - \omega) \end{bmatrix} \tag{9}$$

Clearly, $|M|$ is proportional to $\omega(1 - \omega)$, and it follows the D-optimal design is the balanced design with $\omega = \frac{1}{2}$ (i.e., $n_1 = n_2$).

To demonstrate differences here between the 95% Wald and likelihood-based confidence regions for nonlinear models given in the last section, consider the Fieller-Creasy setup with $n_1 = 3$ observations in the first group and responses $y = 3, 4, 5$, and $n_2 = 8$ observations in the second group with responses $y = 6, 6, 7, 8, 8, 9, 10, 10$. The LSE/MLE parameter estimates are then $\hat{\theta}_1 = \bar{y}_1 = 4$ and $\hat{\theta}_2 = \bar{y}_2 / \bar{y}_1 = 8/4 = 2$, and this point is plotted in Fig. 1 along with the 95% Wald (dashed ellipse) and 95% likelihood-based (solid) confidence regions. It is useful to note that the spans of these confidence regions in the horizontal direction (corresponding to θ_1) are the same since θ_1 is a linear parameter here, and so the corresponding confidence intervals, obtained by projection, are identical. On the

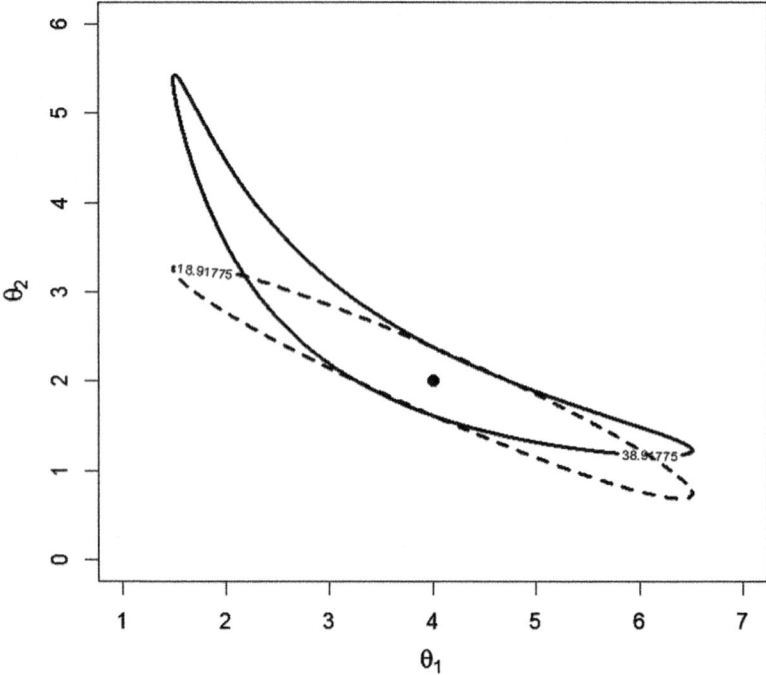

Fig. 1 Wald (dashed) and likelihood-based (solid) 95% confidence region

other hand, the projected intervals for θ_2 are quite different: the 95% Wald interval is $(0.982, 3.018)$, whereas the 95% likelihood-based interval is $(1.302, 3.943)$.

Key advantages of likelihood-based regions and intervals for nonlinear models are given in Bates and Watts (1988), Seber and Wild (1989), and Pawitan (2013). The difference between these regions and intervals is important in the current context since although D-optimal designs minimize the volume of Wald regions, curvature-adjusted methods suggested in Sect. 4.1 minimize a second-order volume approximation to the likelihood region. Furthermore, this example provides the opportunity to highlight the connection between differences between Wald and likelihood-based confidence intervals and measures of curvature or nonlinearity. Detailed discussions of curvature are given in Beale (1960), Ratkowsky (1983), Clarke (1987), Bates and Watts (1988), and Seber and Wild (1989). The model function maps the p-dimensional parameter space onto a p-dimensional expectation surface (i.e., manifold) in the n-dimensional sample space (with $n \geq p$). Intrinsic curvature measures the deviation of the expectation surface from a p-dimensional hyperplane, whereas parameter-effects curvatures measure the degree to which straight, parallel, equi-spaced lines in the parameter space remain straight, parallel and equally spaced on the expectation surface.

Table 1 Data used in dose–response illustration	x	0.5	1.0	2.0	3.0	4.0	5.0
	y	1.076	0.944	0.939	0.743	0.627	0.463

3.2   The Two-Parameter Log-Logistic Model

Dose–response models provide a useful tool in biostatistical modeling, and popular two-parameter dose–response model functions are the log-logistic (*LL2*) and the Weibull (*WEIB2*) functions given by the respective expressions

$$\eta_{LL2}(x, \boldsymbol{\theta}) = \frac{1}{1 + (x/\theta_1)^{\theta_2}} \quad \eta_{WEIB2}(x, \boldsymbol{\theta}) = e^{-(x/\theta_1)^{\theta_2}} \tag{10}$$

Here, the response variables are assumed independent and normally distributed, although extensions to correlated data and other distributions are straightforward. The *LL2* model function is often preferred since the LD_{50}, θ_1, is a model parameter and is easily interpreted by the practitioner since the point $(\theta_1, \frac{1}{2})$ can readily be estimated by a plot of the data. The data used here is given in Table 1 and plotted in Fig. 2.

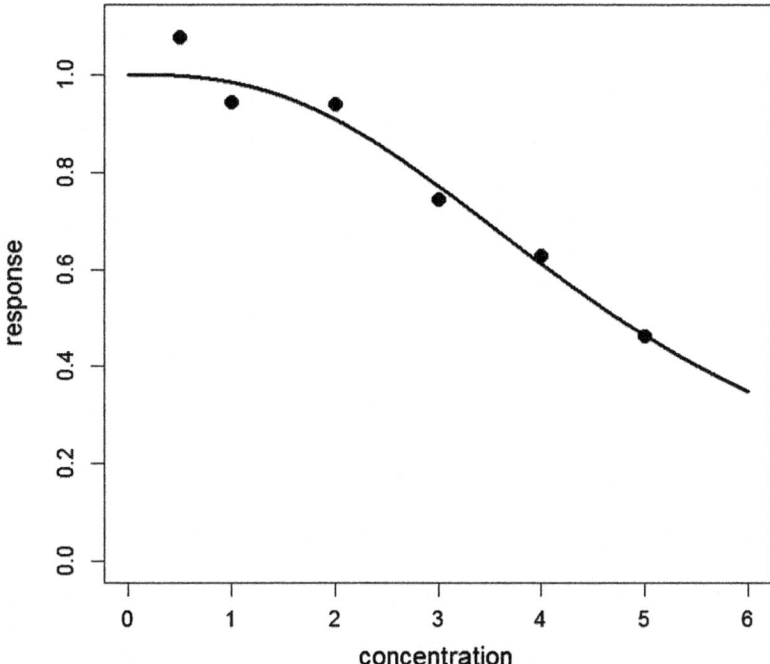

Fig. 2 Log-logistic (*LL2*) model fit to dose–response data

Although both model functions in (10) provide very similar fits to these data, we model these data using the *LL2* function, and in this case the parameter estimates are $\theta_1 = 4.7475$ and $\theta_2 = 2.6544$. In this instance, the columns of the 6×2 Jacobian matrix correspond to $\frac{\partial \eta}{\partial \theta_1} = \frac{\theta_2}{\theta_1} \frac{t}{(1+t)^2}$ and $\frac{\partial \eta}{\partial \theta_2} = -\frac{1}{\theta_2} \frac{t \log(t)}{(1+t)^2}$ for $t = (x/\theta_1)^{\theta_2}$ evaluated at each of the six observations. The fitted model function is also plotted in Fig. 2.

This latter example illustrates that since several (nonlinear) model functions may exist to adequately model a given process, researchers desire designs to estimate model parameters and to distinguish the best-fitting model; these issues are addressed in Sect. 4.2.

4 Robust Design Approaches

A common shortcoming of optimal designs for models involving p model parameters is that these designs may contain only p support points, thereby providing no ability to test for model goodness-of-fit. For example, for the *LL2* model function and parameter estimates given in Sect. 3.2, the D-optimal design contains only the two support points, $x = 3.204$ and $x = 7.034$. Here, we introduce several means to obtain so-called robust (near-optimal) designs which are efficient for parameter estimation yet which provide extra support points to test for model adequacy. Key references include Atkinson (1972) and O'Brien (1994) and those given below. Suggestions on which robust design criterion to choose are given in Sect. 5.

The design measure in (3) is expressed in terms of the sample size (n), and whenever each $\omega_i = 1/n$, the design is called a discrete design; when this condition is not met for at least one ω_i, the design is called continuous. Sample size notwithstanding, optimal designs often have a fewer number of actual support points (with some replication), and the design measure in (3) can also be expressed in terms of these $r \leq n$ support points (s_i) as

$$\xi = \left\{ \begin{array}{cccc} s_1 & s_2 & \cdots & s_r \\ \lambda_1 & \lambda_2 & \cdots & \lambda_r \end{array} \right\} \tag{11}$$

Thus, in terms of (11), discrete designs are those for which $\lambda_i = n_i/n$ for all λ_i. Thus, continuous designs are more general than discrete designs (i.e., discrete designs are a special case of continuous designs).

4.1 A Quadratic Design Strategy

In terms of discrete designs and a given nonlinear model, D-optimal designs minimize the volume of the Wald confidence region given in Sect. 2, and thus the first-order approximation to volume of the actual likelihood-based region. A second-order approximation to this likelihood region volume, given in Hamilton and Watts (1985), is

$$vol_2 = c \left| V^T V \right|^{-1/2} |D|^{-1/2} \left\{ 1 + k^2 \times tr\left(D^{-1} C \right) \right\} \tag{12}$$

In this expression, c and k are constants relative to the design, C (of size $p \times p$) is a function of the parameter-effects curvature, and D (also of size $p \times p$) measures the intrinsic curvature in the direction $= I_p - B$ with $B = L^T [e^T][W]L$, and bracket multiplication of arrays is discussed in Seber and Wild (1989).

Claiming that this volume approximation cannot be used as a design criterion since the residual vector (e) is not known at the design stage, Hamilton and Watts (1985) suggests obtaining designs to minimize the modified volume approximation, $vol_2' = c \left| V^T V \right|^{-1/2} \left\{ 1 + k^2 \times tr(C) \right\}$, obtained by assuming $e = 0$. These designs are called Q'-optimal here. Also, these authors also note that for all their examples, the (local) Q'-optimal designs have only $n = p$ support points—thereby providing no ability to test for lack-of-fit.

This claim notwithstanding, since the residual vector is always orthogonal to the tangent plane at the least-squares estimate, O'Brien (1992) points out that the residual vector can be written as $e = N\alpha$ since N spans this orthogonal space. From the QR decomposition, we write $V = QR = [U|N]R = UL^{-1}$. It follows that when the sample size (n) equals $p + 1$, α is a scalar, and we can then restrict the expected squared length of e to equal σ^2. The design procedure given in O'Brien (1992) yields so-called Q-optimal designs or designs which minimize the original second-order volume approximation of Hamilton and Watts (1985) given in (12).

To illustrate, we use the two-parameter intermediate product (IP2) model function used in pharmacokinetic modeling and given by the expression

$$\eta(x, \boldsymbol{\theta}) = \frac{\theta_1}{\theta_1 - \theta_2} \left\{ e^{-\theta_2 x} - e^{-\theta_1 x} \right\} \tag{13}$$

For initial parameter estimates $\theta_1 = 0.70, \theta_2 = 0.20, \sigma = 0.10$, the (local) D-optimal design comprises the two points $x = 1.23, 6.86$, whereas the three-point (local) Q-optimal design consists of the support points $x = 1.02, 4.72, 6.81$.

These results are extended in O'Brien et al. (2010) by allowing both for discrete designs with $n = p + s$ (for $s > 1$) support points and for continuous designs. In these cases, Q-optimal designs minimize the expected volume, $E(vol_2)$, and use polar or spherical coordinates. For the IP2 case given above, the four-point discrete Q-optimal design has support points $x = 1.00, 1.23, 5.35, 6.72$, and the continuous Q-optimal design associates the weights $\lambda = 0.46, 0.28, 0.26$ with the respective support points $s = 1.06, 5.02, 6.78$. The additional support points given by these designs provide the means to test for model misspecification.

The second example demonstrating the Q-optimality procedure involves the Fieller-Creasy (ratio of two normal means) illustration discussed in Sect. 3.1, where it was noted that regardless of the parameter values, the D-optimal design is the balanced design with $\omega = 1/2$; here, ω is the proportion of the participants in the first treatment group. Further, since by definition the design is a two-point design (so $n = p$), $B = 0$ and $D = I_p$ in (12). O'Brien et al. (2010) demonstrate that due to the curvature captured in the C term in (12), the Q-optimal value of ω varies between 0.50 and 0.75 as the noise (σ) increases from zero to infinity. Note that the original design used in this illustration employed $\omega = \frac{3}{11} = 0.27$. However, with fixed sample size of $n = 11$, the optimal sample size of participants given the first treatment group should have been closer to $n_1 = 6, 7$ or 8—depending upon the anticipated level of noise.

4.2 An Estimation–Discrimination Strategy

Suppose that several model functions, $\eta_1(x, \theta_1), \eta_2(x, \theta_2) \ldots \eta_m(x, \theta_m)$, can be used to describe a given process, each with associated information matrices, $M_1, M_2 \ldots M_m$. The parameter vectors (θ_i) contain p_i model parameters, respectively. A combined estimation measure is $E(\xi) = \sum (\pi_k/p_k) log|M_k(\xi)|$; the weights ($\pi_1, \pi_2 \ldots \pi_m$, which sum to one) here control the emphasis placed upon each of the m rival model functions. O'Brien and Rawlings (1996) use this and an analogous discrimination measure, $D(\xi)$, useful to highlight which model best fits the data, by combining these two terms into a single estimation/discrimination design objective function:

$$B(\xi) = \alpha E(\xi) + (1 - \alpha)D(\xi) \qquad (14)$$

In this expression, $\alpha \in [0, 1]$ controls relative weight placed on estimation versus discrimination, and designs which maximize the criteria function $B(\xi)$ are called (locally) D_B-optimal.

To illustrate this estimation/discrimination technique, consider again the $LL2$ model function discussed in Sect. 3.2, given in Eq. (10), and graphed in Fig. 2; for the given data, the estimated parameter vector is $\hat{\theta}_1 = \begin{pmatrix} 4.7475 \\ 2.6544 \end{pmatrix}$. Given these preliminary results, a researcher wishes to obtain a robust near-optimal design where it is felt the data follow the $LL2$ model, but a possible rival function is the two-parameter Weibull ($WEIB2$) model function given in (10). For these data, the estimated parameters for the $WEIB2$ function are $\hat{\theta}_2 = \begin{pmatrix} 5.5999 \\ 2.2008 \end{pmatrix}$, and we now use the D_B-optimality criterion in (14). To emphasize the $LL2$ model function over the $WEIB2$ function, we choose $\pi_1 = 0.95$ (and so $\pi_2 = 0.05$), and to emphasize estimation over discrimination, we choose $\alpha = 0.75$. This gives the four-point design

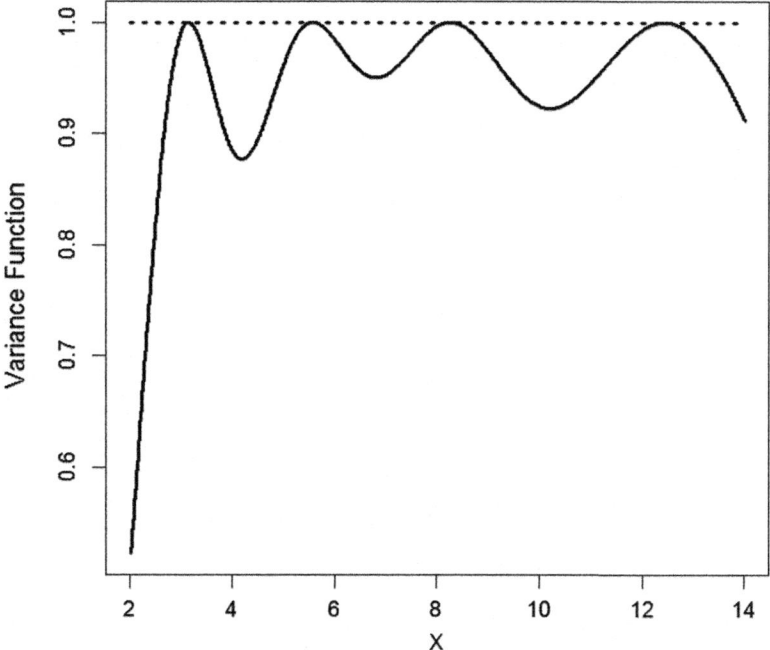

Fig. 3 Plot of variance function for D_B-optimal design

(ξ^*) which associates the weights $\omega = 0.344, 0.245, 0.251, 0.161$, respectively, with the support points $x = 3.137, 5.584, 8.269, 12.441$. Continuous designs of this sort can be approximated by discrete designs using the rounding technique given in Pukelsheim and Rieder (1992). Furthermore, O'Brien and Rawlings (1996) provide a variation of the General Equivalence Theorem and variance function, and the plot of this function in Fig. 3 demonstrates optimality since the variance function does not exceed the cut line $y = 1$ and is equal to unity at the support points of the D_B-optimal design—thereby confirming D_B-optimality.

Although helpful in terms of checking for model misspecification, the addition of two extra design support points comes at a loss in terms of estimating the *LL2* model parameters, and this is captured in the D-efficiency:

$$\left(\frac{|\boldsymbol{M}_1(\xi^*)|}{|\boldsymbol{M}_1(\xi_D)|}\right)^{1/2} = 0.8601 \tag{15}$$

As noted in Atkinson et al. (2007) and as used here, D-efficiency serves to measure the distance (and information loss) between the D_B-optimal design ξ^* and the D-optimal design ξ_D (as noted above comprises the two support points $x = 3.204$ and $x = 7.034$); note that the exponent in (15) is $1/p$ where p is the number of model parameters in the base model function. Thus, the loss in choosing the D_B-optimal design over the D-optimal design here is approximately 14%, and the gain is the information regarding discrimination/estimation departures from the *LL2* model function in the direction of the *WEIB2* model function as captured in the two additional design support points. Clearly, in general, the derived D_B-optimal design and D-efficiency can be adjusted by choosing the tuning parameters $(\alpha, \pi_1, \pi_2 \ldots \pi_m)$ in (14) to suit a specific situation.

4.3   Model Nesting and Geometric Designs

In the context of linear models, one can envisage that the true model for a given situation is $\eta_1(x, \theta_1) = X_1\theta_1$, but we wish to check for departures in the direction of the larger model—the so-called supermodel—$\eta_2(x, \theta) = X_1\theta_1 + X_2\theta_2$, with $\theta = \begin{pmatrix} \theta_1 \\ \theta_2 \end{pmatrix}$. We seek a design that is efficient for η_1 but which can be used to check for departures in the direction of the larger η_2. To illustrate this in the context of response surface modeling, let $X_1\theta_1$ represent the hyperplane, whereas $X_2\theta_2$ includes the additional interaction and quadratic terms. In this case and based on the partition given in Eq. (2) where θ_1 and θ_2 contain p_1 and p_2 parameters, respectively, Atkinson (1972) introduced and illustrated the compound nesting design criterion function

$$\phi(\xi, \theta) = \frac{\kappa}{p_1} log|M_{11}| + \frac{1 - \kappa}{p_2} log|M_{22} - M_{21}M_{11}^{-1}M_{12}| \qquad (16)$$

In Eq. (16), $M_{ij} = X_i^T \Omega X_j$ for $i, j = 1, 2$. Also, $\kappa \in [0, 1]$ controls the emphasis placed on the original parameters versus the additional parameters. As in previously discussed situations, D_ϕ-optimality is confirmed by plotting the corresponding variance function plot and confirming that this graph does not exceed the horizontal cut line.

Since the case of nonlinear models is more complicated, as in O'Brien (1994), we restrict our attention here to sigmoidal models, and these models are typically chosen from the Richards, Weibull, and log-logistic classes or families. A supermodel which generalizes and connects the *LL2* and *WEIB2* model functions given in Eq. (10) is our so-called three-parameter Eclectic (EC3) model function

$$\eta_{EC3}(x, \theta) = \frac{1}{\left(1 + \frac{(x/\theta_1)^{\theta_2}}{\theta_3}\right)^{\theta_3}} \qquad (17)$$

The *LL2* model function is obtained in this expression when $\theta_3 = 1$, and the *WEIB2* model function results for $\theta_3 \to \infty$, thereby demonstrating that *EC3* generalizes both these two-parameter model functions. Other important general-izations of these model functions and Richards, Weibull, and log-logistic dose–response families are given in O'Brien (1994). When for example it is felt that the *LL2* model function is the true model function, but a design is obtained by nesting the *LL2* model in the *EC3* model, the researcher is then protected against departures from the *LL2* model in the direction of all other models in the larger *EC3* class.

To illustrate, consider again the *LL2* model function and the parameter estimates given in the previous section, recalling that the D-optimal design comprises only the two support points $x = 3.204$ and $x = 7.034$ (thereby providing no ability to check for goodness-of-fit). When this model is nested in the *EC3* model using the nesting criteria function in (16) with $\kappa = 0.93$, we obtain the design that associates the weights $\omega = 0.420, 0.258, 0.322$ with the support points $x = 2.978, 5.369, 8.308$. Since the D-efficiency of this latter design exceeds 93.7%, the nested design represents only a modest loss in efficiency (6.3%) but provides the extra support point.

In the spirit of O'Brien et al. (2009), this nesting procedure is further extended here by allowing for (discrete) geometric nesting designs where the design support points have the form $x = a, ab, ab^2 \ldots ab^{K-1}$. For a given value of K, the values of a and b in this expression are chosen to optimize the criteria function in (16). For the *LL2* model, the geometric design support points are $x = 2.716, 3.998, 5.883, 8.659$, for $K = 4$ and $x = 2.626, 3.568, 4.847, 6.586, 8.947$ for $K = 5$; the respective D-efficiencies are indeed quite high—90.4% and 90.1%, respectively. Additional applications of geometric and uniform robust designs are given in O'Brien (2016) in the context of assessing relative potency of similar compounds.

4.4 A General Departure Procedure

The final robust design strategy discussed here is the general departure criterion originally introduced and illustrated in O'Brien (1995), which is akin to so-called space-filling designs popular in software packages such as JMP®. For a given model function with p model parameters, this method entails obtaining the D-optimal design (ξ_D) and corresponding variance function, and adding to the D-optimal design the t additional points where the variance function intersects the cut line $y = p\left\{\left[\frac{(p+1)\delta}{p}\right]^p - 1\right\}$ with say $\delta = 0.90$; here $\delta \in \left[0, \frac{p}{p+1}2^{1/p}\right]$ is chosen to control the final efficiency.

To illustrate, for the *IP2* model function given in (13) discussed in Sect. 4.1 and initial parameter values $\theta_1 = 0.70$ and $\theta_2 = 0.20$, the (local) D-optimal design comprises the two points $x = 1.23$ and $x = 6.86$. The variance function is plotted in Fig. 4 along with (dashed) cut line $y = 2$ to demonstrate D-optimality of this design; taking $\delta = 0.90$, also plotted in this figure is the (dotted) cut line $y = 2\left\{\left[\frac{3(0.90)}{2}\right]^2 - 1\right\} = 1.645$. In this case, the $t = 4$ additional intersection points are $x = 0.761, 1.909, 4.890, 9.366$.

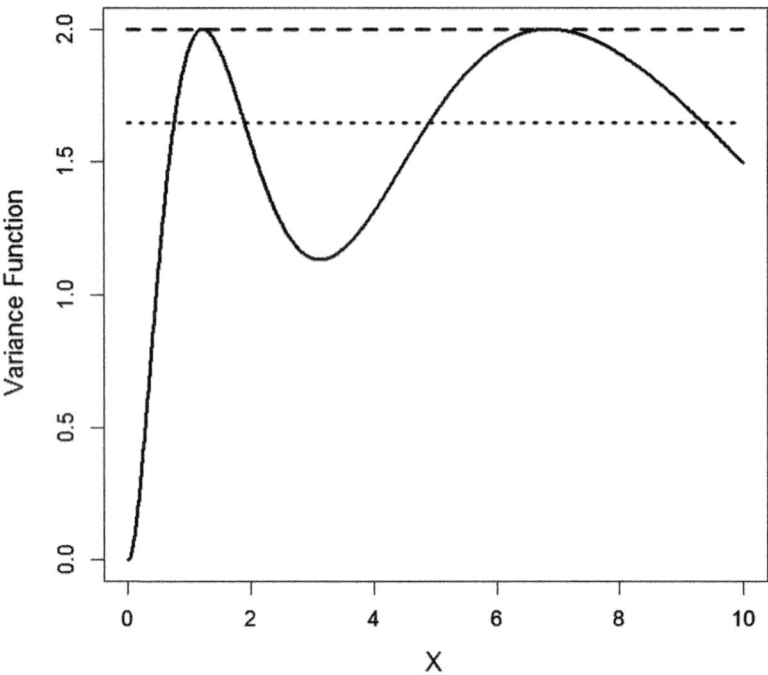

Fig. 4 Plot of variance function for D-optimal design and IP2 model function

To motivate the above procedure, let ξ_x denote the one-point design which puts all mass at the single support point x. Then, the design $\xi_N = \frac{p}{p+1}\xi_D + \frac{1}{p+1}\xi_x$ associates the weight $1/(p+1)$ with each of the p D-optimal support points and x. Then, the D-efficiency $(\delta = \left[\frac{|M(\xi_N)|}{|M(\xi_D)|}\right]^{1/p})$ is equal to $\frac{p}{p+1}\left[1 + \frac{1}{p}d(x,\xi_D)\right]^{1/p}$; when this is solved for $d(x,\xi_D)$, we get the above cut line. If the final design includes r_1 replicates of the D-optimal design and r_2 replicates of the additional t intersection points, the final D-efficiency is $DE_F = \frac{(r_1 p)^{1-t/p}}{r_1 p + r_2 t}|r_1 p I_t + r_2 \mathbf{D}(x,\xi_D)|^{1/p}$.

For the IP2 illustration, with $r_1 = 1$ replicate of the D-optimal design and $r_2 = 1$ replicate of the additional $t = 4$ intersection points (i.e., for the six-point design $x = 1.229, 6.858; 0.761, 1.909, 4.890, 9.366$), the final D-efficiency is 87.9%. With $r_1 = 3$ replicates of the D-optimal design and $r_2 = 1$ replicates of the additional $t = 4$ intersection points (i.e., a 10-point design $x = 3 \times 1.229, 3 \times 6.858; 0.761, 1.909, 4.890, 9.366$), the final D-efficiency increases to 92.8%.

5 Summary

Govaerts (1996), O'Brien and Funk (2003), and others have highlighted the inadequacy of theoretical optimal designs especially those comprised of only p support points for models containing p model parameters. This has motivated our search for viable robust near-optimal design strategies such as those given here. Of the four methods given in Sect. 4, choice of the specific design approach should be dictated by the researcher's belief in the assumed model function. For example, when this belief is quite high and an additional support point or two is desired to augment the usual design, the Q-optimality procedure is recommended. When two or three model functions form a reasonable set of contending model functions, the estimation–discrimination strategy can be used. Model nesting is useful when protection is desired from the assumed model function in the direction of a whole class of model functions—such as sigmoidal dose–response functions useful in bioassay modeling. Finally, the general departures procedure is useful when one wishes to guard against departures from the assumed function in all directions.

All designs discussed here have been obtained using SAS/IML software, and optimality of the respective design was confirmed using the corresponding variance function (when appropriate). A challenge for the practicing statistician is to make these designs and design algorithms available to the practitioner. Providing statistical consulting clients with robust optimal designs affords statisticians the opportunity to engage in consulting at the planning/design phase with an eye to more efficient use of scarce resources.

Acknowledgements The author expresses his appreciation to the J. William Fulbright Foreign Scholarship Board for ongoing grant support and to Chiang Mai University (Thailand), Gadjah Mada University and Islamic University (Indonesia), Kathmandu University (Nepal), and Vietnam National University (Hanoi) for kind hospitality during recent research visits.

References

Atkinson, A.C.: Planning experiments to detect inadequate regression models. Biometrika **59**, 275–293 (1972)

Atkinson, A.C., Donev, A.N., Tobias, R.D.: Optimum Experimental Designs, with SAS. Oxford, New York (2007)

Bates, D.M., Watts, D.G.: Nonlinear Regression Analysis and Its Applications. Wiley, New York (1988)

Beale, E.M.L.: Confidence regions in non-linear estimation (with Discussion). J. R. Stat. Soc. B **22**, 41–88 (1960)

Clarke, G.P.Y.: Marginal curvatures and their usefulness in the analysis of nonlinear regression models. J. Am. Stat. Assoc. **82**, 844–850 (1987)

Clarke, G.P.Y., Haines, L.M.: Optimal design for models incorporating the richards function. In: Seeber, G.U.H., Francis, B.J., Hatzinger, R., Steckel-Berger, G. (eds.) Statistical Modelling. Springer, New York (1995)

Cook, R.D., Witmer, J.A.: A note on parameter-effects curvature. J. Am. Stat. Assoc. **80**, 872–878 (1985)

Dette, H., O'Brien, T.E.: Optimality criteria for regression models based on predicted variance. Biometrika **86**, 93–106 (1999)

Govaerts, B.: Discussion of the papers by Atkinson, and Bates et al. J. R. Stat. Soc. B **58**, 95–111 (1996)

Haines, L.M., O'Brien, T.E., Clarke, G.P.Y.: Kurtosis and curvature measures for nonlinear regression models. Stat. Sinica **14**, 547–570 (2004)

Hamilton, D.C., Watts, D.G.: A quadratic design criterion for precise estimation in nonlinear regression models. Technometrics **27**, 241–250 (1985)

Kiefer, J.: General equivalence theory for optimal designs (approximation theory). Ann. Stat. **2**, 849–879 (1974)

Kiefer, J., Wolfowitz, J.: The equivalence of two extremum problems. Can. J. Math. **12**, 363–366 (1960)

O'Brien, T.E.: A note on quadratic designs for nonlinear regression models. Biometrika **79**, 847–849 (1992)

O'Brien, T.E.: A new robust design strategy for sigmoidal models based on model nesting. In: Dutter, R., Grossmann, W. (eds.) CompStat 1994, pp. 97–102. Physica-Verlag, Heidelberg (1994)

O'Brien, T.E.: Optimal design and lack of fit in nonlinear regression models. In: Seeber, G.U.H., Francis, B.J., Hatzinger, R., Steckel-Berger, G. (eds.) Statistical Modelling, pp. 201–206. Springer, New York (1995)

O'Brien, T.E.: Efficient experimental design strategies in toxicology and bioassay. Stat. Optim. Inf. Comput. **4**, 99–106 (2016)

O'Brien, T.E., Funk, G.M.: A gentle introduction to optimal design for regression models. Am. Stat. **57**, 265–267 (2003)

O'Brien, T.E., Rawlings, J.O.: A non-sequential design procedure for parameter estimation and model discrimination in nonlinear regression models. J. Stat. Plann. Infer. **55**, 77–93 (1996)

O'Brien, T.E., Chooprateep, S., Homkham, N.: Efficient geometric and uniform design strategies for sigmoidal regression models. S. Afr. Stat. J. **43**, 49–83 (2009)

O'Brien, T.E., Jamroenpinyo, S., Bumrungsup, C.: Curvature measures for nonlinear regression models using continuous designs with applications to optimal design. Involve J. Math. **3**, 317–332 (2010)

Pawitan, Y.: In All Likelihood: Statistical Modelling and Inference Using Likelihood. Oxford University Press, Oxford (2013)

Pukelsheim, F.: Optimal Design of Experiments. Wiley, New York (1993)

Pukelsheim, F., Rieder, S.: Efficient rounding of approximate designs. Biometrika **79**, 763–770 (1992)

Ratkowsky, D.A.: Nonlinear Regression Modeling: A Unified Practical Approach. Marcel Dekker, New York (1983)

Seber, G.A.F., Wild, C.J.: Nonlinear Regression. Wiley, New York (1989)

Silvey, S.D.: Optimal Design: An Introduction to the Theory for Parameter Estimation. Chapman and Hall, London (1980)

Studden, W.J.: Some robust-type D-optimal designs in polynomial regression. J. Am. Stat. Assoc. **77**, 916–921 (1982)

White, L.V.: An extension of the general equivalence theorem to nonlinear models. Biometrika **60**, 345–348 (1973)

Alternative Approaches for the Use of Uncertain Prior Information to Overcome the Rank-Deficiency of a Linear Model

Burkhard Schaffrin, Kyle Snow and Xing Fang

Abstract The rank-deficiency of a linear model indicates some information deficit that may be covered by "prior information" (p.i.) in spite of its uncertainty. There are several ways of introducing such p.i., which may be characterized as hierarchical or simultaneous. Here, three hierarchical methods will be compared with four simultaneous methods; in particular, the question of rescaling the p.i. itself or only its dispersion matrix will be investigated. A small (surveying) leveling network will serve as a numerical example for the comparison.

Keywords Hierarchical versus simultaneous methods · Rank-deficiency Rescaling prior information itself versus its dispersion matrix · Uncertain prior information

1 Introduction

Oftentimes, the linear(ized) model for a weighted least-squares adjustment problem turns out to be *rank-deficient*, in which case additional information has to be introduced to guarantee a unique solution for the parameter estimates. This can be done by a certain "*datum choice*" following Baarda (1967), or by integrating (stochastic) prior information (p.i.), which may lead to a *Mixed Model* in accordance with Moritz (1970) or, alternatively, to the *Extended Gauss-Markov Model* proposed by Wolf (1977).

In both of the latter cases, *scaling* the prior information may be advised, either directly by applying a scale factor to the (given) expected p.i. vector, or indirectly by allowing a different variance component to govern the uncertainty of the p.i. While

B. Schaffrin · K. Snow (✉) · X. Fang
Geodetic Science Program, School of Earth Sciences, The Ohio State University,
Columbus, OH, USA
e-mail: ksnow6378137@gmail.com

K. Snow
Topcon Positioning Systems, Inc., Columbus, OH, USA

X. Fang
School of Geodesy and Geomatics, Wuhan University, Wuhan, People's Republic of China

© Springer International Publishing AG 2018   181
M. Tez and D. von Rosen (eds.), *Trends and Perspectives
in Linear Statistical Inference*, Contributions to Statistics,
https://doi.org/10.1007/978-3-319-73241-1_12

the *direct approach* has been investigated in much detail by Schaffrin (1985, 1986) under the name of "*robust collocation*," the *indirect approach* may be traced as far back as Helmert (1907), with modern presentations by Schaffrin (1983) and in a particularly comprehensive fashion by Rao and Kleffe (1988), resp. Searle et al. (1992).

In a paper by Schaffrin (1987) that was concerned with the estimation of point heights from leveling data, four different methods were compared on a purely theoretical basis; two of them can be characterized as "*hierarchical*" (fixed datum vs. inner datum), whereas the other two turn out to be "*simultaneous*" in nature (least-squares collocation or inhomBLIP vs. robust collocation or homBLUP). The use of variance component estimation (i.e., the "indirect approach") had not been part of the investigation at the time, nor did the *BLIMPBE (Best LInear Minimum Partial Bias Estimate)* principle exist then, which would be another *hierarchical* procedure of choice; for more details we refer to Schaffrin and Iz (2002) and to Snow and Schaffrin (2007) where it is emphasized that, in general, the BLIMPBE does not belong to the LEast-Squares Solutions (LESS). Further options are provided by Schaffrin (2003), Fok et al. (2009), and Schaffrin and Navratil (2012).

The present contribution is so organized that we begin with a *typical geodetic case study* involving leveling data without any absolute height information. We found this example in the book by Niemeier (2008) and shall introduce it in the following Sect. 2. Afterwards, the three relevant *hierarchical* procedures will be analyzed in Sect. 3, while Sect. 4 will be devoted to the four simultaneous procedures that we believe to be most relevant. Finally, we shall present both an overview of *leveling networks* in general and our *numerical results* in Sect. 5 before drawing some conclusions and providing an outlook on further research.

2   The Problem at Hand: A Leveling Network (Four Loops)

The following problem of a leveling network with four loops was found in the book by Niemeier (2008). There, in Fig. 1, the four loops are sketched out which would allow a "free adjustment" by using four *condition equations*, one for each loop.

Since we are interested in the study of the impact that uncertain (and sometimes unreliable) prior information may have on the adjustment, the use of a *Gauss-Markov Model (GMM) with stochastic constraints* appears to be suitable. The model is defined by:

$$y = \underset{n \times m}{A}\, \xi + e, \quad e \sim (\mathbf{0}, \mathbf{\Sigma}), \qquad\qquad \text{"new data,"} \qquad\qquad (2.1a)$$

$$z_0 = \underset{l \times m}{K}\, \xi + e_0, \quad e_0 \sim (\mathbf{0}, \mathbf{\Sigma}_0), \qquad \text{"prior information,"} \qquad (2.1b)$$

$$C\{e_0, e\} = \mathbf{0}, \qquad\qquad\qquad\qquad\quad \text{"no correlation,"} \qquad\quad (2.1c)$$

$$l := \text{rk}\, K \le m = \text{rk}\left[A^T \mid K^T\right] < n, \qquad \text{"rank relations,"} \qquad (2.1d)$$

$$r := n - q, \quad \text{for}\, q := \text{rk}\, A \qquad\qquad \text{"redundancy."} \qquad\qquad (2.1e)$$

The variables are defined so that

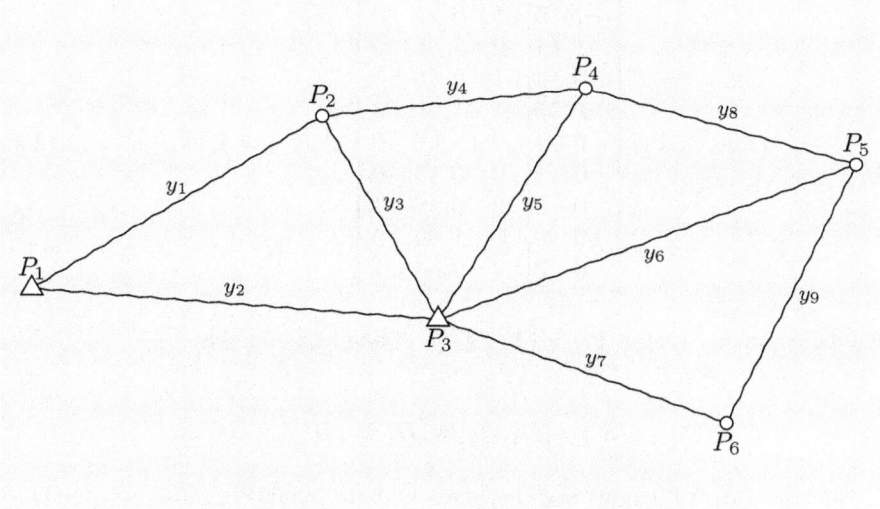

Fig. 1 A 4-loop leveling network with two fiducial points (P_1 and P_3)

y denotes the $n \times 1$ vector of observed height differences;
A is the $n \times m$ coefficient matrix with $q := \mathrm{rk}\, A < m < n$;
ξ is the (unknown) $m \times 1$ parameter vector of heights;
e is the (unknown) $n \times 1$ vector of random observational noise;
Σ is the (given) symmetric positive-definite $n \times n$ dispersion matrix of e.

Similarly,

z_0 denotes the $l \times 1$ vector of previously estimated heights;
$K := [I_l \mid 0]$ is the (given) constraining matrix with $l := \mathrm{rk}\, K$;
e_0 is the (unknown) $l \times 1$ vector of random estimation noise;
Σ_0 is the (given) symmetric nonnegative-definite $l \times l$ dispersion matrix of e_0;
$C\{e_0, e\} = 0$, assuming no correlation between e and e_0.

Two fiducial points were added to Niemeier's network, indicated by triangle symbols in Fig. 1. Fiducial points come with previous estimates, often of poorer quality, along with variances and covariances.

Thus, $l = 2 < m = 6 < n = 9$, and $r = n - q = 9 - 5 = 4$, in view of the coefficient matrix $A := [\underset{9 \times 2}{A_1} \mid \underset{9 \times 4}{A_2}]$ defined by

$$
A = \begin{bmatrix}
-1 & 0 & +1 & 0 & 0 & 0 \\
-1 & +1 & 0 & 0 & 0 & 0 \\
0 & +1 & -1 & 0 & 0 & 0 \\
0 & 0 & -1 & +1 & 0 & 0 \\
0 & -1 & 0 & +1 & 0 & 0 \\
0 & -1 & 0 & 0 & +1 & 0 \\
0 & -1 & 0 & 0 & 0 & +1 \\
0 & 0 & 0 & -1 & +1 & 0 \\
0 & 0 & 0 & 0 & -1 & +1
\end{bmatrix}, \quad \mathrm{rk}\, A = 5, \tag{2.2a}
$$

and the parameter vector $\underset{6\times 1}{\boldsymbol{\xi}} := [\underset{1\times 2}{\boldsymbol{\xi}_1^T} \mid \underset{1\times 4}{\boldsymbol{\xi}_2^T}]^T$ of heights defined by

$$
\boldsymbol{\xi} := \left[H_1, H_3 \mid H_2, H_4, H_5, H_6 \right]^T. \tag{2.2b}
$$

Note the order of heights in $\boldsymbol{\xi}$ that starts with the heights of those two points, P_1 and P_3, for which prior estimates are available, namely

$$
\boldsymbol{z}_0 := \begin{bmatrix} z_1^0 \\ z_2^0 \end{bmatrix} = \boldsymbol{\xi}_1 + \boldsymbol{e}_0 = \left[I_2 \mid 0 \right] \begin{bmatrix} \boldsymbol{\xi}_1 \\ \boldsymbol{\xi}_2 \end{bmatrix} + \boldsymbol{e}_0, \tag{2.2c}
$$

so that $K = [I_2 \mid 0]$ with $\mathrm{rk}\, K = 2 = l$.

Also note that

$$
\mathrm{rk}\, A_1 = 2, \ \mathrm{rk}\, A_2 = 4, \ \text{but}\, \mathrm{rk}\, A = 5 < m \tag{2.2d}
$$

$$
\text{since } \boxed{A\boldsymbol{\tau} = 0} \text{ for } \underset{m\times 1}{\boldsymbol{\tau}} := [1, \ldots, 1]^T. \tag{2.2e}
$$

For the data themselves (both $\boldsymbol{y}$ and $\boldsymbol{z}_0$) as well as for their dispersion matrices ($\boldsymbol{\Sigma}$ and $\boldsymbol{\Sigma}_0$), we refer to the later Sect. 5.

3 Three Hierarchical Procedures

Here we distinguish between two cases, depending on the perceived quality of the prior information; it may either be assumed *superior* (case 1) or *inferior* (cases 2a and 2b) in comparison to the new data.

3.1 Case 1: Prior Information Superior to New Dataset

In this case, highest priority is given to the prior information by the objective

$$\min_{\xi_1} = (z_0 - \xi_1)^T \Sigma_0^{-1}(z_0 - \xi_1),\tag{3.1a}$$

which immediately leads to the "*reproducing estimate*"

$$\boxed{\hat{\xi}_1 = z_0, \quad E\{\hat{\xi}_1\} = \xi_1, \quad D\{\hat{\xi}_1\} = \Sigma_0.}\tag{3.1b}$$

In a second step, the new data are adjusted subject to (3.1a) and (3.1b) by following the objective

$$(y - A_1 z_0 - A_2 \xi_2)^T P(y - A_1 z_0 - A_2 \xi_2) = \min_{\xi_2} \quad \text{for } P := \sigma_0^2 \cdot \Sigma^{-1},\tag{3.2a}$$

which provides the (weighted) least-squares solution

$$\boxed{\hat{\xi}_2 = N_{22}^{-1}(c_2 - N_{21}z_0)} \quad \text{where } \left[N_{21}|N_{22}|c_2\right] := A_2^T P\left[A_1|A_2|y\right],\tag{3.2b}$$

with

$$E\{\hat{\xi}_2\} = N_{22}^{-1}\left(N_{21}\xi_1 + N_{22}\xi_2 - N_{21}\xi_1\right) = \xi_2,\tag{3.2c}$$

and

$$\boxed{D\{\hat{\xi}_2\} = \sigma_0^2 \cdot N_{22}^{-1} + N_{22}^{-1}N_{21}\Sigma_0 N_{21}^T N_{22}^{-1}} \quad \text{since } C\{y, z_0\} = 0.\tag{3.2d}$$

If also $P_0 := \sigma_0^2 \cdot \Sigma_0^{-1}$ is introduced with a *common* variance component σ_0^2, then:

$$(n - m + l)\cdot\hat{\sigma}_0^2 = \tilde{e}_0^T P_0 \tilde{e}_0 + \tilde{e}^T P\tilde{e}\tag{3.3a}$$

for

$$\tilde{e}_0 = z_0 - \hat{\xi}_1 = 0 \quad \text{and } \tilde{e} = y - A_1 z_0 - A_2 \hat{\xi}_2,\tag{3.3b}$$

respectively

$$\boxed{\hat{\sigma}_0^2 = (n - m + l)^{-1} \cdot (y - A_1 z_0)^T P_{\text{red}}(y - A_1 z_0)}\tag{3.4a}$$

with the "*reduced weight matrix*"

$$P_{\text{red}} := P - PA_2 N_{22}^{-1}A_2^T P,\tag{3.4b}$$

which happens to be singular.

The above solution was employed to define the *National Geodetic Vertical Datum of 1929* (U.S. and Canada), which was used for over half a century. Nevertheless, it does not even represent the *"optimal reproducing estimate"* of type repro-BLUUE as was eventually found out by Schaffrin and Navratil (2012).

3.2 Case 2a: Dataset Superior to Prior Information (S-weighted MINOLESS)

In this case the objective functions from case 1 are applied in reverse order, starting with the objective:

$$(y - A\xi)^T P(y - A\xi) = \min_{\xi} \ \text{for} \ P := \sigma_0^2 \cdot \Sigma^{-1}, \tag{3.5a}$$

which produces the (singular) *"normal equations"*

$$N\hat{\xi} = \begin{bmatrix} N_{11} & N_{12} \\ N_{21} & N_{22} \end{bmatrix} \cdot \begin{bmatrix} \hat{\xi}_1 \\ \hat{\xi}_2 \end{bmatrix} = \begin{bmatrix} c_1 \\ c_2 \end{bmatrix} = c \tag{3.5b}$$

with a multitude of (weighted) least-squares solutions. Uniqueness is now achieved by introducing the objective:

$$(z_0' - \hat{\xi})^T S(z_0' - \hat{\xi}) = \min_{\hat{\xi}} \ \text{s.t.} \ N(\hat{\xi} - z_0') = c - Nz_0', \tag{3.6a}$$

with

$$z_0' := [z_1^0, z_2^0, 0, \ldots, 0]^T, \quad P_0 := \sigma_0^2 \cdot \Sigma_0^{-1}, \ \text{and} \tag{3.6b}$$

$$S := K^T P_0 K = \begin{bmatrix} P_0 & 0 \\ 0 & 0 \end{bmatrix} \ \text{as "selection matrix,"} \tag{3.6c}$$

leading to the system of equations

$$\begin{bmatrix} S & N \\ N & 0 \end{bmatrix} \begin{bmatrix} \hat{\xi} - z_0' \\ \hat{\lambda} \end{bmatrix} = \begin{bmatrix} 0 \\ c - Nz_0' \end{bmatrix}. \tag{3.7}$$

By adding the rows of (3.7) together, while assuming the invertibility of $(S+N)$, we obtain:

$$\hat{\xi} - z_0' = (S + N)^{-1}(c - Nz_0') - (S + N)^{-1}N \cdot \hat{\lambda} \tag{3.8a}$$

and with the second row of (3.7) again:

$$c - Nz'_0 = N(\hat{\bar{\xi}} - z'_0) = N(S+N)^{-1}(c - Nz'_0) - N(S+N)^{-1}N \cdot \hat{\lambda} \qquad (3.8b)$$

from which the vector $-N\hat{\lambda}$ can be derived via

$$
\begin{aligned}
-N\hat{\lambda} &= -N[N(S+N)^{-1}N]^- \cdot N(S+N)^{-1}N \cdot \hat{\lambda} = \\
&= N[N(S+N)^{-1}N]^- \cdot [(c - Nz'_0) - N(S+N)^{-1}N \cdot N^- \cdot (c - Nz'_0)] = \qquad (3.8c) \\
&= N[N(S+N)^{-1}N]^- \cdot (c - Nz'_0) - (c - Nz'_0)
\end{aligned}
$$

using various *g-inverse* matrices. By implementing (3.8c) into (3.8a), the final solution of type S-weighted MINOLESS can be represented by:

$$\boxed{\hat{\bar{\xi}} = z'_0 + (S+N)^{-1}N[N(S+N)^{-1}N]^- \cdot (c - Nz'_0).} \qquad (3.8d)$$

It is *not* an unbiased estimate since the "normal equations" (3.5b) happen to be *singular*, even though both diagonal blocks N_{11} and N_{22} are invertible.

Now, by simple error propagation and exploiting the fact that $C\{y, z'_0\} = 0$, the dispersion matrix of $\hat{\bar{\xi}}$ can be obtained via

$$\boxed{\begin{aligned}
D\{\hat{\bar{\xi}}\} = {}& \sigma_0^2 \cdot (S+N)^{-1} \cdot N \cdot [N(S+N)^{-1}N]^- N[N(S+N)^{-1}N]^- N(S+N)^{-1} \\
& + \{I_m - (S+N)^{-1}N[N(S+N)^{-1}N]^- N\} \cdot \begin{bmatrix} \Sigma_0 & 0 \\ 0 & 0 \end{bmatrix} \\
& \cdot \{I_m - (S+N)^{-1}N[N(S+N)^{-1}N]^- N\}^T
\end{aligned}} \qquad (3.9)$$

and the estimated variance component via

$$\boxed{\hat{\sigma}_0^2 = (n-q)^{-1} \cdot (\tilde{e}^T P \tilde{e})} \quad \text{for } \tilde{e} = y - A_1 \hat{\bar{\xi}}_1 - A_2 \hat{\bar{\xi}}_2. \qquad (3.10)$$

Unfortunately, the S-weighted MINOLESS does *not* minimize the bias uniformly, in contrast to the full (I_m-weighted) MINOLESS, which is the only LESS that does so according to Snow and Schaffrin (2007). Thus, for better bias control, we may prefer to follow the approach proposed by Schaffrin and Iz (2002), called BLIMPBE (which stands for Best LInear Minimum Partial Biased Estimate).

3.3 Case 2b: Dataset Superior to Prior Information ($\bar{S}$-BLIMPBE)

According to the acronym of BLIMPBE, now the estimator has to be linear, should minimize the relevant part of the bias, and in this class show the smallest Mean Squared Error (MSE). This means that, firstly,

$$\bar{\xi} = \underset{m \times n}{L} y, \quad \text{where } L \text{ is to be determined,} \tag{3.11}$$

and, secondly, a certain part of the bias vector

$$\beta := E\{\bar{\xi}\} - \xi = -(I_m - LA)\xi =: B \cdot \xi \tag{3.12a}$$

is to be minimized, with B as *"bias matrix."* Thus, the squared norm $\beta^T \beta = \text{tr}(B\xi\xi^T B)$ is replaced by

$$\text{tr}(B\bar{S}B^T) = \text{tr}(LA\bar{S}A^T L^T) - 2\,\text{tr}(LA\bar{S}) + \text{tr}\,\bar{S} = \min_{L^T} \tag{3.12b}$$

with

$$\bar{S} := \begin{bmatrix} P_0^{-1} & 0 \\ 0 & 0 \end{bmatrix} = S^+ \quad \text{(i.e., the "Moore-Penrose inverse" of } S\text{),} \tag{3.12c}$$

leading to the constraints

$$\boxed{A\bar{S}A^T \cdot L^T = A\bar{S}.} \tag{3.12d}$$

Thirdly, under the constraints (3.12d), we minimize the trace of the Mean Squared Error matrix of $\bar{\xi}$, namely

$$\text{tr}\,\text{MSE}\{\bar{\xi} = Ly\}/\sigma_0^2 \approx \text{tr}(LP^{-1}L^T) + \text{tr}(B\bar{S}B^T) = \min_{L^T}, \tag{3.13a}$$

where $\text{tr}(B\bar{S}B^T)$ is now *constant*, as a consequence of (3.12b), so that (3.13a) can be replaced by

$$\text{tr}(LP^{-1}L^T) = \min_{L^T} \text{ s.t. } A\bar{S}A^T \cdot L^T = A\bar{S}, \tag{3.13b}$$

which ultimately leads to the *unique solution*

$$\boxed{L^T = PA \cdot \bar{S}N(N\bar{S}N\bar{S}N)^- N\bar{S}} \tag{3.13c}$$

respectively

$$\boxed{\bar{\xi} = Ly = \bar{S}N(N\bar{S}N\bar{S}N)^- N\bar{S} \cdot c.} \tag{3.14a}$$

Obviously

$$\bar{\xi} \neq \bar{S}N(N\bar{S}N)^- \cdot c \quad \text{unless} \quad \text{rk}\,\bar{S}N = q = \text{rk}\,N, \tag{3.14b}$$

and, thus, the $\bar{S}$-BLIMPBE $\bar{\xi}$ will *not belong* to the LEast-Squares Solutions (LESS), in general.

For the same reason, we obtain

$$E\{\bar{\xi}\} = \bar{S}N(N\bar{S}N\bar{S}N)^-N\bar{S}N \cdot \xi \neq \xi \quad \text{(even if rk } \bar{S}N = q < m), \tag{3.15a}$$

thus establishing a certain (unavoidable) bias for $\bar{\xi}$, with

$$B := \bar{S}N(N\bar{S}N\bar{S}N)^-N\bar{S}N - I_m \tag{3.15b}$$

as the "bias matrix" for $\bar{\xi}$. The corresponding dispersion matrix now results in

$$\boxed{D\{\bar{\xi}\} = \sigma_0^2 \cdot \bar{S}N(N\bar{S}N\bar{S}N)^-N\bar{S}} \tag{3.16}$$

with a still unknown variance component σ_0^2. An *ad-hoc estimate* may use the *residual vector*

$$\bar{e} = y - A\bar{\xi} = y - A_1\bar{\xi}_1 - A_2\bar{\xi}_2, \tag{3.17a}$$

which will, however, *not* be (weakly) unbiased according to

$$E\{\bar{e}\} = A\xi - A \cdot E\{\bar{\xi}\} = -A \cdot B\xi \neq 0 \quad \text{unless} \quad \text{rk } \bar{S}N = q = \text{rk } N. \tag{3.17b}$$

Nonetheless, σ_0^2 may be estimated in analogy to (3.10) by

$$\boxed{\bar{\sigma}_0^2 = (n - q)^{-1} \cdot (\bar{e}^T P\bar{e})} \tag{3.18}$$

before more rigorous estimates are developed.

4   Four Simultaneous Procedures

In this section, the original Gauss-Markov Model with stochastic constraints in (2.1) is given the *equivalent* form of a *Mixed Linear Model*, using "Helmert's knack" (Helmert 1907). It consists of stripping z_0 of its randomness while preserving its numerical values, which can be formalized by writing

$$\underset{l\times 1}{\kappa_0} := z_0 - \underline{0} = (\xi_1 - \underline{0}) + e_0, \quad e_0 \sim (0, \Sigma_0 = \sigma_0^2 P_0^{-1}), \tag{4.1a}$$

where $\underline{0}$ is an $l \times 1$ "*stochastic zero vector*" with

$$\underline{0} \sim (K\xi - \kappa_0, \Sigma_0 = \sigma_0^2 P_0^{-1}), \quad C\{e, \underline{0}\} = 0, \tag{4.1b}$$

and

$$x_1 := \xi_1 - \underline{0} = \kappa_0 - e_0 \sim (\kappa_0, \Sigma_0 = \sigma_0^2 P_0^{-1}), \quad C\{e, e_0\} = 0, \tag{4.2a}$$

as a new (unknown) $l \times 1$ vector of "*random effects*." By further introducing the new observation vector

$$\underset{\sim}{y} := y - A_1 \cdot \underline{0} = A_1 x_1 + A_2 \xi_2 + e, \quad e \sim (0, \Sigma = \sigma_0^2 P^{-1}), \tag{4.2b}$$

the *Mixed Linear Model* (4.2) is formed, which can be transformed into the *equivalent* model of condition equations (or Gauss-Helmert Model)

$$\underset{\sim}{y} - A_1 \kappa_0 = A_2 \xi_2 + (e - A_1 e_0) =: A_2 \xi_2 + e' \tag{4.3a}$$

with

$$e' \sim (0, \Sigma + A_1 \Sigma_0 A_1^T = \sigma_0^2 (P^{-1} + A_1 P_0^{-1} A_1^T)). \tag{4.3b}$$

The models (4.2) and (4.3) are the basis for the first simultaneous procedure that, through the weighted least-squares principle, will generate an optimal estimate of ξ_2 (BLUUE) and an optimal prediction of x_1 (inhomBLIP).

On the other hand, the prior information in (4.2a) may not only be uncertain, but also *unreliable*. In such a case, the introduction of (unknown) *scale factors* for both κ_0 and Σ_0 may be advised, either separately as in the cases 2 and 3, or together as in case 4.

Thus case 2 will result in the *Weak Mixed Model*, defined by the condition equations

$$\underset{\sim}{y} = (A_1 \kappa_0) \cdot \omega + A_2 \xi_2 + e', \quad e' \sim (0, \sigma_0^2 (P^{-1} + A_1 P_0^{-1} A_1^T)), \tag{4.4a}$$

with ω as an additional scale factor. After a weighted least-squares adjustment, optimal estimates of type BLUUE are obtained for ξ_2 and ω, which will lead to an optimal prediction of x_1 of type homBLUP via

$$\tilde{x}_1 = \kappa_0 \cdot \hat{\omega} - \tilde{\tilde{e}}_0. \tag{4.4b}$$

In contrast, a scale factor for Σ_0, as in case 3, will lead to a *Variance Component Model (VCM)*, where the characterization (4.3b) for e' in (4.3a) is replaced by

$$e' \sim (0, \sigma_0^2 (P^{-1} + \alpha \cdot A_1 P_0^{-1} A_1^T) = \sigma_0^2 P^{-1} + \sigma_1^2 \cdot A_1 P_0^{-1} A_1^T). \tag{4.5}$$

The weighted least-squares solution will be *unbiased* if e' follows a symmetric p.d.f., but will not represent BLUUE (as it is a nonlinear estimate).

Finally, cases 2 and 3 can be combined to the *Weak Mixed Model with Variance Components*, represented by a combination of (4.4a) with (4.5). However, for this model to work the number l of components in the vector κ_0 has to be 3 or larger. Since this is not fulfilled here, no numerical results will be presented for case 4 in Sect. 5.

4.1 Case 1: The Mixed Linear Model, resp. Condition Equations

Here, the weighted least-squares approach as applied to the Mixed Linear Model (4.3) will lead to the *normal equations*

$$\begin{bmatrix} N_{11} + P_0 \, N_{12} \\ N_{21} \quad\;\; N_{22} \end{bmatrix} \begin{bmatrix} \tilde{x}_1 \\ \hat{\xi}_2 \end{bmatrix} = \begin{bmatrix} c_1 + P_0 \kappa_0 \\ c_2 \end{bmatrix}, \tag{4.6a}$$

where $\hat{\xi}_2$ represents the BLUUE (Best Linear Uniformly Unbiased Estimate) of ξ_2 and $\tilde{x}_1$ the nhomBLIP (Best inhomogeneously LInear Prediction) of x_1. Since both turn out to be unbiased, due to

$$E\{\hat{\xi}_2\} = \xi_2 \;\text{ for all }\; \xi_2 \in \mathbb{R}^{m-l}, \;\text{ and }\; E\{\tilde{x}_1\} = \kappa_0 = E\{x_1\}, \tag{4.6b}$$

the Mean Squared Error matrix can be represented by

$$\text{MSE}\{ \begin{bmatrix} \tilde{x}_1 \\ \hat{\xi}_2 \end{bmatrix} \} = D\{ \begin{bmatrix} \tilde{x}_1 - x_1 \\ \hat{\xi}_2 \end{bmatrix} \} = \sigma_0^2 \begin{bmatrix} N_{11} + P_0 \, N_{12} \\ N_{21} \quad\;\; N_{22} \end{bmatrix}^{-1}, \tag{4.6c}$$

where the variance component σ_0^2 may be estimated unbiasedly through

$$\hat{\sigma}_0^2 = (n - m + l)^{-1} (\tilde{e}_0^T P_0 \tilde{e}_0 + \tilde{e}^T P \tilde{e}) \tag{4.7a}$$

for

$$\tilde{e}_0 := \kappa_0 - \tilde{x}_1, \;\text{ and }\; \tilde{e} := \underset{\sim}{y} - A_1 \tilde{x}_1 - A_2 \hat{\xi}_2. \tag{4.7b}$$

Alternatively, model (4.4) may be exploited in which case the weighted least-squares principle

$$(e')^T (\Sigma + A_1 \Sigma_0 A_1^T)^{-1} e' \cdot \sigma_0^2 =$$
$$= (e')^T [P - PA_1(P_0 + N_{11})^{-1} A_1^T P] e' = \min_{e'} \tag{4.8}$$

leads to the *normal equations*

$$\left[A_2^T (P^{-1} + A_1 P_0^{-1} A_1^T)^{-1} A_2 \right] \cdot \hat{\xi}_2 = A_2^T (P^{-1} + A_1 P_0^{-1} A_1^T)^{-1} (\underset{\sim}{y} - A_1 \kappa_0) \tag{4.9a}$$

with $\hat{\xi}_2$ as the very same BLUUE of ξ_2 as in (4.6a) and with the same dispersion, resp. MSE-matrix

$$\text{MSE}\{\hat{\hat{\xi}}_2\} = D\{\hat{\hat{\xi}}_2 - \xi_2\} = D\{\hat{\hat{\xi}}_2\} = \sigma_0^2 \left[A_2^T (P^{-1} + A_1 P_0^{-1} A_1^T)^{-1} A_2 \right]^{-1} =$$
$$= \sigma_0^2 \left[N_{22} - N_{21}(P_0 + N_{11})^{-1} N_{12} \right]^{-1} \qquad (4.9b)$$

as in (4.6c), followed by the *residual vectors*

$$\begin{bmatrix} \tilde{e} \\ \tilde{e}_0 \end{bmatrix} = \begin{bmatrix} C\{e, e'\} \\ C\{e_0, e'\} \end{bmatrix} \cdot [D\{e'\}]^{-1} \cdot (\underline{y} - A_1 \kappa_0 - A_2 \hat{\hat{\xi}}_2), \qquad (4.10a)$$

and

$$\tilde{x}_1 = \kappa_0 - \tilde{e}_0 \qquad (4.10b)$$

as inhomBLIP of x_1 with

$$\text{MSE}\{\tilde{x}_1\} = \sigma_0^2 (P_0 + N_{11} - N_{12} N_{22}^{-1} N_{21})^{-1} \qquad (4.10c)$$

as its MSE-matrix, and with $\hat{\sigma}_0^2$ from (4.7a) as its variance component estimate.

4.2   Case 2: Condition Equations for the Weak Mixed Model

Here, the relevant model is defined by (4.4a) and shows an additional (unknown) scale factor for κ_0. Consequently, the weighted least-squares principle (4.8) as applied to (4.4a) will generate the "*normal equations*"

$$\begin{bmatrix} N_{11} + P_0 & N_{12} \\ N_{21} & N_{22} \end{bmatrix} \begin{bmatrix} g_1 & \gamma_1 \\ g_2 & \gamma_2 \end{bmatrix} = \begin{bmatrix} c_1 & N_{11}\kappa_0 \\ c_2 & N_{21}\kappa_0 \end{bmatrix}, \qquad (4.11)$$

from which the estimated scale factor $\hat{\omega}$, the homBLUP (Best Homogeneous Linear (weakly) Unbiased Predictor) $\tilde{x}_1$ of x_1, and the corresponding unbiased estimate $\hat{\hat{\xi}}_2$ of ξ_2 will result as follows:

$$\hat{\omega} = (\kappa_0^T P_0 g_1)/(\kappa_0^T P_0 \gamma_1) \sim (\omega, \sigma_0^2/(\kappa_0^T P_0 \gamma_1)), \qquad (4.12a)$$
$$\tilde{x}_1 = \kappa_0 \cdot \hat{\omega} + (g_1 - \gamma_1 \cdot \hat{\omega}), \quad \hat{\hat{\xi}}_2 = g_2 - \gamma_2 \cdot \hat{\omega}, \qquad (4.12b)$$

along with the MSE-matrix of $\tilde{x}_1$

$$\mathrm{MSE}\{\tilde{\tilde{x}}_1\} = D\{\tilde{\tilde{x}}_1 - x_1\} = \sigma_0^2(P_0 + N_{11} - N_{12}N_{22}^{-1}N_{21})^{-1}+$$
$$+ (P_0 + N_{11} - N_{12}N_{22}^{-1}N_{21})^{-1}P_0\kappa_0\cdot(\kappa_0^T P_0\gamma_1)^{-1}\cdot\kappa_0^T P_0(P_0 + N_{11} - N_{12}N_{22}^{-1}N_{21})^{-1},$$

$$(4.12c)$$

and the variance component estimate

$$\boxed{\hat{\sigma}_0^2 = (n - m + l - 1)^{-1} \cdot (\tilde{\tilde{e}}_0^T P_0\tilde{\tilde{e}}_0 + \tilde{e}^T P\tilde{e})} \tag{4.13a}$$

for

$$\tilde{\tilde{e}}_0 = \kappa_0 \cdot \hat{\omega} - \tilde{\tilde{x}}_1 = -(g_1 - \gamma_1 \cdot \hat{\omega}) \tag{4.13b}$$

and

$$\tilde{e} = \underset{\sim}{y} - A_1\tilde{\tilde{x}}_1 - A_2\hat{\tilde{\xi}}_2 \tag{4.13c}$$

as residual vectors; for more details, see Schaffrin (1985).

4.3   Case 3: The Variance Component Model (VCM)

In this model, defined by (4.3a) in conjunction with (4.5), the vector $\boldsymbol{\vartheta} := [\vartheta_1, \vartheta_2]^T = [\sigma_0^2, \sigma_1^2]^T$ needs to be estimated along with both x_1 and $\boldsymbol{\xi}_2$. For this we rewrite (4.5) as

$$e' \sim (0, \sigma_0^2 P^{-1} + \sigma_1^2 \cdot A_1 P_0^{-1}A_1^T =: \vartheta_1 \cdot Q_1 + \vartheta_2 \cdot Q_2 =: \Sigma'(\boldsymbol{\vartheta})) \tag{4.14}$$

and find the *repro-BIQUUE* (reproducing Best Invariant Quadratic Uniformly Unbiased Estimate) of the vector $\boldsymbol{\vartheta}$ from the *(nonlinear) "normal equations"*

$$\boxed{\begin{bmatrix} \mathrm{tr}(Q_1\hat{W}Q_1\hat{W}) & \mathrm{tr}(Q_1\hat{W}Q_2\hat{W}) \\ \mathrm{tr}(Q_2\hat{W}Q_1\hat{W}) & \mathrm{tr}(Q_2\hat{W}Q_2\hat{W}) \end{bmatrix} \cdot \begin{bmatrix} \hat{\vartheta}_1 \\ \hat{\vartheta}_2 \end{bmatrix} = \begin{bmatrix} (\underset{\sim}{y}-A_1\kappa_0)^T\hat{W}Q_1\hat{W}(\underset{\sim}{y}-A_1\kappa_0) \\ (\underset{\sim}{y}-A_1\kappa_0)^T\hat{W}Q_2\hat{W}(\underset{\sim}{y}-A_1\kappa_0) \end{bmatrix}} \tag{4.15a}$$

with

$$\hat{W} := (\Sigma'(\hat{\boldsymbol{\vartheta}}))^{-1} - (\Sigma'(\hat{\boldsymbol{\vartheta}}))^{-1}A_2[A_2^T \cdot (\Sigma'(\hat{\boldsymbol{\vartheta}}))^{-1} \cdot A_2]^{-1}A_2^T \cdot (\Sigma'(\hat{\boldsymbol{\vartheta}}))^{-1}, \tag{4.15b}$$

followed by the—oftentimes unbiased—estimate

$$\boxed{\hat{\tilde{\xi}}_2 = [A_2^T(\Sigma'(\hat{\boldsymbol{\vartheta}}))^{-1}A_2]^{-1} \cdot A_2^T(\Sigma'(\hat{\boldsymbol{\vartheta}}))^{-1}(\underset{\sim}{y} - A_1\kappa_0)} \tag{4.16a}$$

of $\boldsymbol{\xi}_2$, the *residual vectors*

$$\begin{bmatrix} \tilde{\tilde{e}} \\ \tilde{\tilde{e}}_0 \end{bmatrix} = \begin{bmatrix} \hat{\vartheta}_1 \cdot P^{-1} \\ -\hat{\vartheta}_2 \cdot P_0^{-1} A_1^T \end{bmatrix} \cdot (\Sigma'(\hat{\vartheta}))^{-1}(\underset{\sim}{y} - A_1 \kappa_0 - A_2 \hat{\bar{\xi}}_2), \qquad (4.16b)$$

and the corresponding prediction of x_1, namely

$$\boxed{\tilde{\tilde{x}}_1 = \kappa_0 - \tilde{\tilde{e}}_0.} \qquad (4.16c)$$

For more details, we refer to Schaffrin (1983).

4.4 Case 4: The Weak Mixed Model with Variance Components

Here, the two cases 2 and 3 are combined by considering (4.4a) in conjunction with (4.5). This can be done successfully as soon as the prior information vector κ_0 has *more than three* components ($l \geq 3$). We decided to present the necessary formulas although, in our example, we only have $l = 2$.

After introducing (4.14) for $\vartheta := [\sigma_0^2, \sigma_1^2 = \sigma_0^2 \cdot \alpha]^T$, the *repro-BIQUUE* of ϑ can now be taken from the (nonlinear) *normal equations*

$$\boxed{\begin{bmatrix} \mathrm{tr}(Q_1 \hat{W} Q_1 \hat{W}) \; \mathrm{tr}(Q_1 \hat{W} Q_2 \hat{W}) \\ \mathrm{tr}(Q_2 \hat{W} Q_1 \hat{W}) \; \mathrm{tr}(Q_2 \hat{W} Q_2 \hat{W}) \end{bmatrix} \begin{bmatrix} \hat{\vartheta}_1 \\ \hat{\vartheta}_2 \end{bmatrix} = \begin{bmatrix} \underset{\sim}{y}^T \hat{W} Q_1 \hat{W} \underset{\sim}{y} \\ \underset{\sim}{y}^T \hat{W} Q_2 \hat{W} \underset{\sim}{y} \end{bmatrix}} \qquad (4.17a)$$

with

$$\hat{W} := (\Sigma'(\hat{\vartheta}))^{-1} - (\Sigma'(\hat{\vartheta}))^{-1}[A_1\kappa_0 \,|\, A_2] \cdot$$

$$\cdot \begin{bmatrix} \kappa_0^T A_1^T \cdot (\Sigma'(\hat{\vartheta}))^{-1} \cdot A_1\kappa_0 & \kappa_0^T A_1^T \cdot (\Sigma'(\hat{\vartheta}))^{-1} \cdot A_2 \\ A_2^T \cdot (\Sigma'(\hat{\vartheta}))^{-1} \cdot A_1\kappa_0 & A_2^T \cdot (\Sigma'(\hat{\vartheta}))^{-1} \cdot A_2 \end{bmatrix}^{-1} \begin{bmatrix} \kappa_0^T A_1^T \\ A_2^T \end{bmatrix} (\Sigma'(\hat{\vartheta}))^{-1} \quad (4.17b)$$

followed by the—oftentimes unbiased—estimates

$$\begin{bmatrix} \hat{\omega} \\ \hat{\bar{\xi}}_2 \end{bmatrix} = \begin{bmatrix} \kappa_0^T A_1^T \cdot (\Sigma'(\hat{\vartheta}))^{-1} \cdot A_1\kappa_0 & \kappa_0^T A_1^T \cdot (\Sigma'(\hat{\vartheta}))^{-1} \cdot A_2 \\ A_2^T \cdot (\Sigma'(\hat{\vartheta}))^{-1} \cdot A_1\kappa_0 & A_2^T \cdot (\Sigma'(\hat{\vartheta}))^{-1} \cdot A_2 \end{bmatrix}^{-1} \begin{bmatrix} \kappa_0^T A_1^T \\ A_2^T \end{bmatrix} \cdot (\Sigma'(\hat{\vartheta}))^{-1} \cdot \underset{\sim}{y}$$

$$(4.18a)$$

of ω and ξ_2, the *residual vectors*

$$\begin{bmatrix} \bar{\bar{e}} \\ \bar{\bar{e}}_0 \end{bmatrix} = \begin{bmatrix} \hat{\vartheta}_1 \cdot P^{-1} \\ -\hat{\vartheta}_2 \cdot P_0^{-1} A_1^T \end{bmatrix} \cdot (\Sigma'(\hat{\vartheta}))^{-1}(\underset{\sim}{y} - A_1\kappa_0 \cdot \hat{\omega} - A_2\bar{\bar{\xi}}_2), \qquad (4.19)$$

and the corresponding prediction of x_1, namely

$$\boxed{\bar{\bar{x}}_1 = \kappa_0 \cdot \hat{\bar{\omega}} - \bar{\bar{e}}_0.} \tag{4.20}$$

As in case 3, the various dispersion, resp. MSE-matrices still need to be developed in full detail. But, unlike case 3, case 4 could not be investigated numerically in our small-size example from Sect. 2.

5   Numerical Results

Some comments on leveling networks
A leveling network in surveying and geodesy is a set of points that have been "connected" by a series of spirit-level observations. When such a series of observations returns back to the starting point, it is said to form a circuit (or closed loop). Such observations are then "adjusted" by one of the methods described above, for instance, in order to estimate orthometric heights of the benchmarks, where the model redundancy is equal to the number of observational circuits in the network. However, spirit-leveling only provides information about height differences between benchmarks, and thus some other source of data (e.g., prior information on one or more benchmark heights) must also be specified in the model (e.g., via (2.1b)) to avoid a rank-deficiency (of one).

Because spirit-leveling is a relatively time consuming (but precise) way of determining height differences between points, now-a-days relative GPS positioning methods are also employed to supply height information for one or more points within the leveling network. Depending on the approach used, GPS-derived heights have the added advantage of supplying a "height datum" for the model (e.g., via (2.1b)), thereby overcoming the rank-deficiency mentioned in the preceding paragraph.

In the numerical experiments that follow, we use both orthometric heights derived from GPS (and geoid undulations), e.g., H_A for height at point A, and orthometric height differences from spirit-leveling (and gravity data), e.g., $H_{AB} = H_B - H_A$, as observational data in model (2.1).

5.1   Description of New Data and Prior Information

In the following experiments we use height differences from spirit leveling as "new data" and GPS-derived heights of two points as prior information. The spirit-leveling data are the same as that used by Niemeier (2008). The GPS-derived heights are simulated.

Table 1 Spirit-leveling data corresponding to Fig. 1

y_i	P_j to P_k	H_{jk}(m)	D_i (km)	$\sigma_i^2 \cdot \sigma_0^2$ (mm^2)
y_1	P_1 to P_2	-8.206	0.62	5.58
y_2	P_1 to P_3	-5.734	1.20	10.8
y_3	P_2 to P_3	2.481	0.45	4.05
y_4	P_2 to P_4	-4.433	0.80	7.20
y_5	P_3 to P_4	-6.909	1.00	9.00
y_6	P_3 to P_5	18.872	1.10	9.90
y_7	P_3 to P_6	4.035	0.44	3.96
y_8	P_4 to P_5	11.962	0.72	6.48
y_9	P_5 to P_6	22.904	0.83	7.47

Table 2 Prior information for points P_1 and P_2 shown in Fig. 1

Point	z_0(m)	Element of Diag Σ_0 (mm^2)
P_1	$H_1 = 68.951$	$\sigma_{H_1}^2 = 50^2 \cdot \sigma_0^2$
P_3	$H_3 = 63.201$	$\sigma_{H_3}^2 = 10^2 \cdot \sigma_0^2$

It is common to weight spirit-leveling data as a function of the length of path associated with the observed height difference H_{jk}, $j, k \in \{1, \ldots, 6\}$. Instead of weights, we rather compute observational variances σ_i^2, $i = 1, \ldots, 9$, using the formula $\sigma_i^2 = \sigma_0^2 \cdot (0.003\text{m})^2 \cdot D_{jk}$, where D_{jk} is the distance of the path between points P_j and P_k. The weight matrix used in the model is then computed as a scaled inverse covariance matrix viz. $\boldsymbol{P} := \sigma_0^2 \Sigma^{-1} = \sigma_0^2 [\text{Diag}(\sigma_1^2, \ldots, \sigma_9^2)]^{-1}$. The observed H_{jk} and their associated variances σ_i^2 are listed in Table 1, and the leveling network is depicted in Fig. 1.

As mentioned earlier, the heights derived from "GPS leveling" are simulated for this study. The values adopted are shown in the second column of Table 2. The empirical standard deviations of heights derived from relative GPS positioning often range in the level of ± 1 to ± 5 cm depending on the length of the observation session and the observational standards and specifications adhered to. The values adopted for this study are shown in the last column of Table 2, which indicate that the standard deviation of the height H_1 is assumed to be five times greater than that of H_3.

The data from Tables 1 and 2 were used in all the numerical estimations discussed herein; the results are presented in the following sections.

5.2 Description of Parameter Estimates and Predictions

The following quantities are shown in the tables below for each of the six cases investigated:

- Estimated and predicted parameters (orthometric heights)
- Prior information residuals
- Measurement residuals
- Model redundancy
- Estimated variance component(s)

The number of significant digits shown in Tables 3, 4, 5, 6, 7 and 8 is greater than what the precision of the data would warrant. We only show so many digits for the benefit of readers who might want to replicate our work.

Hierarchical—Case 1: Prior information reproduced
The parameter estimates determined according to Sect. 3.1 are listed in Table 3. As explained in Sect. 3.1, the prior information will be reproduced, which can be confirmed by comparing the numerical values for heights H_1 and H_3 in Tables 2 and 3. As noted earlier, this reproducing estimator is not optimal, which appears to be reflected in its relatively large estimated variance component $\hat{\sigma}_0^2 = 18.196779$ shown in the table. Thus we would not recommend this estimator in practice, but merely have included it here for comparative purposes.

Hierarchical—Case 2a: S-weighted MINOLESS
The parameter estimates determined according to Sect. 3.2 are listed in Table 4. As explained in Sect. 3.2, the estimates are of type S-weighted MINOLESS, which has the following properties:

1. The change in prior information is minimized (in terms of the S-weighted L_2-norm) for the selected parameters (here H_1 and H_3).

Table 3 Hierarchical—Case 1: ξ_1 reproduced, $r = 5$, $\hat{\sigma}_0^2 = 18.196779$

$\hat{\xi}$ (m)	$\tilde{e}_0$ (mm)	$\tilde{e}$ (mm)
H_1 : 68.9510	0.0	15.3
H_3 : 63.2010	0.0	16.0
H_2 : 60.7297		9.7
H_4 : 56.2940		2.6
H_5 : 44.3311		−2.0
H_6 : 67.2357		−2.1
		0.3
		0.9
		−0.6

Table 4 Hierarchical—Case 2a: S-weighted MINOLESS, $r = 4$, $\hat{\tilde{\sigma}}_0^2 = 1.282825$

$\hat{\tilde{\xi}}$ (m)	Δz_0 (mm)	$\tilde{\tilde{e}}$ (mm)
H_1 : 68.9315	19.5	2.2
H_3 : 63.2018	−0.8	−4.3
H_2 : 60.7233		2.5
H_4 : 56.2918		−1.6
H_5 : 44.3306		0.9
H_6 : 67.2360		−0.8
		0.8
		−0.7
		−1.4

2. The trace of the respective subblock of the dispersion matrix $D\{\hat{\tilde{\xi}}\}$, shown in (3.9), is minimized.
3. The residuals $\tilde{\tilde{e}}$, shown in (3.10), are unique, i.e., they are not affected by the specification of the selection matrix S.
4. The parameter estimates $\hat{\tilde{\xi}}$ are biased by the selection matrix S.

Note that in Table 4 the term Δz_0 denotes the change in prior information and is defined by $\Delta z_0 := z_0 - \hat{\tilde{\xi}}_1$. It turns out that the prior information for height H_1 changed more significantly than that for H_3, which is somewhat expected considering the larger variance of H_1 shown in Table 2.

Hierarchical—Case 2b: $\bar{S}$-BLIMPBE

The parameter estimates determined according to Sect. 3.3 are listed in Table 5. As explained in Sect. 3.3, the estimates are of type $\bar{S}$-BLIMPBE. The numerical algorithm linearizes the problem such that $c_i = A^T P(y - A \cdot \bar{\xi}_{i-1})$ for the i-th iteration, thereby requiring an initial approximation $\bar{\xi}_0$ for the parameters. It turns out that for this particular implementation of BLIMPBE, the "non-selected" parameters (H_2, H_4, H_5, and H_6) are reproduced. This is consistent with the fact that the $(I_m - \bar{S})$-BLIMPBE coincides with the standard least-squares solution in the overconstrained model as shown in Snow and Schaffrin (2007, Theorem 9).

Note that in Table 5 the term Δz_0 denotes the change in prior information and is defined by $\Delta z_0 := z_0 - \bar{\xi}_1$.

Simultaneous—Case 1: BLUUE of ξ_2 and inhomBLIP of x_1

The parameter estimates determined according to Sect. 4.1 are listed in Table 6. As explained in Sect. 4.1, the estimates and predictions are BLUUE of ξ_2 and inhomBLIP of x_1, respectively.

Comparison of Tables 4 and 6 show identical results for the parameters and residuals to the precisions listed. However, this is suspected to be an artifact of the dataset rather than indicative of the general case, as the two different estimators associated with these solutions have different statistical properties in general.

Table 5 Hierarchical—Case 2b: $\bar{S}$-BLIMPBE, $r = 4$, $\bar{\sigma}_0^2 = 1.283321$

$\bar{\xi}$ (m)	Δz_0 (mm)	$\tilde{\bar{e}}$ (mm)
H_1 : 68.9234	27.6	2.2
H_3 : 63.1937	7.3	−4.3
H_2 : 60.7152		2.5
H_4 : 56.2838		−1.6
H_5 : 44.3226		0.9
H_6 : 67.2280		−0.9
		0.7
		−0.8
		−1.4

Table 6 Simultaneous—Case 1: inhomBLIP of x_1, BLUUE of ξ_2, $r = 5$, $\hat{\sigma}_0^2 = 1.057875$

$\tilde{x}_1, \hat{\xi}_2$ (m)	$\tilde{e}_0$ (mm)	$\tilde{e}$ (mm)
H_1 : 68.9315	19.5	2.2
H_3 : 63.2018	−0.8	−4.3
H_2 : 60.7233		2.5
H_4 : 56.2918		−1.6
H_5 : 44.3306		0.9
H_6 : 67.2360		−0.8
		0.8
		−0.7
		−1.4

Simultaneous—Case 2: BLUUE of ξ_2 and homBLUP of x_1

The parameter estimates determined according to Sect. 4.2 are listed in Table 7. As explained in Sect. 4.2, the estimates and predictions are BLUUE of ξ_2 and homBLUP of x_1, respectively.

The table lists both the formal residual vector $\tilde{\bar{e}}_0 = \kappa_0 \cdot \hat{\omega} - \tilde{\bar{x}}_1$ (Eq. (4.13b)) for the prior information and the change in prior information $\Delta\kappa_0 := \kappa_0 - \tilde{\bar{x}}_1$. Apparently the estimated scale factor $\hat{\omega}$ results in an unexpectedly large shift in the prior information, though the observation residuals $\tilde{\bar{e}}$ turn out to be the same as the residuals listed in Table 4, which may be due to an artifact of the dataset.

Simultaneous—Case 3: repro-BIQUUE of ϑ

The parameter estimates determined according to Sect. 4.3 are listed in Table 8. As explained in Sect. 4.3, the estimates of the variance components are repro-BIQUUE. The fact that the estimate of the first variance component $\hat{\vartheta}_1$ turns out to be greater than 1, suggests that the variances of the prior information (from Table 2) might have been too small. Likewise, the fact that the estimate of the second variance component $\hat{\vartheta}_2$ turns out to be less than 1, suggests that the variances of the observations (from

Table 7 Simultaneous—Case 2: homBLUP of x_1, BLUUE of ξ_2, $\hat{\omega} = 0.99647103$, $r = 4$, $\hat{\sigma}_0^2 = 1.282825$

$\tilde{x}_1, \hat{\tilde{\xi}}_2$ (m)	$\tilde{e}_0, \Delta\kappa_0$ (mm)	$\tilde{\tilde{e}}$ (mm)
H_1 : 68.7077	0.0, 243.3	2.2
H_3 : 62.9780	0.0, 223.0	−4.3
H_2 : 60.4995		2.5
H_4 : 56.0680		−1.6
H_5 : 44.1068		0.9
H_6 : 67.0122		−0.8
		0.8
		−0.7
		−1.4

Table 8 Simultaneous—Case 3: repro-BIQUUE of ϑ, $\hat{\vartheta}_1 = 1.28282530$, $\hat{\vartheta}_2 = 0.15599792$, $r = 5$, $\hat{\alpha} = \hat{\sigma}_0^2/\hat{\sigma}_1^2 = 8.22334888$

$\tilde{x}_1, \hat{\tilde{\xi}}_2$ (m)	$\tilde{e}_0$ (mm)	$\tilde{\tilde{e}}$ (mm)
H_1 : 68.9318	19.2	2.4
H_3 : 63.2018	−0.8	−4.0
H_2 : 60.7234		2.6
H_4 : 56.2919		−1.5
H_5 : 44.3306		0.9
H_6 : 67.2360		−0.8
		0.8
		−0.7
		−1.4

Table 1) might have been too large. Nevertheless, the ratio $\hat{\alpha}$ turned out to be greater than 8, which is consistent with our assumption that the leveling observations are more precise than the GPS-derived heights comprising the prior information.

6   Conclusions and Outlook

Our research has investigated various ways to combine prior information on a subset of unknown parameters with newer observational data to be used for estimating the full set of unknown parameters within certain linear models.

While the particular application of our work is from geodetic science and surveying, it is expected that the theory can be applied to a variety of estimation problems in other fields as well.

The relative superiority or inferiority of the prior information versus that of the observational data must be considered when choosing which one of the various models and associated least-squares solutions are best to be used.

The simultaneous case 4 described in Sect. 4.4 could not be included in our numerical example due to the prior information vector $\boldsymbol{\kappa}_0$ only having two elements. However, studies are planned for the near future that involve larger networks with more prior information. These studies will provide opportunities to experiment with case 4, too, and the findings are expected to be published in the geodetic literature once the experiments are completed.

References

Baarda, W.: Statistical Concepts in Geodesy. Netherlands Geodetic Commission, New Series 2, No. 4, Delft/NL (1967)

Fok, H., Baki Iz, H., Schaffrin, B.: Comparison of four geodetic network densification solutions. Surv. Rev. **41**(311), 44–56 (2009)

Helmert, F.R.: Adjustment Computations by the Method of Least Squares (in German), 2nd edn. Teubner, Leipzig, Germany (1907)

Moritz, H.: A generalized least-squares model. Studia Geophysica et Geodaetica **14**(2), 353–362 (1970)

Niemeier, W.: Adjustment Computations (in German), 2nd edn. de Gruyter, Berlin, New York (2008)

Rao, C.R., Kleffe, J.: Estimation of Variance Components and Applications, North-Holland, Amsterdam, New York, Oxford, Tokyo (1988)

Schaffrin, B.: Variance covariance component estimation for the adjustment of heterogeneous replicated observations (in German). Ph.D. thesis (1983). Publication of the German Geodetic Community C-282, Munich

Schaffrin, B.: The geodetic datum with stochastic prior information (in German). Habilitation thesis (1985). Publication of the German Geodetic Community C-313, Munich

Schaffrin, B.: On robust collocation. In: First Hotine–Marussi Symposium on Mathematical Geodesy (Rome, 1985), pp. 343–361, Milan (1986)

Schaffrin, B.: Merging gauge registrations of minor accuracy into a first order levelling network. In: Pelzer, H., Niemeier, W. (eds.) Determination of Heights and Height Changes, pp. 397–401. Dümmler, Bonn (1987)

Schaffrin, B.: Reproducing estimates via least-squares: an optimal alternative to the Helmert transformation. In: Grafarend, E., Krumm, F.W., Schwarze, V.S. (eds.) Geodesy-The Challenge of the 3rd Millennium, pp. 387–392. Springer, Berlin (2003)

Schaffrin, B., Iz, H.B.: BLIMPBE and its geodetic applications. In: Adam, J., Schwarz, K. (eds.) Vistas for Geodesy in the New Millenium, vol. 125, Springer Series, IAG-Symp., pp. 377–381. Springer, Berlin (2002)

Schaffrin, B., Navratil, G.: On reproducing linear estimators within the Gauss-Markov model with stochastic constraints. Commun. Stat.-Theory Methods **41**(13–14), 2570–2587 (2012)

Searle, S., Casella, G., McCulloch, C.: Variance Components. Wiley Interscience, Hoboken, New Jersey (1992)

Snow, K., Schaffrin, B.: GPS network analysis with BLIMPBE: an alternative to least-squares adjustment for better bias control. J. Surv. Eng. **133**(3), 114–122 (2007)

Wolf, H.: Zur Grundlegung der Kollokationsmethode. Z. für Vermessungswesen **102**, 237–239 (1977)

Exact Likelihood-Based Point and Interval Estimation for Lifetime Characteristics of Laplace Distribution Based on hybrid Type-I and Type-II Censored Data

Feng Su, N. Balakrishnan and Xiaojun Zhu

Abstract In this chapter, we first derive explicit expressions for the Maximum likelihood estimators (MLEs) of the parameters of Laplace distribution-based on a hybrid Type-I censored sample (Type-I HCS). We then derive the conditional moment generating functions (MGF) of the MLEs, and then use them to obtain the means, variances, and covariance of the MLEs. From the conditional MGFs, we also derive the exact conditional distributions of the MLEs, which are then used to develop exact conditional confidence intervals (CIs) for the parameters. Proceeding similarly, we obtain the MLEs of quantile, reliability, and cumulative hazard functions, and discuss the construction of exact CIs for these functions as well. By using the relationships between Type-I, Type-II, Type-I HCS, and hybrid Type-II censored samples (Type-II HCS), we develop exact inferential methods based on a Type-II HCS as well. Then, a Monte Carlo simulation study is carried out to evaluate the performance of the developed inferential results. Finally, a numerical example is presented to illustrate the point and interval estimation methods developed here under both Type-I HCS and Type-II HCS.

Keywords Bias · Confidence interval · Cumulative hazard function · Hybrid Type-I censoring · Hybrid Type-II censoring · Laplace distribution · Maximum likelihood estimators · Moment generating function · P-P plot · Q-Q plot Quantile · Reliability function · Type-I censoring · Type-II censoring · Variance

F. Su
Department of Basic Courses, Guangzhou Maritime College,
Guangzhou 510725, Guangdong, China
e-mail: sufeng40@hotmail.com

N. Balakrishnan (✉)
Department of Mathematics and Statistics, McMaster University,
Hamilton, ON L8S 4K1, Canada
e-mail: bala@mcmaster.ca

X. Zhu
Department of Mathematical Sciences, Xi'an Jiaotong-Liverpool University,
215123 Suzhou, People's Republic of China
e-mail: Xiaojun.Zhu@xjtlu.edu.cn

© Springer International Publishing AG 2018
M. Tez and D. von Rosen (eds.), *Trends and Perspectives in Linear Statistical Inference*, Contributions to Statistics,
https://doi.org/10.1007/978-3-319-73241-1_13

1 Introduction

The Laplace (μ, σ) distribution has its cumulative distribution function (CDF) as

$$F(x) = \begin{cases} \frac{1}{2} e^{-\frac{\mu-x}{\sigma}}, & x < \mu, \\ 1 - \frac{1}{2} e^{-\frac{x-\mu}{\sigma}}, & x \geq \mu, \end{cases} \tag{1}$$

where μ and σ are the location and scale parameters, respectively. The probability density function (PDF) corresponding to (1) is

$$f(x) = \frac{1}{2\sigma} e^{-\frac{|x-\mu|}{\sigma}}, \qquad -\infty < x < \infty. \tag{2}$$

One may refer to Johnson et al. (1995) and Kotz et al. (2001) for detailed overviews on several inferential results for Laplace distribution based on the complete and censored data. For the case of complete sample, Bain and Engelhardt (1973), and Kappenman (1975, 1977) derived approximate CIs, tolerance intervals, and conditional CIs. For the case of censored data, Balakrishnan and Cutler (1995) first derived the MLEs based on general Type-II censored samples in closed form. These explicit expressions were then used by Childs and Balakrishnan (1996, 1997, 2000) to construct conditional inferential results based on the Type-II and progressively Type-II censored samples. Recently, several exact likelihood inferential procedures have been developed. For example, Iliopoulos and Balakrishnan (2011) developed exact likelihood inference and exact distributions of some pivotal quantities. Iliopoulos and MirMostafaee (2014) developed prediction intervals based on the exact distributions of the MLEs under Type-II censored samples. Zhu and Balakrishnan (2016, 2017) developed exact MLE-based inferential procedures for the parameters as well as quantile, reliability, and cumulative hazard functions based on the Type-I and Type-II censored samples. However, no work has been developed for the case when the life-testing experiment is of a hybrid form resulting in a hybrid censored sample. We, therefore, focus our attention here on this situation and derive the MLEs of Laplace parameters based on the Type-I HCS and Type-II HCS. For a detailed review of this form of censoring and associated developments, interested readers may refer to Balakrishnan and Kundu (2013).

The rest of this chapter is organized as follows. In Sect. 2, we first derive explicit expressions for the MLEs based on the Type-I HCS. In Sect. 3, we derive the conditional joint MGF of the MLEs, and use it to determine the conditional means, variances, and covariance of the MLEs. From this conditional joint MGF, we also derive the exact conditional marginal and joint density functions of the MLEs, which are then used to develop exact conditional CIs for the parameters μ and σ. In Sect. 4, we derive the exact conditional distribution of the MLE of a quantile, which is then used to develop exact conditional CIs for population quantiles. In Sects. 5 and 6, we briefly discuss the construction of exact conditional CIs for reliability and cumulative hazard functions, respectively. In Sect. 7, by using the relationships between Type-II HCS, Type-I HCS, and Type-I and Type-II censoring schemes, we develop exact inference for a Type-II HCS. A Monte Carlo simulation study is then carried out in Sect. 8 to evaluate the performance of the MLEs. In Sect. 9, we present an example to illustrate all the methods of inference developed here. Finally, some concluding comments are made in Sect. 10.

2 MLEs from Type-I HCS

Epstein (1954) proposed Type-I HCS as a compromise between Type-I and Type-II censoring schemes. This life-test would get terminated at the kth failure or a pre-fixed time T, whichever occurs first; that is, the termination time is $T^* = \min\{X_{k:n}, T\}$, where k $(2 \leq k \leq n)$ is a fixed value. Now, let D denote the number of failures up to time T^*. Obviously, the MLEs of μ and σ exist only when $D \geq 1$, and so all subsequent results developed here are based on this condition of observing at least one failure. We then have the following lemma.

Lemma 2.1 *The probability mass function (PMF) of D is*

$$
P(D = d) = \begin{cases}
\frac{1}{1-p_0} \frac{n!}{d!(n-d)!} \left(1 - \frac{1}{2}e^{\frac{\mu-T}{\sigma}}\right)^d \left(\frac{1}{2}e^{\frac{\mu-T}{\sigma}}\right)^{n-d} & T \geq \mu, 1 \leq d \leq k-1 \\
\frac{1}{1-p_0} \sum_{i=k}^{n} \frac{n!}{i!(n-i)!} \left(1 - \frac{1}{2}e^{\frac{\mu-T}{\sigma}}\right)^i \left(\frac{1}{2}e^{\frac{\mu-T}{\sigma}}\right)^{n-i} & T \geq \mu, d = k, \\
\frac{1}{1-q_0} \frac{n!}{d!(n-d)!} \left(\frac{1}{2}e^{\frac{T-\mu}{\sigma}}\right)^d \left(1 - \frac{1}{2}e^{\frac{T-\mu}{\sigma}}\right)^{n-d} & T < \mu, 1 \leq d \leq k-1, \\
\frac{1}{1-q_0} \sum_{i=k}^{n} \frac{n!}{i!(n-i)!} \left(\frac{1}{2}e^{\frac{T-\mu}{\sigma}}\right)^i \left(1 - \frac{1}{2}e^{\frac{T-\mu}{\sigma}}\right)^{n-i} & T < \mu, d = k,
\end{cases} \tag{3}
$$

where $p_0 = \left(\frac{1}{2}e^{\frac{\mu-T}{\sigma}}\right)^n$ and $q_0 = \left(1 - \frac{1}{2}e^{\frac{T-\mu}{\sigma}}\right)^n$.
Moreover, we have

$$
P(T^* = T) = \sum_{d=1}^{k-1} P(D = d). \tag{4}
$$

Proof By considering the fact that the number of failures up to time T follows a $B(n, F(T))$, we readily obtain this lemma. ∎

The likelihood function in this case is given by (see Balakrishnan and Cohen 1991; Arnold et al. 1992)

$$
L = \frac{n!}{(n-d)!} \prod_{i=1}^{d} f(x_{i:n}) \left[1 - F(T^*)\right]^{n-d}, \qquad -\infty < x_{1:n} < \cdots < x_{d:n} \leq T^*. \tag{5}
$$

Theorem 2.1 *By maximizing the likelihood function in (5), the MLEs of μ and σ are obtained as*

$$
\hat{\mu} = \begin{cases}
[X_{m:n}, X_{m+1:n}], & n = 2m, d \geq m+1, \\
X_{m+1:n}, & n = 2m+1, d \geq m+1, \\
[X_{m:n}, T^*], & n = 2m, d = m, \\
T^* + \hat{\sigma} \log(\frac{n}{2d}), & d < \frac{n}{2};
\end{cases}
\tag{6}
$$

$$
\hat{\sigma} = \begin{cases}
\frac{1}{d}\left[(n-d)T^* + \sum_{i=m+1}^{d} X_{i:n} - \sum_{i=1}^{m} X_{i:n}\right], & n = 2m, d \geq m, \\
\frac{1}{d}\left[(n-d)T^* + \sum_{i=m+2}^{d} X_{i:n} - \sum_{i=1}^{m} X_{i:n}\right], & n = 2m+1, d \geq m+1, \\
\frac{1}{d}\sum_{i=1}^{d}\left(T^* - X_{i:n}\right), & d < \frac{n}{2}.
\end{cases}
\tag{7}
$$

Proof The proof can be provided by proceeding along the lines of Zhu and Balakrishnan (2017) for the Type-I censoring case and is therefore omitted for the sake of brevity. ∎

We observe that in (6), in some cases, $\hat{\mu}$ can be any value in a specific interval with equal likelihood. In these cases, as done usually, we take the midpoints of the intervals to obtain

$$
\hat{\mu} = \begin{cases}
\frac{1}{2}(X_{m:n} + X_{m+1:n}) & n = 2m, d \geq m+1, \\
X_{m+1:n} & n = 2m+1, d \geq m+1, \\
\frac{1}{2}(X_{m:n} + T^*) & n = 2m, d = m, \\
T^* + \hat{\sigma} \log(\frac{n}{2d}) & d < \frac{n}{2}.
\end{cases}
\tag{8}
$$

3 Exact Conditional MGF and Density Function of the MLEs

To derive the exact joint density function of the MLEs, we first need the following lemma.

Lemma 3.1 *The expectation of any function $g(\mathbf{X})$, conditioned on $D > 0$, where*

$$
g(\mathbf{X}) = \begin{cases}
g_1(\mathbf{X}), & X_{k:n} < T, \\
g_2(\mathbf{X}), & X_{k:n} > T,
\end{cases}
$$

based on a Type-I HCS can be readily obtained from the expectation of Type-I censored samples along with $1 \leq d \leq k-1$ and a Type-I HCS along with $X_{k:n} < T$, that is,

$$E[g(\mathbf{X})|D > 0] = \sum_{d=1}^{k-1} E[g_2(\mathbf{X}), D = d, X_{k:n} > T|D > 0]$$

$$+E[g_1(\mathbf{X}), D = k, X_{k:n} < T|D > 0]. \tag{9}$$

Proof The proof is straightforward and is therefore omitted. ∎

For obtaining the exact conditional joint distribution of the MLEs from the conditional joint MGF, we also need the following lemma.

Lemma 3.2 *Let* Y_1, Y_2, Z_1 *and* Z_2 *be independent random variables, where* $Y_1 \sim \Gamma(\alpha_1, \beta_1)$, $Y_2 \sim N\Gamma(\alpha_2, \beta_2)$, $Z_1 \sim E(1)$ *and* $Z_2 \sim E(1)$, *with* β_1, $\beta_2 > 0$. *Here,* $N\Gamma$ *denotes the negative gamma distribution, i.e., if* $Y \sim N\Gamma(\alpha, \beta)$ *then* $-Y \sim \Gamma(\alpha, \beta)$. *Now, let* $W_1 = Y_1 + Y_2 + a_1^* Z_1 + a_2^* Z_2$ *and* $W_2 = b_1^* Z_1 + b_2^* Z_2$, *where* a_1^*, a_2^*, b_1^* *and* b_2^* *are any real values. Then, the joint MGF of* W_1 *and* W_2 *is given by*

$$E\left(e^{tW_1+sW_2}\right) = \left(1 - t\beta_1\right)^{-\alpha_1} \left(1 + t\beta_2\right)^{-\alpha_2} \left(1 - a_1^* t - b_1^* s\right)^{-1} \left(1 - a_2^* t - b_2^* s\right)^{-1}. \tag{10}$$

Here and in what follows, we will denote

$$W_1 \stackrel{d}{=} \Gamma(\alpha_1, \beta_1) + N\Gamma(\alpha_2, \beta_2) + a_1 E_1 + a_2 E_2,$$

$$W_2 \stackrel{d}{=} b_1 E_1 + b_2 E_2.$$

Proof The required result is readily obtained from well-known properties of exponential and gamma distributions; see Johnson et al. (1994). Zhu and Balakrishnan (2017) have provided explicit expressions for the exact joint and marginal CDFs of W_1 and W_2. ∎

Moreover, Zhu and Balakrishnan (2017) have derived the exact joint and marginal conditional distributions of $\hat{\mu}$ and $\hat{\sigma}$. By using their results, we directly arrive at an expression for $\sum_{d=1}^{k-1} E[e^{t\hat{\sigma}+s\hat{\mu}}, D = d, X_{k:n} > T|D > 0]$ (say, E_1) as given in the following lemma.

Lemma 3.3 *If the sample size is even, i.e.,* $n = 2m$, *then*

$$E_1 = 1_{\{T>\mu, k\le m\}} \left\{ \sum_{d=1}^{k-1} \sum_{j=0}^{d} \sum_{l=0}^{d-j} p_1 M_{Z_{p1}^{(1)}, Z_{p1}^{(2)}}(t, s) \right\}$$

$$+1_{\{T>\mu, k>m+1\}} \left\{ \sum_{d=1}^{m-1} \sum_{j=0}^{d} \sum_{l=0}^{d-j} p_1 M_{Z_{p1}^{(1)}, Z_{p1}^{(2)}}(t, s) + \sum_{j=0}^{m-1} \sum_{l=0}^{m-1-j} \left[p_{4,a,e} M_{Z_{p4,a,e}^{(1)}, Z_{p4,a,e}^{(2)}}(t, s) \right. \right.$$

$$\left. + p_{4,b,e} M_{Z_{p4,b,e}^{(1)}, Z_{p4,b,e}^{(2)}}(t, s) \right] + p_{5,e} M_{Z_{p5,e}^{(1)}, Z_{p5,e}^{(2)}}(t, s)$$

$$+ \sum_{d=m+1}^{k-1} \sum_{j=0}^{m-1} \sum_{l_1=0}^{m-1-j} \sum_{l_2=0}^{d-m-1} \left[p_{10,a,e} M_{Z_{p10,a,e}^{(1)}, Z_{p10,a,e}^{(2)}}(t, s) + p_{10,b,e} M_{Z_{p10,b,e}^{(1)}, Z_{p10,b,e}^{(2)}}(t, s) \right.$$

$$+ p_{10,c,e} M_{Z^{(1)}_{p10,c,e}, Z^{(2)}_{p10,c,e}}(t,s) + p_{10,d,e} M_{Z^{(1)}_{p10,d,e}, Z^{(2)}_{p10,d,e}}(t,s) \Big]$$

$$+ \sum_{d=m+1}^{k-1} \sum_{l=0}^{d-m-1} \Big[p_{11,a,e} M_{Z^{(1)}_{p11,a,e}, Z^{(2)}_{p11,a,e}}(t,s) + p_{11,b,e} M_{Z^{(1)}_{p11,b,e}, Z^{(2)}_{p11,b,e}}(t,s) \Big]$$

$$+ \sum_{d=m+1}^{k-1} \sum_{j=m+1}^{d} \sum_{l_1=0}^{j-m-1} \sum_{l_2=0}^{d-j} p_{12,e} M_{Z^{(1)}_{p12,e}, Z^{(2)}_{p12,e}}(t,s) \Bigg\}$$

$$+ 1_{\{T \le \mu, k \le m\}} \left\{ \sum_{d=1}^{k-1} q_1 M_{Z^{(2)}_{q1}, Z^{(2)}_{q1}}(t,s) \right\}$$

$$+ 1_{\{T \le \mu, k \ge m+1\}} \left\{ \sum_{d=1}^{m-1} q_1 M_{Z^{(2)}_{q1}, Z^{(2)}_{q1}}(t,s) + q_{3,e} M_{Z^{(1)}_{q3,e}, Z^{(2)}_{q3,e}}(t,s) \right.$$

$$\left. + \sum_{d=m+1}^{k-1} \sum_{l=0}^{d-m-1} q_5 M_{Z^{(1)}_{q5,e}, Z^{(2)}_{q5,e}}(t,s) \right\}.$$

If the sample size is odd, i.e., $n = 2m+1$, then

$$E_1 = 1_{\{T > \mu\}} \left\{ \sum_{d=1}^{\min(m,k-1)} \sum_{j=0}^{d} \sum_{l=0}^{d-j} p_1 M_{Z^{(1)}_{p1}, Z^{(2)}_{p1}}(t,s) \right.$$

$$+ \sum_{d=m+1}^{k-1} \sum_{j=0}^{m} \sum_{l_1=0}^{m-j} \sum_{l_2=0}^{d-m-1} \Big[p_{4,a,o} M_{Z^{(1)}_{p4,a,o}, Z^{(2)}_{p4,a,o}}(t,s) + p_{4,b,o} M_{Z^{(1)}_{p4,b,o}, Z^{(2)}_{p4,b,o}}(t,s) \Big]$$

$$\left. + \sum_{d=m+1}^{k-1} \sum_{j=m+1}^{d} \sum_{l_1=0}^{j-m-1} \sum_{l_2=0}^{d-j} p_{5,o} M_{Z^{(1)}_{p5,o}, Z^{(2)}_{p5,o}}(t,s) \right\}$$

$$+ 1_{\{T \le \mu, k \le m+1\}} \left\{ \sum_{d=1}^{m} q_1 M_{Z^{(1)}_{q1}, Z^{(2)}_{q1}}(t,s) \right\}$$

$$+ 1_{\{T \le \mu, k > m+1\}} \left\{ \sum_{d=1}^{m} q_1 M_{Z^{(1)}_{q1}, Z^{(2)}_{q1}}(t,s) + q_{3,o} M_{Z^{(1)}_{q3,o}, Z^{(2)}_{q3,o}}(t,s) \right.$$

$$\left. + \sum_{d=m+2}^{k-1} \sum_{l=0}^{d-m-1} q_5 M_{Z^{(1)}_{q5,o}, Z^{(2)}_{q5,o}}(t,s) \right\},$$

where all the coefficients and variables involved here are presented in the Appendix. Here and in the rest of this chapter, we will use $\sum_{i=1}^{n} P_{X_i} X_i$ to denote the generalized mixture of distributions of variables $X_1, \cdots, X_n$ with probabilities $P_{X_1}, \cdots, P_{X_n}$, such that $\sum_{i=1}^{n} P_{X_i} = 1$, but P_{X_i}'s need not necessarily be non-negative.

To derive the joint MGF of $(\hat{\mu}, \hat{\sigma})$, we then only need to derive $E[e^{t\hat{\sigma}+s\hat{\mu}}, D = k, X_{k:n} < T | D > 0]$, (say, E_2). Due to the two forms of the density function in (2), we need to consider the cases of $\mu \geq T^*$ and $\mu < T^*$ separately. In the case, when $\mu < T^*$, we will use the conditional approach and let J $(0 \leq J \leq D)$ denote the number of observations in the Type-I HCS that are smaller than μ.

Lemma 3.4 *If the sample size is even, i.e., $n = 2m$, we have*

$$
\begin{aligned}
E_2 = & \, 1_{\{T>\mu, k<m\}} \left\{ \sum_{j=0}^{k-1} \sum_{l=0}^{k-1-j} \left[p_{2,a} M_{Z_{p2,a}^{(1)}, Z_{p2,a}^{(2)}}(t,s) + p_{2,b} M_{Z_{p2,b}^{(1)}, Z_{p2,b}^{(2)}}(t,s) \right] \right. \\
& \left. + \sum_{l=0}^{n-k} p_3 M_{Z_{p3}^{(1)}, Z_{p3}^{(2)}}(t,s) \right\} \\
& + 1_{\{T>\mu, k=m\}} \left\{ \sum_{j=0}^{k-1} \sum_{l=0}^{k-1-j} \left[p_{2,a} M_{Z_{p2,a}^{(1)}, Z_{p2,a}^{(3)}}(t,s) + p_{2,b} M_{Z_{p2,b}^{(1)}, Z_{p2,b}^{(3)}}(t,s) \right] \right. \\
& \left. + \sum_{l=0}^{n-k} p_3 M_{Z_{p3}^{(1)}, Z_{p3}^{(3)}}(t,s) \right\} \\
& + 1_{\{T>\mu, k=m+1\}} \left\{ \sum_{j=0}^{m-1} \sum_{l=0}^{m-1-j} \left[p_{6,a,e} M_{Z_{6,a,e}^{(1)}, Z_{6,a,e}^{(2)}}(t,s) + p_{6,b,e} M_{Z_{6,b,e}^{(1)}, Z_{p6,b,e}^{(2)}}(t,s) \right. \right. \\
& + p_{6,c,e} M_{Z_{p6,c,e}^{(1)}, Z_{p6,c,e}^{(2)}}(t,s) + p_{6,d,e} M_{Z_{p6,d,e}^{(1)}, Z_{p6,d,e}^{(2)}}(t,s) \Big] + \left[p_{7,a,e} M_{Z_{p7,a,e}^{(1)}, Z_{p7,a,e}^{(2)}}(t,s) \right. \\
& \left. \left. + p_{7,b,e} M_{Z_{7,b,e}^{(1)}, Z_{7,b,e}^{(2)}}(t,s) \right] + \sum_{l=0}^{n-k} p_{9,e} M_{Z_{p9,e}^{(1)}, Z_{p9,e}^{(2)}}(t,s) \right\} \\
& + 1_{\{T>\mu, k>m+1\}} \left\{ \sum_{j=0}^{m-1} \sum_{l_1=0}^{m-1-j} \sum_{l_2=0}^{d-m-2} \left[p_{13,a} M_{Z_{p13,a}^{(1)}, Z_{p13,a}^{(2)}}(t,s) + p_{13,b} M_{Z_{p13,b}^{(1)}, Z_{p13,b}^{(2)}}(t,s) \right. \right. \\
& + p_{13,c} M_{Z_{p13,c}^{(1)}, Z_{p13,c}^{(2)}}(t,s) + p_{13,d} M_{Z_{p13,d}^{(1)}, Z_{p13,d}^{(2)}}(t,s) + p_{13,e} M_{Z_{p13,e}^{(1)}, Z_{p13,e}^{(2)}}(t,s) \\
& \left. + p_{13,f} M_{Z_{p13,f}^{(1)}, Z_{p13,f}^{(2)}}(t,s) + p_{13,g} M_{Z_{p13,g}^{(1)}, Z_{p13,g}^{(2)}}(t,s) + p_{13,h} M_{Z_{p13,h}^{(1)}, Z_{p13,h}^{(2)}}(t,s) \right] \\
& + \sum_{l=0}^{k-m-2} \left[p_{14,a} M_{Z_{p14,a}^{(1)}, Z_{p14,a}^{(2)}}(t,s) + p_{14,b} M_{Z_{p14,b}^{(1)}, Z_{p14,b}^{(2)}}(t,s) + p_{14,c} M_{Z_{p14,c}^{(1)}, Z_{p14,c}^{(2)}}(t,s) \right. \\
& \left. + p_{14,d} M_{Z_{p14,d}^{(1)}, Z_{p14,d}^{(2)}}(t,s) \right] + \sum_{j=m+1}^{k-1} \sum_{l_1=0}^{j-m-1} \sum_{l_2=0}^{k-1-j} \left[p_{15,a} M_{Z_{p15,a}^{(1)}, Z_{p15,a}^{(2)}}(t,s) \right. \\
& \left. + p_{15,b} M_{Z_{p15,b}^{(1)}, Z_{p15,b}^{(2)}}(t,s) \right] \\
& \left. + \sum_{l_1=0}^{k-m-2} \sum_{l_2=0}^{n-k} \left[p_{16,a} M_{Z_{p16,a}^{(1)}, Z_{p16,a}^{(2)}}(t,s) + p_{16,b} M_{Z_{p16,b}^{(1)}, Z_{p16,b}^{(2)}}(t,s) \right] \right\}
\end{aligned}
$$

$$+1_{\{T\leq\mu,k<m\}}\left\{\sum_{l=0}^{n-k}q_2 M_{Z^{(1)}_{q2,a},Z^{(2)}_{q2,a}}(t,s)\right\}+1_{\{T\leq\mu,k=m\}}\left\{\sum_{l=0}^{n-k}q_2 M_{Z^{(1)}_{q2,b},Z^{(2)}_{q2,b}}(t,s)\right\}$$

$$+1_{\{T\leq\mu,k\leq m+1\}}\left\{\sum_{l=0}^{n-k}q_{4,e} M_{Z^{(1)}_{q4,e},Z^{(2)}_{q4,e}}(t,s)\right\}$$

$$+1_{\{T\leq\mu,k>m+1\}}\left\{\sum_{l_1=0}^{k-m-2}\sum_{l_2=0}^{n-k}q_6 M_{Z^{(2)}_{q6,e},Z^{(2)}_{q6,e}}(t,s)\right\}.$$

If the sample size is odd, i.e., $n=2m+1$, we have

$$E_2=1_{\{T>\mu,k<m+1\}}\left\{\sum_{j=0}^{k-1}\sum_{l=0}^{k-1-j}\left[p_{2,a}M_{Z^{(1)}_{p2,a},Z^{(2)}_{p2,a}}(t,s)+p_{2,b}M_{Z^{(1)}_{p2,b},Z^{(2)}_{p2,b}}(t,s)\right]\right.$$

$$\left.+\sum_{l=0}^{n-k}p_3 M_{Z^{(1)}_{p3},Z^{(2)}_{p3}}(t,s)\right\}$$

$$1_{\{T>\mu,k=m+1\}}\left\{\sum_{j=0}^{k-1}\sum_{l=0}^{k-1-j}\left[p_{2,a}M_{Z^{(1)}_{p2,a},Z^{(3)}_{p2,a}}(t,s)+p_{2,b}M_{Z^{(1)}_{p2,b},Z^{(3)}_{p2,b}}(t,s)\right]\right.$$

$$\left.+\sum_{l=0}^{n-k}p_3 M_{Z^{(1)}_{p3},Z^{(2)}_{p3}}(t,s)\right\}$$

$$1_{\{T>\mu,k>m+1\}}\left\{\sum_{j=0}^{m}\sum_{l_1=0}^{m-j}\sum_{l_2=0}^{d-m-2}\left[p_{6,a,o}M_{Z^{(1)}_{p6,a,o},Z^{(2)}_{p6,a,o}}(t,s)+p_{6,b,o}M_{Z^{(1)}_{p6,b,o},Z^{(2)}_{p6,b,o}}(t,s)\right.\right.$$

$$\left.+p_{6,c,o}M_{Z^{(1)}_{p6,c,o},Z^{(2)}_{p6,c,o}}(t,s)+p_{6,d,o}M_{Z^{(1)}_{p6,d,o},Z^{(2)}_{p6,d,o}}(t,s)\right]$$

$$+\sum_{j=m+1}^{k-1}\sum_{l_1=0}^{j-m-1}\sum_{l_2=0}^{k-1-j}\left[p_{7,a,o}M_{Z^{(1)}_{p7,a,o},Z^{(2)}_{p7,a,o}}(t,s)+p_{7,b,o}M_{Z^{(1)}_{p7,b,o},Z^{(2)}_{p7,b,o}}(t,s)\right]$$

$$\left.+\sum_{l_1=0}^{k-m-2}\sum_{l_2=0}^{n-k}\left[p_{8,a,o}M_{Z^{(1)}_{p8,a,o},Z^{(2)}_{p8,a,o}}(t,s)+p_{8,b,o}M_{Z^{(1)}_{p8,b,o},Z^{(2)}_{p8,b,o}}(t,s)\right]\right\}$$

$$+1_{\{T\leq\mu,k<m+1\}}\left\{\sum_{l=0}^{n-k}q_2 M_{Z^{(1)}_{q2,a},Z^{(2)}_{q2,a}}(t,s)\right\}$$

$$+1_{\{T\leq\mu,k=m+1\}}\left\{\sum_{l=0}^{n-k}q_2 M_{Z^{(1)}_{q2,b},Z^{(2)}_{q2,b}}(t,s)\right\}$$

$$+1_{\{T\leq\mu,k>m+1\}}\left\{\sum_{l_1=0}^{k-m-2}\sum_{l_2=0}^{n-k}q_6 M_{Z^{(2)}_{q6,o},Z^{(2)}_{q6,o}}(t,s)\right\},$$

where the coefficients and variables involved are all as presented in the Appendix.

Proof See the Appendix.   ∎

Theorem 3.1 *The exact conditional joint MGF of $(\hat{\mu}, \hat{\sigma})$ is given by*

$$E\left[e^{t\hat{\sigma}+s\hat{\mu}}|D > 0\right] = E_1 + E_2. \tag{11}$$

Proof It is straightforward with the use of Lemma 3.1.   ∎

Remark 3.1 Balakrishnan and Zhu (2016) have derived an explicit expression for the exact CDF of the variable having the form $\Gamma\left(\alpha_1, \beta_1\right) + N\Gamma(\alpha_2, \beta_2) + E(\theta)$. Here, all the conditional mixture marginal distributions of $\hat{\mu}$ and most of the conditional mixture marginal distributions of $\hat{\sigma}$ take on this form. We can then directly apply the results of Balakrishnan and Zhu (2016). The only variable that does not have this form is $Z_{p16,a}$ corresponding to the case when $X_{k:n} < \mu < T, k > m + 1$ and $n = 2m$. If we are only interested in the distribution of $\hat{\sigma}$, we can then replace $p_{16,a}Z_{p16,a} + p_{16,b}Z_{p16,b}$ by $p_{16}^* Z_{p16}^*$, which follows this form. If we are only interested in finding the exact distribution of one of the MLEs, then some simplifications are possible; for example,

$$\sum_{j=0}^{m-1}\sum_{l_1=0}^{m-1-j}\sum_{l_2=0}^{d-m-2}\left[p_{13,a}Z_{p13,a}^{(1)} + p_{13,b}Z_{p13,b}^{(1)} + p_{13,c}Z_{p13,c}^{(1)} + p_{13,d}Z_{p13,d}^{(1)} + p_{13,e}Z_{p13,e}^{(1)}\right.$$

$$\left.+p_{13,f}Z_{p13,f}^{(1)} + p_{13,g}Z_{p13,g}^{(1)} + p_{13,h}Z_{p13,h}^{(1)}\right]$$

$$= \sum_{j=0}^{m}\sum_{l_1=0}^{m-j}\sum_{l_2=0}^{k-2-m}\left[p_{13,a}^*Z_{p13,a}^* + p_{13,b}^*Z_{p13,b}^* + p_{13,c}^*Z_{p13,c}^* + p_{13,d}^*Z_{p13,d}^*\right]. \tag{12}$$

Here, we do not list all the simplifications for the sake of brevity. From the exact conditional distributions of $\hat{\mu}$ and $\hat{\sigma}$, we can readily obtain the exact conditional CIs for μ and σ.

Remark 3.2 From Theorem 3.1, we can also readily obtain the conditional moments of $\hat{\mu}$ and $\hat{\sigma}$ as well as the correlation coefficient between $\hat{\mu}$ and $\hat{\sigma}$; but, we refrain from presenting them here.

4 Conditional MLE of Population Quantile and Its Exact Distribution

The MLE of the q-quantile of a standard Laplace distribution $(0, 1)$ is given by

$$\hat{Q}_q = q\hat{\sigma} + \hat{\mu}. \tag{13}$$

We then readily obtain the exact conditional MGF of $\hat{Q}_q$ to be

$$M(t) = E\left(e^{t\hat{Q}_q}\right) = E\left(e^{tq\hat{\sigma}+t\hat{\mu}}\right),\tag{14}$$

where the exact conditional MGF of $(\hat{\mu}, \hat{\sigma})$ is as presented in Theorem 3.1. So, from the exact conditional joint distribution $\hat{\mu}$ and $\hat{\sigma}$ derived above, we can readily obtain the exact conditional distribution of $\hat{Q}_q$, which can then be used to develop exact conditional CI for any particular quantile of interest. These CIs can be then used as bounds in a Q–Q plot; see the details for the example presented later.

5 Conditional MLE of Reliability and Associated Confidence Interval

Balakrishnan and Chandramouleeswaran (1996) discussed the BLUE of reliability function. Zhu and Balakrishnan (2016, 2017) developed exact CIs based on the Type-II and Type-I censored samples for the reliability function. Here, we consider the MLE of the reliability function based on a Type-I HCS sample. A natural estimator for the reliability at mission time t is

$$\hat{S}(t) = \begin{cases} 1 - \frac{1}{2}e^{-\left(\frac{\hat{\mu}-t}{\hat{\sigma}}\right)} & \text{if } t < \hat{\mu}, \\ \frac{1}{2}e^{-\left(\frac{t-\hat{\mu}}{\hat{\sigma}}\right)} & \text{if } t \geq \hat{\mu}. \end{cases}\tag{15}$$

The distribution function of $\hat{S}(t)$ can be obtained as

$$P\left(\hat{S}(t) \leq s\right) = \begin{cases} P(\hat{\mu} + \log(2(1-s))\hat{\sigma} \leq t) & \text{if } t < \hat{\mu}, s < \frac{1}{2}, \\ P(\hat{\mu} - \log(2s)\hat{\sigma} \leq t) & \text{if } t \geq \hat{\mu}, s \geq \frac{1}{2}. \end{cases}\tag{16}$$

To construct an exact equi-tailed $100(1-\alpha)\%$ CI for $S(t)$ is equivalent to find a s such that

$$\begin{cases} P(\hat{\mu} + \log(2(1-s))\hat{\sigma} \leq t) = \frac{\alpha}{2} & \text{if } t < \hat{\mu}, s < \frac{1}{2}, \\ P(\hat{\mu} - \log(2s)\hat{\sigma} \leq t) = \frac{\alpha}{2} & \text{if } t \geq \hat{\mu}, s \geq \frac{1}{2}, \end{cases}\tag{17}$$

and

$$\begin{cases} P(\hat{\mu} + \log(2(1-s))\hat{\sigma} \leq t) = 1 - \frac{\alpha}{2} & \text{if } t < \hat{\mu}, s < \frac{1}{2}, \\ P(\hat{\mu} - \log 2s\hat{\sigma} \leq t) = 1 - \frac{\alpha}{2} & \text{if } t \geq \hat{\mu}, s \geq \frac{1}{2}. \end{cases}\tag{18}$$

To obtain the exact distribution of $\hat{\mu} + \log(2(1-s))\hat{\sigma}$ and $\hat{\mu} - \log(2s)\hat{\sigma}$, we can use the results of the exact distribution of the quantile by setting $q = \log(2(1-s))$ or

$q = -\log(2s)$. These CIs can then be used as bounds in the P–P plot or Kaplan–Meier curve.

6 Conditional MLE of Cumulative Hazard and Associated Exact Confidence Interval

The cumulative hazard function, denoted by Λ, is defined as

$$\Lambda(t) = -\ln(S(t)). \tag{19}$$

As mission time t, a natural estimator for the cumulative hazard function is

$$\hat{\Lambda}(t) = -\ln(\hat{S}(t)), \tag{20}$$

where $\hat{S}(t)$ is as defined in (15). Now, the distribution function of $\hat{\Lambda}(t)$ can be expressed as

$$P(\hat{\Lambda}(t) \le h) = P(\hat{S}(t) \ge e^{-h}) = 1 - P(\hat{S}(t) < e^{-h}). \tag{21}$$

So, if an exact equi-tailed $100(1 - \alpha)\%$ CI for $S(t)$ is (s_l, s_u), then an exact equi-tailed $100(1 - \alpha)\%$ CI for $\Lambda(t)$ is simply $(-\log(s_u), -\log(s_l))$.

7 MLEs from Type-II HCS

Epstein (1954) and Childs et al. (2003) proposed the Type-II HCS in which the life-test would be terminated at $T^* = \max\{X_{k:n}, T\}$, where k $(2 \le k \le n)$ and T are pre-fixed.

In this case, the MLEs have the same expression as presented in (6) and (8).

Remark 7.1 When $k = 1$, the MLEs exist only if $X_{k:n} < T$. For convenience, we will consider here only the case when $k \ge 2$.

To develop the exact inference based on $(\hat{\mu}, \hat{\sigma})$, we need the following lemma.

Lemma 7.1 *The expectation of any function $g(\mathbf{X})$, where*

$$g(\mathbf{X}) = \begin{cases} g_1(\mathbf{X}), & X_{k:n} > T, \\ g_2(\mathbf{X}), & X_{k:n} < T, \end{cases}$$

based on a Type-II HCS, can be readily obtained from the expectation from a Type-II censored sample based on k observations, Type-I HSC together with $X_{k:n} < T$, and Type-I censored sample together with $d \ge k$, as

$$E[g(\mathbf{X})] = E[g_1(\mathbf{X}), D = k, X_{k:n} < \infty] - E[g_1(\mathbf{X}), D = k, X_{k:n} < T]$$
$$+ \sum_{d=k}^{n} E[g_2(\mathbf{X}), D = d].$$

Proof We have

$$E[g(\mathbf{X})] = E[g_1(\mathbf{X}), D = k, X_{k:n} > T] + \sum_{d=k}^{n} E[g_2(\mathbf{X}), D = d]$$
$$= E[g_1(\mathbf{X}), D = k, X_{k:n} < \infty] - E[g_1(\mathbf{X}), D = k, X_{k:n} < T]$$
$$+ \sum_{d=k}^{n} E[g_2(\mathbf{X}), D = d],$$

as required.   ∎

By using this lemma, we readily obtain the joint and marginal distributions of $\hat{\mu}$ and $\hat{\sigma}$. The results of Type-I HSC along with $X_{k:n} < T$ is presented in Lemma 3.4. Note that, in this case, we don't need to condition on $D > 0$, and so all the mixture probabilities presented in the Appendix should be multiplied by $(1 - p_0)$ and $(1 - q_0)$ corresponding to the cases when $T > \mu$ and $T < \mu$, respectively.

For $\sum_{d=k}^{n} E[g_2(\mathbf{X}), D = d]$ (say E_1) and $E[g_1(\mathbf{X}), D = k, X_{k:n} < \infty]$ (say E_3), we can directly apply the results of Balakrishnan and Zhu (2016), Zhu and Balakrishnan (2016, 2017). Here, we give their results for the sake of completeness.

Lemma 7.2 *If the sample size is even, i.e., $n = 2m$, then*

$$E_1 = 1_{\{T > \mu, k \le m\}} \left\{ \sum_{d=k}^{m-1} \sum_{j=0}^{d} \sum_{l=0}^{d-j} p_1 M_{Z_{p1}^{(1)}, Z_{p1}^{(2)}}(t, s) + \sum_{j=0}^{m-1} \sum_{l=0}^{m-1-j} \left[p_{4,a,e} M_{Z_{p4,a,e}^{(1)}, Z_{p4,a,e}^{(2)}}(t, s) \right. \right.$$

$$+ p_{4,b,e} M_{Z_{p4,b,e}^{(1)}, Z_{p4,b,e}^{(2)}}(t, s) \Big] + p_{5,e} M_{Z_{p5,e}^{(1)}, Z_{p5,e}^{(2)}}(t, s)$$

$$+ \sum_{d=m+1}^{n} \sum_{j=0}^{m-1} \sum_{l_1=0}^{m-1-j} \sum_{l_2=0}^{d-m-1} \left[p_{10,a,e} M_{Z_{p10,a,e}^{(1)}, Z_{p10,a,e}^{(2)}}(t, s) + p_{10,b,e} M_{Z_{p10,b,e}^{(1)}, Z_{p10,b,e}^{(2)}}(t, s) \right.$$

$$+ p_{10,c,e} M_{Z_{p10,c,e}^{(1)}, Z_{p10,c,e}^{(2)}}(t, s) + p_{10,d} M_{Z_{p10,d,e}^{(1)}, Z_{p10,d,e}^{(2)}}(t, s) \Big]$$

$$+ \sum_{d=m+1}^{k-1} \sum_{l=0}^{d-m-1} \left[p_{11,a,e} M_{Z_{p11,a,e}^{(1)}, Z_{p11,a,e}^{(2)}}(t, s) + p_{11,b,e} M_{Z_{p11,b,e}^{(1)}, Z_{p11,b,e}^{(2)}}(t, s) \right]$$

$$+ \sum_{d=m+1}^{k-1} \sum_{j=m+1}^{d} \sum_{l_1=0}^{j-m-1} \sum_{l_2=0}^{d-j} p_{12,e} M_{Z_{p12,e}^{(1)}, Z_{p12,e}^{(2)}}(t, s) \right\}$$

$$1_{\{T > \mu, k \ge m+1\}} \left\{ \sum_{d=k}^{n} \sum_{j=0}^{m-1} \sum_{l_1=0}^{m-1-j} \sum_{l_2=0}^{d-m-1} \left[p_{10,a,e} M_{Z_{p10,a,e}^{(1)}, Z_{p10,a,e}^{(2)}}(t, s) \right. \right.$$

$$
\begin{aligned}
&+ p_{10,b,e} M_{Z^{(1)}_{p10,b,e}, Z^{(2)}_{p10,b,e}}(t,s) + p_{10,c,e} M_{Z^{(1)}_{p10,c,e}, Z^{(2)}_{p10,c,e}}(t,s) + p_{10,d,e} M_{Z^{(1)}_{p10,d,e}, Z^{(2)}_{p10,d,e}}(t,s) \Bigg] \\
&+ \sum_{d=m+1}^{k-1} \sum_{l=0}^{d-m-1} \left[p_{11,a,e} M_{Z^{(1)}_{p11,a,e}, Z^{(2)}_{p11,a,e}}(t,s) + p_{11,b,e} M_{Z^{(1)}_{p11,b,e}, Z^{(2)}_{p11,b,e}}(t,s) \right] \\
&+ \sum_{d=m+1}^{k-1} \sum_{j=m+1}^{d} \sum_{l_1=0}^{j-m-1} \sum_{l_2=0}^{d-j} p_{12,e} M_{Z^{(1)}_{p12,e}, Z^{(2)}_{p12,e}}(t,s) \Bigg\} \\
&+ 1_{\{T \leq \mu, k \leq m\}} \left\{ \sum_{d=k}^{m-1} q_1 M_{Z^{(2)}_{q1}, Z^{(2)}_{q1}}(t,s) + q_3 M_{Z^{(1)}_{q3,e}, Z^{(2)}_{q3,e}}(t,s) \right. \\
&\left. + \sum_{d=m+1}^{n} \sum_{l=0}^{d-m-1} q_5 M_{Z^{(1)}_{q5,e}, Z^{(2)}_{q5,e}}(t,s) \right\} \\
&+ 1_{\{T \leq \mu, k \geq m+1\}} \left\{ \sum_{d=k}^{n} \sum_{l=0}^{d-m-1} q_5 M_{Z^{(1)}_{q5,e}, Z^{(2)}_{q5,e}}(t,s) \right\}.
\end{aligned}
\tag{22}
$$

If the sample size is odd, i.e., $n = 2m + 1$, then

$$
\begin{aligned}
E_1 = 1_{\{T > \mu\}} &\left\{ \sum_{d=k}^{m} \sum_{j=0}^{d} \sum_{l=0}^{d-j} p_1 M_{Z^{(1)}_{p1}, Z^{(2)}_{p1}}(t,s) \right. \\
&+ \sum_{d=\max(m+1,k)}^{n} \sum_{j=0}^{m} \sum_{l_1=0}^{m-j} \sum_{l_2=0}^{d-m-1} \left[p_{4,a,o} M_{Z^{(1)}_{p4,a,o}, Z^{(2)}_{p4,a,o}}(t,s) + p_{4,b,o} M_{Z^{(1)}_{p4,b,o}, Z^{(2)}_{p4,b,o}}(t,s) \right] \\
&\left. + \sum_{d=\max(m+1,k)}^{n} \sum_{j=m+1}^{d} \sum_{l_1=0}^{j-m-1} \sum_{l_2=0}^{d-j} p_{5,o} M_{Z^{(1)}_{p5,o}, Z^{(2)}_{p5,o}}(t,s) \right\} \\
&+ 1_{\{T \leq \mu, k \leq m+1\}} \left\{ \sum_{d=k}^{m} q_1 M_{Z^{(1)}_{q1}, Z^{(2)}_{q1}}(t,s) + q_{3,o} M_{Z^{(1)}_{q3,o}, Z^{(2)}_{q3,o}}(t,s) \right. \\
&\left. + \sum_{d=m+2}^{n} \sum_{l=0}^{d-m-1} q_5 M_{Z^{(1)}_{q5,o}, Z^{(2)}_{q5,o}}(t,s) \right\} \\
&+ 1_{\{T \leq \mu, k > m+1\}} \left\{ \sum_{d=k}^{n} \sum_{l=0}^{d-m-1} q_5 M_{Z^{(1)}_{q5,o}, Z^{(2)}_{q5,o}}(t,s) \right\},
\end{aligned}
\tag{23}
$$

where the coefficients and the variables involved are all as presented in the Appendix, but the coefficients need to be multiplied by $(1 - p_0)$ and $(1 - q_0)$ corresponding to the cases $T > \mu$ and $T < \mu$, respectively.

Lemma 7.3 *If the sample size is even, i.e., $n = 2m$, we have*

$$
E_3 = 1_{\{k<m\}}\left\{\sum_{j=0}^{k-1}\sum_{l=0}^{k-1-j}p_{a1}M_{Z_{a1}^{(1)},Z_{a1}^{(2)}}(t,s) + \sum_{l=0}^{n-k}p_{a2}M_{Z_{a2}^{(1)},Z_{a2}^{(2)}}(t,s)\right\}
$$

$$
+1_{\{k=m\}}\left\{\sum_{j=0}^{k-1}\sum_{l=0}^{k-1-j}p_{b1,e}M_{Z_{b1,e}^{(1)},Z_{b1,e}^{(2)}}(t,s) + \sum_{l=0}^{n-k}p_{b2}M_{Z_{b2}^{(1)},Z_{b2}^{(2)}}(t,s)\right\}
$$

$$
+1_{\{k=m+1\}}\left\{\sum_{j=0}^{m-1}\sum_{l=0}^{m-1-j}p_{c1}M_{Z_{c1}^{(1)},Z_{c1}^{(2)}}(t,s) + p_{c2}M_{Z_{c2}^{(1)},Z_{c2}^{(2)}}(t,s)\right.
$$

$$
\left. + \sum_{l=0}^{n-k}p_{c3}M_{Z_{c3}^{(1)},Z_{c3}^{(2)}}(t,s)\right\}
$$

$$
+1_{\{k>m+1\}}\left\{\sum_{j=0}^{m-1}\sum_{l_1=0}^{m-1-j}\sum_{l_2=0}^{k-2-m}p_{d1,e}M_{Z_{d1,e}^{(1)},Z_{d1,e}^{(2)}}(t,s) + \sum_{l=0}^{k-2-m}p_{d2}M_{Z_{d2}^{(1)},Z_{d2}^{(2)}}(t,s)\right.
$$

$$
\left. + \sum_{j=m+1}^{k-1}\sum_{l_1=0}^{j-m-1}\sum_{l_2=0}^{k-j-1}p_{d3,e}M_{Z_{d3,e}^{(1)},Z_{d3,e}^{(2)}}(t,s) + \sum_{l_1=0}^{n-k}\sum_{l_2=0}^{k-m-2}p_{d4}M_{Z_{d4,e}^{(1)},Z_{d4,e}^{(2)}}(t,s)\right\}.
$$

$$\tag{24}$$

Table 1 Simulated values of the first, second, and cross moments of $\hat{\mu}$ and $\hat{\sigma}$ based on the Type-I HCS when $\mu = 0$, $\sigma = 1$ and $n = 20$, while the corresponding exact values are reported within parentheses

k	$F(T)$	$\hat{\mu}$	$\hat{\mu}^2$	$\hat{\sigma}$	$\hat{\sigma}^2$	$\hat{\mu}\hat{\sigma}$
8	0.4	−0.0440	0.1827	0.9269	1.0003	0.0367
		(−0.0425)	(0.1879)	(0.9270)	(1.0012)	(0.0398)
	0.6	−0.1103	0.1109	0.8802	0.8858	−0.0679
		(−0.1106)	(0.1102)	(0.8793)	(0.8840)	(−0.0689)
10	0.4	0.0167	0.1449	0.9753	1.0886	0.0774
		(0.0183)	(0.1504)	(0.9755)	(1.0900)	(0.0809)
	0.6	−0.0521	0.0799	0.9211	0.9447	−0.0353
		(−0.0521)	(0.0795)	(0.9205)	(0.9436)	(−0.0358)
11	0.4	0.0302	0.1415	0.9898	1.1181	0.0876
		(0.0305)	(0.1395)	(0.9884)	(1.1146)	(0.0871)
	0.6	−0.0056	0.0653	0.9450	0.9858	0.0042
		(−0.0055)	(0.0650)	(0.9446)	(0.9850)	(0.0039)
15	0.4	0.0302	0.1415	0.9969	1.1311	0.0838
		(0.0305)	(0.1395)	(0.9954)	(1.1276)	(0.0833)
	0.6	−0.0056	0.0653	0.9700	1.0262	−0.0023
		(−0.0055)	(0.0650)	(0.9695)	(1.0246)	(−0.0026)

Table 2 Simulated values of the first, second and cross moments of $\hat{\mu}$ and $\hat{\sigma}$ based on the Type-I HCS when $\mu = 0$, $\sigma = 1$ and $n = 15$, while the corresponding exact values are reported within parentheses

k	$F(T)$	$\hat{\mu}$	$\hat{\mu}^2$	$\hat{\sigma}$	$\hat{\sigma}^2$	$\hat{\mu}\hat{\sigma}$
6	0.4	−0.0580	0.3071	0.9016	1.0095	0.0781
		(−0.0573)	(0.3017)	(0.9005)	(1.0062)	(0.0761)
	0.6	−0.1463	0.1630	0.8409	0.8496	−0.0831
		(−0.1447)	(0.1632)	(0.8412)	(0.8507)	(−0.0813)
8	0.4	0.0426	0.2187	0.9719	1.1333	0.1381
		(0.0453)	(0.2239)	(0.9733)	(1.1388)	(0.1443)
	0.6	−0.0043	0.0966	0.9116	0.9553	0.0177
		(−0.0018)	(0.0971)	(0.9129)	(0.9584)	(0.0205)
10	0.4	0.0426	0.2187	0.9893	1.1658	0.1281
		(0.0453)	(0.2239)	(0.9901)	(1.1685)	(0.1346)
	0.6	−0.0043	0.0966	0.9504	1.0192	0.0049
		(−0.0018)	(0.0971)	(0.9507)	(1.0200)	(0.0081)

Table 3 Simulated values of the first, second, and cross moments of $\hat{\mu}$ and $\hat{\sigma}$ based on the Type-II HCS when $\mu = 0$, $\sigma = 1$ and $n = 20$, while the corresponding exact values are reported within parentheses

k	$F(T)$	$\hat{\mu}$	$\hat{\mu}^2$	$\hat{\sigma}$	$\hat{\sigma}^2$	$\hat{\mu}\hat{\sigma}$
8	0.4	−0.0398	0.0584	0.9466	1.0070	−0.0294
		(−0.0401)	(0.0580)	(0.9457)	(1.0056)	(−0.0300)
	0.6	−0.0081	0.0616	0.9691	1.0230	−0.0071
		(−0.0081)	(0.0612)	(0.9686)	(1.0213)	(−0.0074)
10	0.4	−0.0467	0.0625	0.9343	0.9651	−0.0450
		(−0.0466)	(0.0624)	(0.9339)	(0.9645)	(−0.0450)
	0.6	−0.0125	0.0590	0.9644	1.0105	−0.0146
		(−0.0122)	(0.0588)	(0.9640)	(1.0094)	(−0.0145)
11	0.4	0.0006	0.0664	0.9465	0.9825	0.0021
		(0.0000)	(0.0666)	(0.9448)	(0.9787)	(0.0019)
	0.6	−0.0003	0.0667	0.9640	1.0076	−0.0013
		(0.0000)	(0.0666)	(0.9638)	(1.0067)	(−0.0011)
15	0.4	−0.0009	0.0667	0.9648	0.9954	−0.0008
		(0.0000)	(0.0666)	(0.9637)	(0.9930)	(0.0000)
	0.6	−0.0003	0.0667	0.9653	0.9961	−0.0005
		(0.0000)	(0.0666)	(0.9647)	(0.9945)	(−0.0003)

Table 4 Simulated values of the first, second, and cross moments of $\hat{\mu}$ and $\hat{\sigma}$ based on the Type-II HCS when $\mu = 0$, $\sigma = 1$ and $n = 15$, while the corresponding exact values are reported within parentheses

k	$F(T)$	$\hat{\mu}$	$\hat{\mu}^2$	$\hat{\sigma}$	$\hat{\sigma}^2$	$\hat{\mu}\hat{\sigma}$
6	0.4	−0.0496	0.0757	0.9275	1.0045	−0.0368
		(−0.0480)	(0.0755)	(0.9275)	(1.0037)	(−0.0347)
	0.6	−0.0099	0.0873	0.9544	1.0207	−0.0082
		(−0.0077)	(0.0872)	(0.9545)	(1.0210)	(−0.0055)
8	0.4	−0.0025	0.0962	0.9181	0.9561	−0.0003
		(0.0000)	(0.0963)	(0.9187)	(0.9562)	(0.0026)
	0.6	−0.0025	0.0962	0.9467	0.9981	−0.0048
		(0.0000)	(0.0963)	(0.9469)	(0.9983)	(−0.0019)
10	0.4	−0.0025	0.0962	0.9404	0.9779	−0.0023
		(0.0000)	(0.0963)	(0.9403)	(0.9782)	(0.0005)
	0.6	−0.0025	0.0962	0.9475	0.9883	−0.0040
		(0.0000)	(0.0963)	(0.9474)	(0.9884)	(−0.0013)

Table 5 Data from an accelerated life test of 16 integrated circuit chips taken from Devore (2015)

11.6	26.5	82.8	179.7	204.6	212.6	229.9	242.0
244.8	304.3	307.8	359.5	366.7	379.1	502.5	558.9

If the sample size is odd, i.e., $n = 2m + 1$, we have

$$
\begin{aligned}
E_3 = 1_{\{k \le m\}} & \left\{ \sum_{j=0}^{k-1} \sum_{l=0}^{k-1-j} p_{a1} M_{Z_{a1}^{(1)}, Z_{a1}^{(2)}}(t, s) + \sum_{l=0}^{n-k} p_{a2} M_{Z_{a2}^{(1)}, Z_{a2}^{(2)}}(t, s) \right\} \\
+ 1_{\{k = m+1\}} & \left\{ \sum_{j=0}^{k-1} \sum_{l=0}^{k-1-j} p_{b1,o} M_{Z_{b1,o}^{(1)}, Z_{b1,o}^{(2)}}(t, s) + \sum_{l=0}^{n-k} p_{b2} M_{Z_{b2}^{(1)}, Z_{b2}^{(2)}}(t, s) \right\} \\
+ 1_{\{k > m+1\}} & \left\{ \sum_{j=0}^{m} \sum_{l_1=0}^{m-j} \sum_{l_2=0}^{k-2-m} p_{d1,o} M_{Z_{d1,o}^{(1)}, Z_{d1,o}^{(2)}}(t, s) + \sum_{j=m+1}^{k-1} \sum_{l_1=0}^{j-m-1} \sum_{l_2=0}^{k-1-j} p_{d3} M_{Z_{d3,o}^{(1)}, Z_{d3,o}^{(2)}}(t, s) \right. \\
& \left. + \sum_{l_1=0}^{n-k} \sum_{l_2=0}^{k-2-m} p_{d4} M_{Z_{d4,o}^{(1)}, Z_{d4,o}^{(2)}}(t, s) \right\},
\end{aligned}
\tag{25}
$$

where the coefficients and the variables involved are all as presented in the Appendix.

Remark 7.2 It is of interest to mention here that we can rewrite one of their results as

Table 6 MLEs of quantiles, estimates of their bias, MSEs, and correlation coefficients from exact formulae and bootstrap method (within brackets) based on data from Table 5

Complete sample

n	$\hat{\mu}$	$\widehat{Bias}(\hat{\mu})$	$\widehat{MSE}(\hat{\mu})$	$\hat{\sigma}$	$\widehat{Bias}(\hat{\sigma})$	$\widehat{MSE}(\hat{\sigma})$	$\widehat{corr}(\hat{\mu},\hat{\sigma})$
16	243.4000	0.0000	1122.8394	114.6188	−3.9161	841.8631	0.0000
		(−0.1485)	(1120.9140)		(−3.8949)	(811.6926)	(−0.0018)

Type-I HCS

T	k	$\hat{\mu}$	$\widehat{Bias}(\hat{\mu})$	$\widehat{MSE}(\hat{\mu})$	$\hat{\sigma}$	$\widehat{Bias}(\hat{\sigma})$	$\widehat{MSE}(\hat{\sigma})$	$\widehat{corr}(\hat{\mu},\hat{\sigma})$
230	6	239.3448	−10.8706	1996.8638	92.9667	−12.5141	1564.2788	0.4527
			(−11.0864)	(1952.7850)		(−12.5738)	(1548.8781)	(0.4449)
	12	242.634	2.1509	1415.5427	94.6143	−1.1469	1409.9987	0.3314
			(2.2996)	(1486.6625)		(−1.1449)	(1408.7212)	(0.3358)
400	6	239.3448	−15.0037	1405.4560	92.9667	−15.3098	1436.3546	0.3008
			(−15.0725)	(1410.0102)		(−15.2810)	(1433.2357)	(0.3008)
	12	243.4000	−0.0034	1272.4762	122.0583	−5.2772	1255.7857	0.0129
			(−0.1627)	(1269.7588)		(−5.2236)	(1255.2068)	(0.0096)

Type-II HCS

T	k	$\hat{\mu}$	$\widehat{Bias}(\hat{\mu})$	$\widehat{MSE}(\hat{\mu})$	$\hat{\sigma}$	$\widehat{Bias}(\hat{\sigma})$	$\widehat{MSE}(\hat{\sigma})$	$\widehat{corr}(\hat{\mu},\hat{\sigma})$
230	6	242.6340	−2.8763	677.3824	94.6143	−4.5107	1246.6015	0.1055
			(−2.9544)	(676.1956)		(−4.5087)	(1241.4338)	(0.1029)
	12	243.4000	0.0000	1273.3306	122.0583	−5.6029	1218.733	−0.0040
			(−0.1935)	(1276.5319)		(−5.5768)	(1217.8716)	(−0.0035)
400	6	243.4000	−0.0015	1077.9086	112.3214	−2.7691	917.7861	0.0130
			(−0.1476)	(1075.7892)		(−2.7364)	(917.5885)	(0.0110)
	12	243.4000	0.0000	1078.2800	112.3214	−2.9736	894.6194	0.0038
			(−0.1455)	(1076.4309)		(−2.9411)	(894.4223)	(0.0017)

Table 7 Exact and bootstrap 95% CIs for μ and σ based on data from Table 5

Complete sample			Exact result		Bootstrap results	
			μ	σ	μ	σ
			(62.5255, 172.4032)	(175.5387, 311.2613)	(175.0376, 311.0353)	(62.2222, 172.4491)
Type-I HCS			Exact result		Bootstrap results	
			μ	σ	μ	σ
T	k					
230	6		(148.9267, 321.4518)	(25.2426, 169.3095)	(148.6992, 319.9853)	(25.3147, 168.5342)
	12		(186.581, 331.2992)	(34.8836, 180.3554)	(186.3366, 331.7197)	(34.9342, 180.6664)
400	6		(148.9268, 284.4663)	(25.2815, 158.9101)	(148.6992,284.3417)	(25.2709, 158.8336)
	12		(171.1340,315.6648)	(58.9833, 195.3864)	(170.6004,315.4253)	(58.6954, 195.1298)
Type-II HCS			Exact result		Bootstrap results	
			μ	σ	μ	σ
T	k					
230	6		(186.5811, 289.4002)	(34.1404,170.3456)	(186.1776, 289.2119)	(33.9493, 170.0125)
	12		(171.1340, 315.6660)	(59.0524, 193.1321)	(170.7054, 315.3186)	(58.9966, 192.4856)
400	6		(176.8989, 309.9009)	(60.5973,178.5442)	(176.4078, 309.6796)	(60.4959, 177.6320)
	12		(176.8989, 309.9011)	(60.5973,176.8232)	(176.4078, 309.6796)	(60.4959, 176.1976)

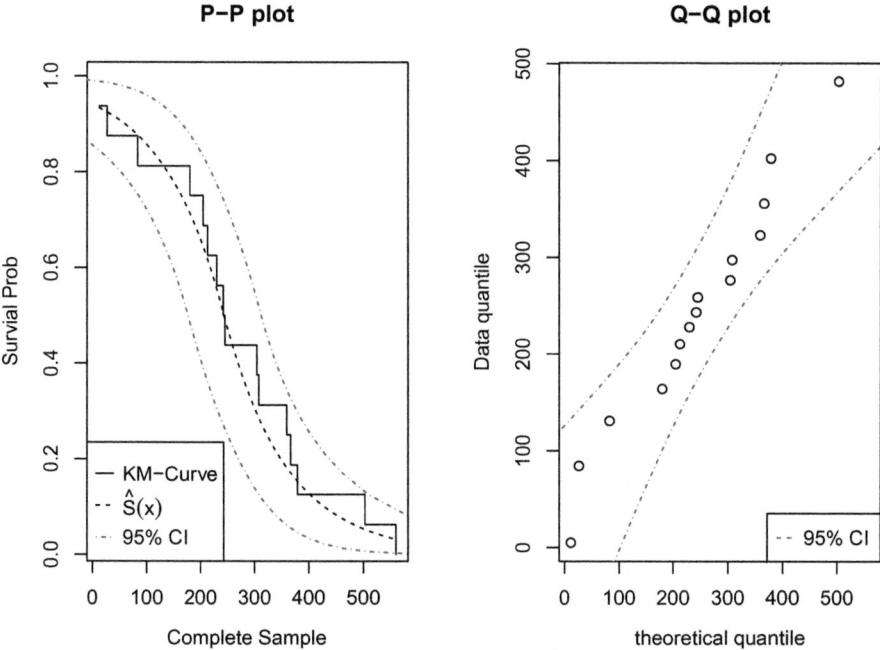

Fig. 1 P–P and Q–Q plots with 95% confidence bounds based on the complete data from Table 5

$$\sum_{l_1=0}^{n-k}\sum_{l_2=0}^{k-m-2} P_{d4}M_{Z_{d4,e}^{(1)},Z_{d4,e}^{(2)}}(t,s) = \sum_{l_1=0}^{k-m-2}\sum_{l_2=0}^{n-k}\left[p_{16,a}^*M_{Z_{p16,a}^{(1)},Z_{p16,a}^{(2)}}(t,s) + p_{16,b}^*M_{Z_{p16,b}^{(1)},Z_{p16,b}^{(2)}}(t,s)\right],$$

$$\sum_{l_1=0}^{n-k}\sum_{l_2=0}^{k-2-m} P_{d4}M_{Z_{d4,o}^{(1)},Z_{d4,o}^{(2)}}(t,s) = \sum_{l_1=0}^{k-m-2}\sum_{l_2=0}^{n-k}\left[p_{8,a,o}^*M_{Z_{p8,a,o}^{(1)},Z_{p8,a,o}^{(2)}}(t,s) + p_{8,b,o}^*M_{Z_{p8,b,o}^{(1)},Z_{p8,b,o}^{(2)}}(t,s)\right],$$

$$(26)$$

where $p_{16,a}^* = (1-p_0)p_{16,a}$, $p_{16,b}^* = (1-p_0)p_{16,b}$, $p_{8,a,o}^* = (1-p_0)p_{8,a,o}$ and $p_{8,b,o}^* = (1-p_0)p_{8,b,o}$. Then, the joint distribution can be easily obtained by the use of Lemma 3.2.

Remark 7.3 From the exact distribution of $(\hat{\mu},\hat{\sigma})$, we can readily obtain the moments of $\hat{\mu}$ and $\hat{\sigma}$ as well as the correlation coefficient between them.

Remark 7.4 Based on $(\hat{\mu},\hat{\sigma})$, we can propose a plug-in estimator of quantile, reliability function $R(t)$, and cumulative hazard function $\Lambda(t)$ at mission time t, as described in Sects. 4–7. These can in turn be used to produce exact $100(1-\alpha)\%$ CIs for population quantile, $R(t)$, and $\Lambda(t)$. However, the details for these are not presented here for the sake of conciseness.

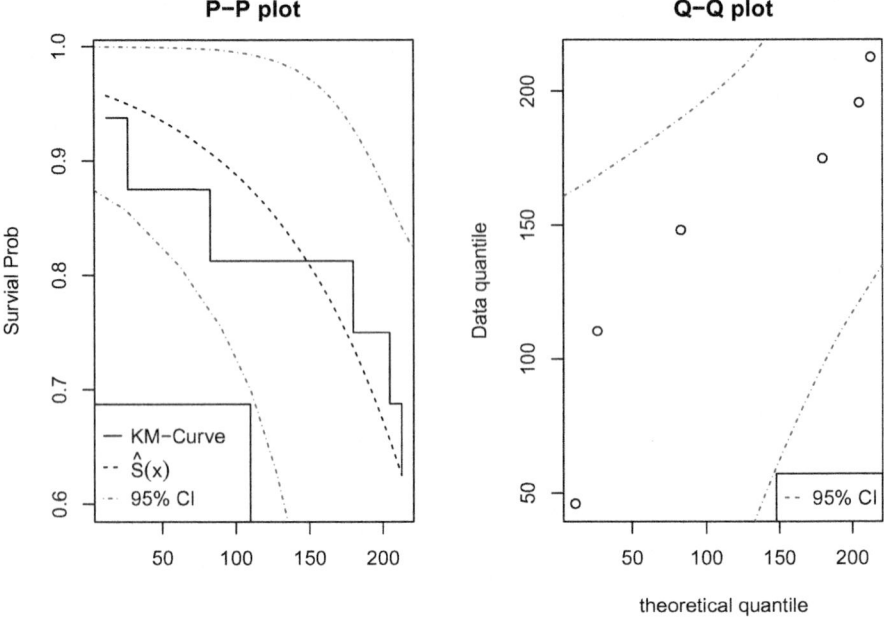

Fig. 2 P–P and Q–Q plots with 95% confidence bounds based on the Type-I HCS with $k = 6$ and $T = 230$ from Table 5

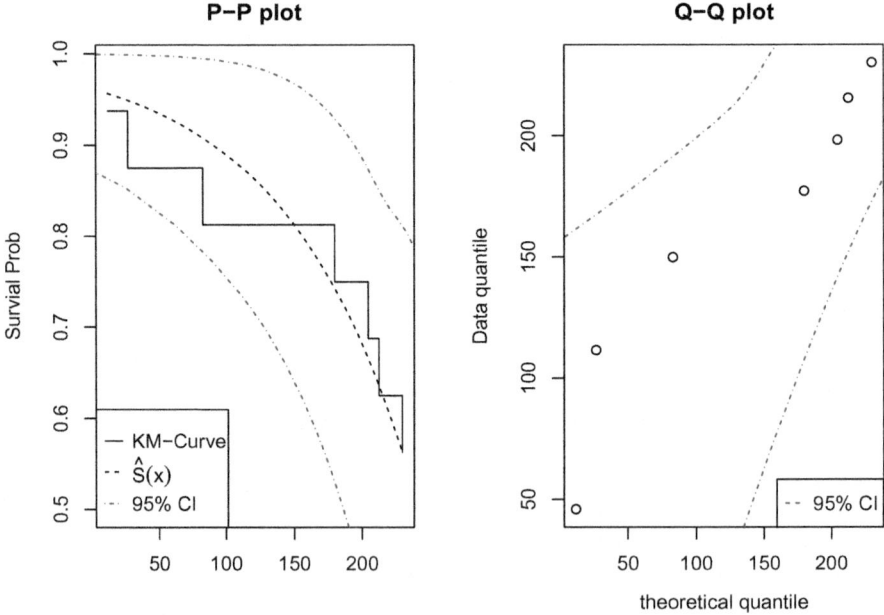

Fig. 3 P–P and Q–Q plots with 95% confidence bounds based on the Type-I HCS with $k = 12$ and $T = 230$ from Table 5

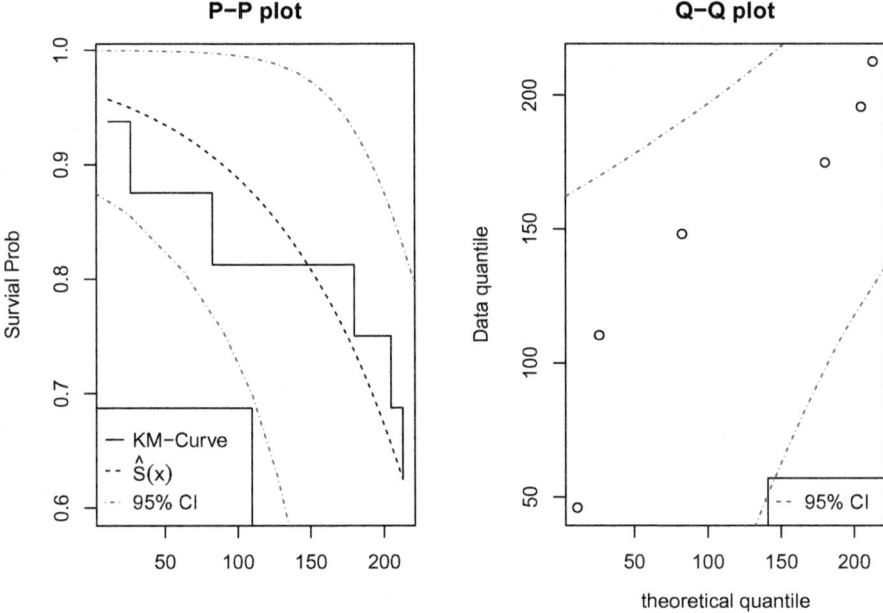

Fig. 4 P–P and Q–Q plots with 95% confidence bounds based on the Type-I HCS with $k = 6$ and $T = 400$ from Table 5

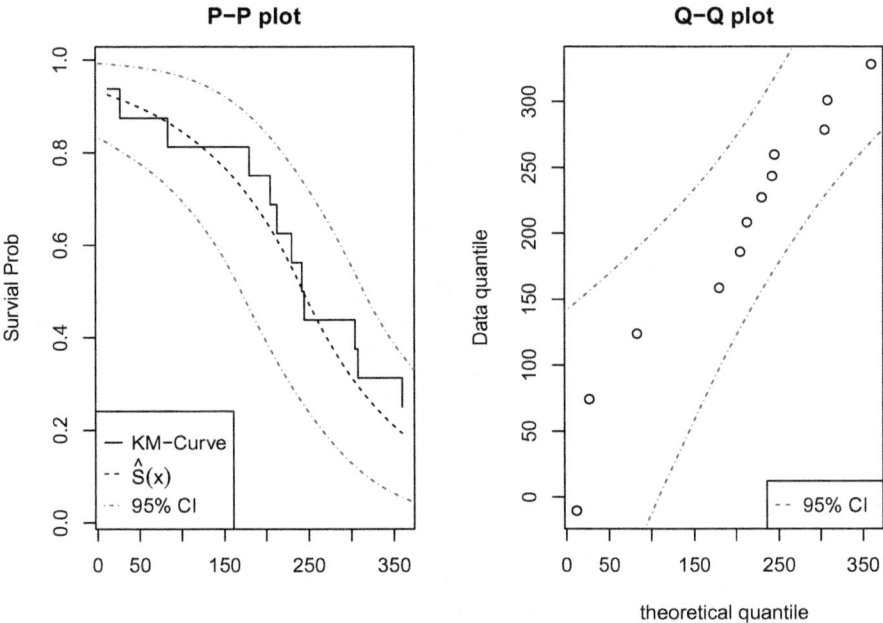

Fig. 5 P–P and Q–Q plots with 95% confidence bounds based on the Type-I HCS with $k = 12$ and $T = 400$ from Table 5

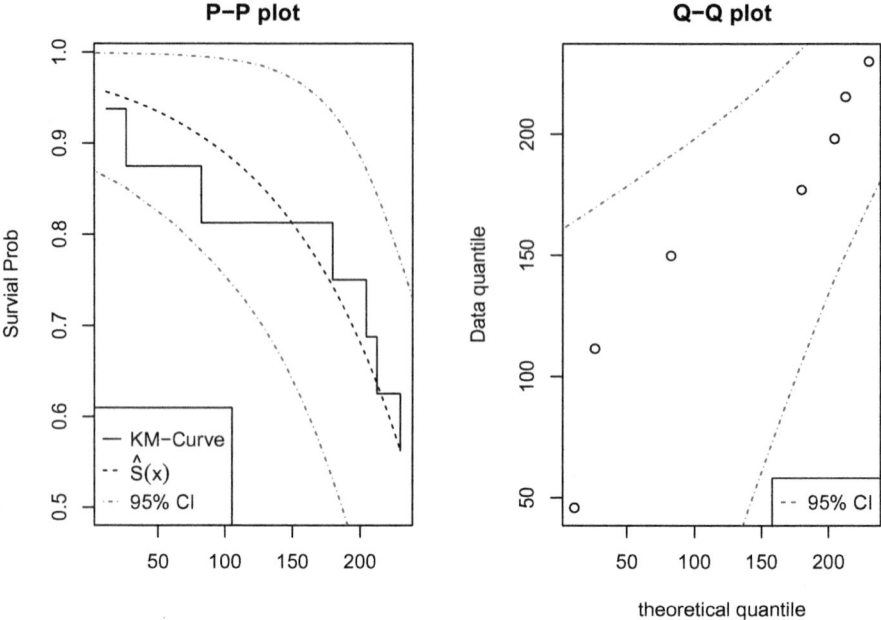

Fig. 6 P–P and Q–Q plots with 95% confidence bounds based on the Type-II HCS with $k = 6$ and $T = 230$ from Table 5

8 Monte Carlo Simulation Study

We carried out a Monte Carlo simulation study for Type-I HCS and Type-II HCS based on $n = 15, 20$ and by taking $\mu = 0$ and $\sigma = 1$, without loss of any generality. The values of k were chosen as 8, 10, 11, 15 for $n = 20$, and 7, 10 for $n = 15$, respectively, while the values of T were chosen so as to make $F(T) = 0.4, 0.6$. We then computed the first, second, and cross moments of $\hat{\mu}$ and $\hat{\sigma}$ through simulation as well as by the use of exact formulas established here. All these results are presented in Tables 1, 2, 3 and 4, and we observe that in all cases, $\hat{\sigma}$ is negatively biased with the bias decreasing as k increases in general. We further observe that the exact values are in close agreement with the corresponding simulated values.

9 Illustrative Example

In this section, we illustrate the results established in the preceding sections with one real dataset taken from the reliability literature.

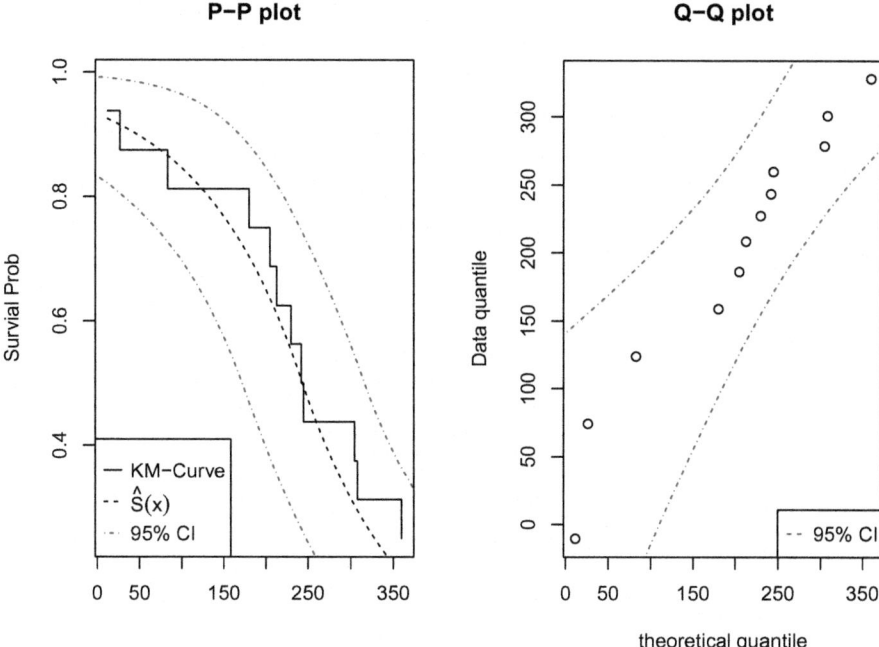

Fig. 7 P–P and Q–Q plots with 95% confidence bounds based on the Type-II HCS with $k = 12$ and $T = 230$ from Table 5

Example 9.1 The following failure time observations (in thousands of hours) are obtained by Devore (2015) from an accelerated life testing of 16 integrated circuit chips. For illustrative purpose, we analyze these data here by assuming a Laplace distribution and compute the MLEs based on complete, Type-I HCS and Type-II HCS for different choices of T and k, and the MSEs and the correlation coefficient of the MLEs based on the exact formulas as well as by the bootstrap method. The corresponding results are all presented in Table 6. In all the cases considered, the bias is negligible, and so all further computations are based on the MLEs without a bias-reduction. The 95% CIs, constructed from the exact conditional CDFs, are presented in Table 7, and are compared with the results obtained from the bootstrap approach. We note that these two sets of results are quite close. Finally, we have presented the Q–Q and P–P plots in Figs. 1, 2, 3, 4, 5, 6, 7, 8, and 9 with corresponding 95% confidence bounds, which do support the assumption of Laplace distribution for these data.

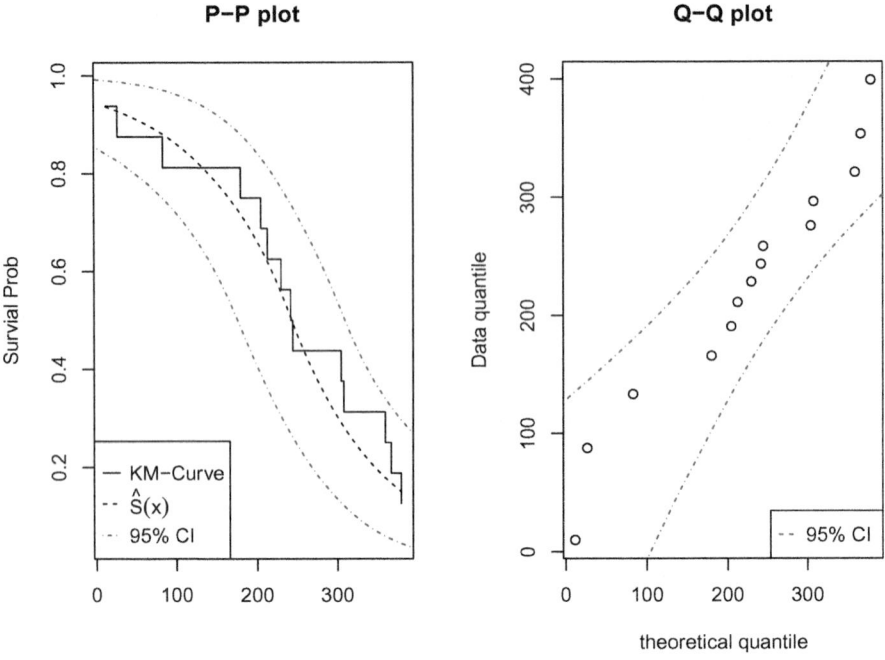

Fig. 8 P–P and Q–Q plots with 95% confidence bounds based on the Type-II HCS with $k = 6$ and $T = 400$ from Table 5

10 Discussion and Concluding Remarks

In this chapter, we have derived the MLEs of the location and scale parameters of the Laplace distribution under Type-I HCS and Type-II HCS. We have then derived the exact distributions of the MLEs, and the estimates of quantile, reliability, and cumulative hazard functions of Laplace (μ, σ) distribution. These exact distributions have then been used to determine the bias and variances of the estimates as well as exact CIs. These exact CIs have been further utilized to develop exact confidence bounds for Q–Q plot and K–M curves. It will naturally be of interest to develop goodness-of-fit tests based on the Type-I HCS and Type-II HCS by utilizing the results established here; see Puig and Stephens (2000) for a discussion on goodness-of-fit tests for the Laplace model based on a complete sample. Work is currently under progress on this problem and we hope to report the findings in a future article.

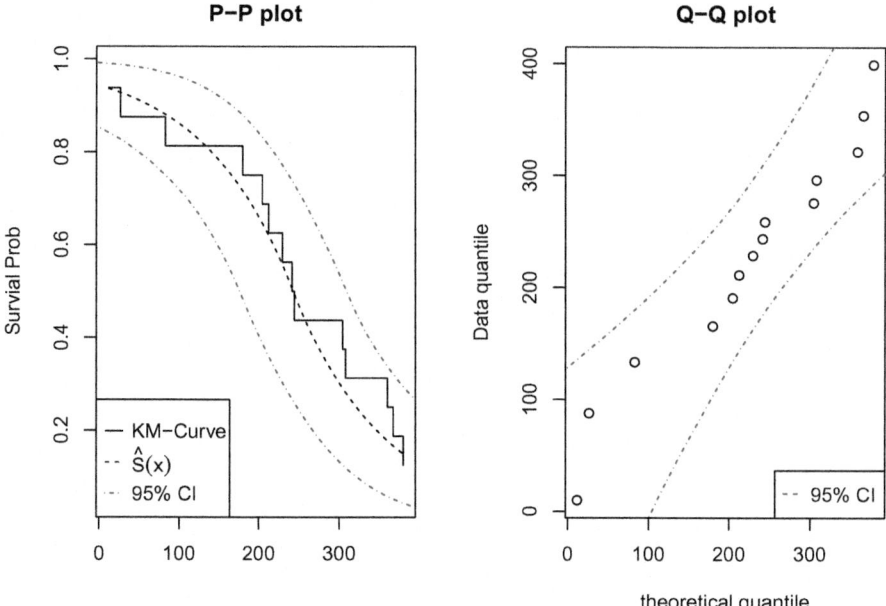

Fig. 9 P–P and Q–Q plots with 95% confidence bounds based on the Type-II HCS with $k = 12$ and $T = 400$ from Table 5

Acknowledgements This research was partially supported by the National Natural Science Foundation of China (No. 11571263) and by the National Sciences and Engineering Research Council of Canada.

Appendix

List of notation used throughout

$$p_1 = \frac{(-1)^l n! e^{-\frac{(T-\mu)(n+l-d)}{\sigma}}}{2^n (n-d)! j! (d-j-l)! l! (1-p_0)},$$

$$p_{2,a} = \frac{(-1)^l n!}{2^n (n-k)! j! (k-j-1-l)! l! (n-k)! (n-k+l+1)(1-p_0)},$$

$$p_{2,b} = -p_{2,a} e^{-\frac{(T-\mu)(n-k+l+1)}{\sigma}},$$

$$p_3 = \frac{(-1)^l n!}{2^{k+l}(k-1)!(n-k-l)!l!(k+l)(1-p_0)},$$

$$p_{4,a,e} = \frac{(-1)^l n!e^{-\frac{(T-\mu)m}{\sigma}}}{2^n j!(m-1-j-l)!(l+1)!m!(1-p_0)},$$

$$p_{4,b,e} = -p_{4,a,e}e^{-\frac{(T-\mu)(l+1)}{\sigma}},$$

$$p_{5,e} = \frac{n!e^{-\frac{(T-\mu)m}{\sigma}}}{2^n(n-m)!m!(1-p_0)},$$

$$p_{6,a,e} = \frac{(-1)^l n!}{2^n j!(m-1-j-l)!m!(l+1)!(1-p_0)},$$

$$p_{6,b,e} = -p_{6,a,e}e^{-\frac{(T-\mu)m}{\sigma}},$$

$$p_{6,c,e} = \frac{(-1)^{l+1} n!}{2^n j!(m-1-j-l)!(m-1)!(l+1)!(m+l+1)(1-p_0)},$$

$$p_{6,d,e} = -p_{6,c,e}e^{-\frac{(T-\mu)(m+l+1)}{\sigma}},$$

$$p_{7,a,e} = \frac{n!}{2^n m!m!(1-p_0)},$$

$$p_{7,b,e} = -p_{7,a,e}e^{-\frac{m(T-\mu)}{\sigma}},$$

$$p_{9,e} = \frac{(-1)^l n!}{m!(n-k-l)!l!2^{k+l}(m+k+l)(1-p_0)},$$

$$p_{10,a,e} = \frac{(-1)^{l_1+d-m-1-l_2} n!e^{-\frac{(T-\mu)(m-1-l_2)}{\sigma}}}{2^n j!(m-1-j-l_1)!(d-m-1-l_2)!(l_2+1)!(n-d)!(l_1+1)!(1-p_0)},$$

$$p_{10,b,e} = -p_{10,a,e}e^{-\frac{(T-\mu)(l_2+1)}{\sigma}},$$

$$p_{10,c,e} = -\frac{p_{10,a,e}}{l_1+l_2+2},$$

$$p_{10,d,e} = -p_{10,c,e}e^{-\frac{(T-\mu)(l_1+l_2+2)}{\sigma}},$$

$$p_{11,a,e} = \frac{(-1)^{d-m-l-1} n!e^{-\frac{(T-\mu)(m-l-1)}{\sigma}}}{2^n m!(d-m-l-1)!(l+1)!(n-d)!(1-p_0)},$$

$$p_{11,b,e} = -p_{11,a,e}e^{-\frac{(T-\mu)(l+1)}{\sigma}},$$

$$p_{12,e} = \frac{(-1)^{l_1+l_2} n!e^{-\frac{(T-\mu)(n-d+l_2)}{\sigma}}}{2^n m!(j-m-l_1-1)!l_1!l_2!(d-j-l_2)!l_2!(n-d)!(m+l_1+1)(1-p_0)},$$

$$p_{13,a} = \frac{(-1)^{l_1+l_2} n!}{2^n j!(m-1-j-l_1)!(l_1+1)!(k-m-2-l_2)!l_2!(n-k)!(n-k+l_2+1)m(1-p_0)},$$

$$p_{13,b} = -p_{13,a}e^{-\frac{(T-\mu)m}{\sigma}},$$

$$p_{13,c} = \frac{(-1)^{l_1+l_2+1} n!e^{-\frac{(T-\mu)(n-k+l_2+1)}{\sigma}}}{2^n j!(m-1-j-l_1)!(l_1+1)!(k-m-l_2-1)!l_2!(n-k)!(n-k+l_2+1)(1-p_0)},$$

$$p_{13,d} = -p_{13,c}e^{-\frac{(T-\mu)(k-m-l_2-1)}{\sigma}},$$

$$p_{13,e} = \frac{(-1)^{l_1+l_2+1} n!}{2^n j!(m-1-j-l_1)!(l_1+1)!(k-m-2-l_2)!l_2!(n-k)!(n-k+l_2+1)(m+l_1+1)(1-p_0)},$$

$$p_{13,f} = -p_{13,e}e^{-\frac{(T-\mu)(m+l_1+1)}{\sigma}},$$

$$p_{13,g} = \frac{(-1)^{l_1+l_2} n!e^{-\frac{n-k+l_2+1}{\sigma}}}{2^n j!(m-1-j-l_1)!(l_1+1)!(k-m-2-l_2)!l_2!(n-k)!(n-k+l_2+1)(k-m-l_2+l_1)(1-p_0)},$$

$$p_{13,h} = -p_{13,g}e^{-\frac{(T-\mu)(k-m-l_2+l_1)}{\sigma}},$$

$$p^*_{13,a} = \frac{(-1)^{l_1+k-m-l_2}n!}{2^n j!(m-j-l_1)!l_1!(k-m-2-l_2)!l_2!(n-k)!(m-l_2-1)(m+l_1)},$$

$$p^*_{13,b} = -p^*_{13,a}e^{-\frac{(T-\mu)(m+l_1)}{\sigma}},$$

$$p^*_{13,c} = \frac{(-1)^{l_1+k-m-1-l_2}n!e^{-\frac{(T-\mu)(m-l_2-1)}{\sigma}}}{2^n j!(m-j-l_1)!l_1!(k-m-2-l_2)!l_2!(n-k)!(m-l_2-1)(l_1+l_2+1)},$$

$$p^*_{13,d} = -p^*_{13,c}e^{-\frac{(T-\mu)(l_1+l_2+1)}{\sigma}},$$

$$p_{14,a} = \frac{(-1)^l n!}{2^n m!(k-m-2-l)!l!(n-k+l+1)m(1-p_0)},$$

$$p_{14,b} = -p_{14,a}e^{-\frac{(T-\mu)m}{\sigma}},$$

$$p_{14,c} = \frac{(-1)^{l+1}n!e^{-\frac{(T-\mu)(n-k+l+1)}{\sigma}}}{2^n m!(k-m-1-l)!l!(n-k+l+1)(1-p_0)},$$

$$p_{14,d} = -p_{14,c}e^{-\frac{(T-\mu)(k-m-1-l)}{\sigma}},$$

$$p_{15,a} = \frac{n!(-1)^{l_1+l_2}(m+l_1+1)^{-1}}{2^n m!(j-m-1-l_1)!l_1!(k-1-j-l_2)!l_2!(n-k)!(n-k+l_2+1)(1-p_0)},$$

$$p_{15,b} = -p_{15,a}e^{-\frac{(T-\mu)(n-k+l_2+1)}{\sigma}},$$

$$p_{16,a} = \frac{(-1)^{l_1+l_2}n!}{2^{l_2+d}m!(k-m-2-l_1)!l_1!(n-k-l_2)!l_2!(k-m-1-l_1+l_2)(m+l_1+1)(1-p_0)},$$

$$p_{16,b} = \frac{(-1)^{l_1+l_2+1}n!}{2^{l_2+d}m!(k-m-2-l_1)!l_1!(n-k-l_2)!l_2!(k-m-1-l_1+l_2)(k+l_2)(1-p_0)},$$

$$q_1 = P(D=d),$$

$$q_2 = \frac{(-1)^l e^{\frac{(T-\mu)(k+l)}{\sigma}}n!}{2^{k+l}(k-1)!l!(n-k-l)!(k+l)(1-q_0)},$$

$$q_{3,e} = P(D=m),$$

$$q_{4,e} = \frac{(-1)^l n!e^{\frac{(T-\mu)(m+l+1)}{\sigma}}}{2^{k+l}m!l!(m-1-l)!(m+l+1)(1-q_0)},$$

$$q_5 = \frac{(-1)^l \left(1-\frac{1}{2}e^{-\frac{\mu-T}{\sigma}}\right)^{n-d}\left(\frac{1}{2}e^{-\frac{(\mu-T)}{\sigma}}\right)^d n!}{(l+m+1)m!l!(d-m-1-l)!(n-d)!(1-q_0)},$$

$$q_6 = \frac{(-1)^{l_1+l_2}e^{\frac{(T-\mu)(k+l_2)}{\sigma}}n!}{2^{k+l_2}(l_1+m+1)(k+l_2)m!l_1!(k-m-2-l_1)!l_2!(n-k-l_2)!(1-q_0)},$$

$$p_{4,a,o} = \frac{(-1)^{l_1+d-m-1-l_2}n!e^{-\frac{(T-\mu)(m-l_2)}{\sigma}}}{(1-p_0)2^n(l_1+l_2+1)l_1!l_2!j!(n-d)!(m-j-l_1)!(d-m-1-l_2)!},$$

$$p_{4,b,o} = -p_{4,a,o}e^{-\frac{(T-\mu)(l_1+l_2+1)}{\sigma}},$$

$$p_{5,o} = \frac{(-1)^{l_1+l_2}n!e^{-\frac{(T-\mu)(n-d+l_2)}{\sigma}}}{2^n m!(j-m-l_1-1)!l_1!(d-j-l_2)!l_2!(n-d)!(m+l_1+1)(1-p_0)},$$

$$p_{6,a,o} = \frac{(-1)^{l_1+l_2}n!}{2^n j!(m-j-l_1)!l_1!(k-m-2-l_2)!l_2!(n-k)!(n-k+l_2+1)(m+l_1+1)(1-p_0)},$$

$$p_{6,b,o} = -p_{6,a,o}e^{-\frac{(T-\mu)(m+l_1+1)}{\sigma}},$$

$$p_{6,c,o} = \frac{(-1)^{l_1+l_2+1}n!e^{-\frac{(T-\mu)(n-d+l_2+1)}{\sigma}}}{(1-p_0)2^n j!(m-j-l_1)!l_1!(k-m-2-l_2)!l_2!(n-k)!(n-k+l_2+1)(k-m-l_2+l_1-1)},$$

$$p_{6,d,o} = -p_{6,c,o}e^{-\frac{(T-\mu)(d-m+l_1-l_2-1)}{\sigma}},$$

$$p_{7,a,o} = \frac{(-1)^{l_1+l_2} n!}{2^n m! (j - m - 1 - l_1)! l_1! (k - 1 - j - l_2)! l_2! (n - k)! (m + l_1 + 1)(n - k + l_2 + 1)(1 - p_0)},$$

$$p_{7,b,o} = -p_{7,a,o} e^{-\frac{(T-\mu)(n-k+l_2+1)}{\sigma}},$$

$$p_{8,a,o} = \frac{(-1)^{k-m-2-l_1+l_2} n!}{2^{k+l_2} m! (k - m - 2 - l_1)! l_1! (n - k - l_2)! l_2! (l_1 + l_2 + 1)(k - l_1 - 1)(1 - p_0)},$$

$$p_{8,b,o} = \frac{(-1)^{k-m-1-l_1+l_2} n!}{2^{k+l_2} m! (k - m - 2 - l_1)! l_1! (n - k - l_2)! l_2! (l_1 + l_2 + 1)(k + l_2)(1 - p_0)},$$

$$q_{3,o} = P(D = m + 1),$$

$$p_{16}^* = \frac{(-1)^{l_1+l_2} n!}{m! (k - m - 2 - l_1)! l_1! (n - k - l_2)! l_2! (m + 1 + l_1)(k + l_2)(1 - p_0)},$$

$$Z_{p1}^{(1)} \stackrel{d}{=} \Gamma\left(j, \frac{\sigma}{d}\right) + N\Gamma\left(d - j, \frac{\sigma}{d}\right) + \frac{(T - \mu)(d - l)}{d},$$

$$Z_{p1}^{(2)} \stackrel{d}{=} \log\left(\frac{n}{2d}\right) Z_{p1}^{(1)} + T,$$

$$Z_{p2,a}^{(1)} \stackrel{d}{=} \Gamma_A\left(j, \frac{\sigma}{k}\right) + N\Gamma_A\left(k - 1 - j, \frac{\sigma}{k}\right) + \frac{(k - l - 1)\sigma}{(n - k + l + 1)k} E_1,$$

$$Z_{p2,a}^{(2)} \stackrel{d}{=} \log\left(\frac{n}{2k}\right)\left[\Gamma_A\left(j, \frac{\sigma}{k}\right) + N\Gamma_A\left(k - 1 - j, \frac{\sigma}{k}\right)\right] + \left[\log\left(\frac{n}{2k}\right)\frac{k - l - 1}{n - k + l + 1} + \frac{k}{n - k + l + 1}\right]\frac{\sigma}{k} E_1 + \mu,$$

$$Z_{p2,a}^{(3)} \stackrel{d}{=} \frac{\sigma}{n - k + l + 1} E_1 + \mu,$$

$$Z_{p2,b}^{(1)} \stackrel{d}{=} \Gamma_A\left(j, \frac{\sigma}{k}\right) + N\Gamma_A\left(k - 1 - j, \frac{\sigma}{k}\right) + \frac{(k - l - 1)\sigma}{(n - k + l + 1)k} E_2 + \frac{(T - \mu)(k - l - 1)}{k},$$

$$Z_{p2,b}^{(2)} \stackrel{d}{=} \log\left(\frac{n}{2k}\right)\left[\Gamma_A\left(j, \frac{\sigma}{k}\right) + N\Gamma_A\left(k - 1 - j, \frac{\sigma}{k}\right) + \frac{(T - \mu)(k - l - 1)}{k}\right]$$

$$+ \left[\log\left(\frac{n}{2k}\right)\frac{k - l - 1}{n - k + l + 1} + \frac{k}{n - k + l + 1}\right]\frac{\sigma}{k} E_1 + T,$$

$$Z_{p2,b}^{(3)} \stackrel{d}{=} \frac{\sigma}{n - k + l + 1} E_1 + T$$

$$Z_{p3}^{(1)} \stackrel{d}{=} \Gamma\left(k - 1, \frac{\sigma}{k}\right),$$

$$Z_{p3}^{(2)} \stackrel{d}{=} \log\left(\frac{n}{2k}\right) Z_{p3}^{(1)} + NE\left(\frac{\sigma}{k + l}\right) + \mu,$$

$$Z_{p3}^{(3)} \stackrel{d}{=} NE\left(\frac{\sigma}{k + l}\right) + \mu,$$

$$Z_{p4,a,e}^{(1)} \stackrel{d}{=} \Gamma\left(j, \frac{\sigma}{m}\right) + N\Gamma\left(m - 1 - j, \frac{\sigma}{m}\right) - \frac{\sigma}{m} E + (T - \mu),$$

$$Z_{p4,a,e}^{(2)} \stackrel{d}{=} \frac{\sigma}{2(l + 1)} E + \frac{T + \mu}{2},$$

$$Z_{p4,b,e}^{(1)} \stackrel{d}{=} \Gamma\left(j, \frac{\sigma}{m}\right) + N\Gamma\left(m - 1 - j, \frac{\sigma}{m}\right) - \frac{\sigma}{m} E + \frac{(T - \mu)(m - l - 1)}{m},$$

$$Z_{p4,b,e}^{(2)} \stackrel{d}{=} \frac{\sigma}{2(l + 1)} E + T,$$

$$Z_{p5,e}^{(1)} \stackrel{d}{=} \Gamma\left(m - 1, \frac{\sigma}{m}\right) + \frac{\sigma}{m} E + (T - \mu),$$

$$Z_{p5,e}^{(2)} \stackrel{d}{=} -\frac{\sigma}{n} E + \frac{T + \mu}{2},$$

$$Z_{p6,a,e}^{(1)} \stackrel{d}{=} \Gamma\left(j, \frac{\sigma}{m + 1}\right) + N\Gamma\left(m - 1 - j, \frac{\sigma}{m + 1}\right) - \frac{\sigma}{m + 1} E_1 + \frac{\sigma}{m + 1} E_2,$$

$$Z_{p6,a,e}^{(2)} \stackrel{d}{=} \frac{\sigma}{2(l + 1)} E_1 + \frac{\sigma}{n} E_2 + \mu,$$

$$Z_{p6,b,e}^{(1)} \overset{d}{=} \Gamma\left(j, \frac{\sigma}{m+1}\right) + N\Gamma\left(m-1-j, \frac{\sigma}{m+1}\right) - \frac{\sigma}{m+1}E_1 + \frac{\sigma}{m+1}E_2 + \frac{m(T-\mu)}{m+1},$$

$$Z_{p6,b,e}^{(2)} \overset{d}{=} \frac{\sigma}{2(l+1)}E_1 + \frac{\sigma}{n}E_2 + \frac{T+\mu}{2},$$

$$Z_{p6,c,e}^{(1)} \overset{d}{=} \Gamma\left(j, \frac{\sigma}{m+1}\right) + N\Gamma\left(m-1-j, \frac{\sigma}{m+1}\right) - \frac{\sigma}{m+1}E_1 + \frac{\sigma(m-l-1)}{(m+l+1)(m+1)}E_2,$$

$$Z_{p6,c,e}^{(2)} \overset{d}{=} \frac{\sigma}{2(l+1)}E_1 + \frac{\sigma}{m+l+1}E_2 + \mu,$$

$$Z_{p6,d,e}^{(1)} \overset{d}{=} \Gamma\left(j, \frac{\sigma}{k}\right) + N\Gamma\left(m-1-j, \frac{\sigma}{m+1}\right) - \frac{\sigma}{m+1}E_1 + \frac{\sigma(m-l-1)}{(m+l+1)(m+1)}E_2 + \frac{(T-\mu)(m-l-1)}{m+1},$$

$$Z_{p6,d,e}^{(2)} \overset{d}{=} \frac{\sigma}{2(l+1)}E_1 + \frac{\sigma}{m+l+1}E_2 + T,$$

$$Z_{p7,a,e}^{(1)} \overset{d}{=} \Gamma\left(m-1, \frac{\sigma}{m+1}\right) + \frac{\sigma}{k}E_1 + \frac{\sigma}{m+1}E_2,$$

$$Z_{p7,a,e}^{(2)} \overset{d}{=} -\frac{\sigma}{n}E_1 + \frac{\sigma}{n}E_2 + \mu,$$

$$Z_{p7,b,e}^{(1)} \overset{d}{=} \Gamma\left(m-1, \frac{\sigma}{m+1}\right) + \frac{\sigma}{m+1}E_1 + \frac{\sigma}{m+1}E_2 + \frac{m(T-\mu)}{m+1},$$

$$Z_{p7,b,e}^{(2)} \overset{d}{=} -\frac{\sigma}{n}E_1 + \frac{\sigma}{n}E_2 + \frac{(T+\mu)}{2},$$

$$Z_{p9,e}^{(1)} \overset{d}{=} \Gamma\left(m-1, \frac{\sigma}{k}\right) + \frac{\sigma}{k}E,$$

$$Z_{p9,e}^{(2)} \overset{d}{=} -\frac{\sigma}{n}E + NE\left(\frac{\sigma}{m+l+1}\right) + \mu,$$

$$Z_{p10,a,e}^{(1)} \overset{d}{=} \Gamma\left(d-m-1+j, \frac{\sigma}{d}\right) + N\Gamma\left(m-1-j, \frac{\sigma}{d}\right) + \frac{\sigma}{d}E_1 - \frac{\sigma}{d}E_2 + \frac{(T-\mu)(m-1-l_2)}{d},$$

$$Z_{p10,a,e}^{(2)} \overset{d}{=} \frac{\sigma}{2(l_2+1)}E_1 + \frac{\sigma}{2(l_1+1)}E_2 + \mu,$$

$$Z_{p10,b,e}^{(1)} \overset{d}{=} \Gamma\left(d-m-1+j, \frac{\sigma}{d}\right) + N\Gamma\left(m-1-j, \frac{\sigma}{d}\right) + \frac{\sigma}{d}E_1 - \frac{\sigma}{d}E_2 + \frac{(T-\mu)m}{d},$$

$$Z_{p10,b,e}^{(2)} \overset{d}{=} \frac{\sigma}{2(l_2+1)}E_1 + \frac{\sigma}{2(l_1+1)}E_2 + \frac{T+\mu}{2},$$

$$Z_{p10,c,e}^{(1)} \overset{d}{=} \Gamma\left(d-m-1+j, \frac{\sigma}{d}\right) + N\Gamma\left(m-1-j, \frac{\sigma}{d}\right) + \frac{(l_2-l_1)\sigma}{(l_1+l_2+2)d}E_1 - \frac{\sigma}{d}E_2 + \frac{(T-\mu)(m-l_2-1)}{d},$$

$$Z_{p10,c,e}^{(2)} \overset{d}{=} \frac{\sigma}{l_1+l_2+2}E_1 + \frac{\sigma}{2(l_1+1)}E_2 + \mu,$$

$$Z_{p10,d,e}^{(1)} \overset{d}{=} \Gamma\left(d-m-1+j, \frac{\sigma}{d}\right) + N\Gamma\left(m-1-j, \frac{\sigma}{d}\right) + \frac{(l_2-l_1)\sigma}{(l_1+l_2+2)d}E_1 - \frac{\sigma}{d}E_2 + \frac{(T-\mu)(m-l_1-1)}{d},$$

$$Z_{p10,d,e}^{(2)} \overset{d}{=} \frac{\sigma}{l_1+l_2+2}E_1 + \frac{\sigma}{2(l_1+1)}E_2 + T,$$

$$Z_{p11,a,e}^{(1)} \overset{d}{=} \Gamma\left(d-2, \frac{\sigma}{d}\right) + \frac{\sigma}{d}E_1 + \frac{\sigma}{d}E_2 + \frac{(T-\mu)(m-l-1)}{d},$$

$$Z_{p11,a,e}^{(2)} \overset{d}{=} -\frac{\sigma}{n}E_1 + \frac{\sigma}{2(l+1)}E_2 + \mu,$$

$$Z_{p11,b,e}^{(1)} \overset{d}{=} \Gamma\left(d-2, \frac{\sigma}{d}\right) + \frac{\sigma}{d}E_1 + \frac{\sigma}{d}E_2 + \frac{(T-\mu)m}{d},$$

$$Z_{p11,b,e}^{(2)} \overset{d}{=} -\frac{\sigma}{n}E_1 + \frac{\sigma}{2(l+1)}E_2 + \frac{T+\mu}{2},$$

$$Z_{p12,e}^{(1)} \overset{d}{=} \Gamma\left(d-j+m-1, \frac{\sigma}{d}\right) + N\Gamma\left(j-m-1, \frac{\sigma}{d}\right) + \frac{\sigma}{d}E_1 + \frac{(m-l_1-1)\sigma}{(m+l_1+1)d}E_2 + \frac{(T-\mu)(n-d+l_2)}{d},$$

$$Z_{p12,e}^{(2)} \overset{d}{=} -\frac{\sigma}{n}E_1 - \frac{\sigma}{m+l_1+1}E_2 + \mu,$$

$$Z_{p13,a}^{(1)} \overset{d}{=} \Gamma\left(k-m-1+j,\frac{\sigma}{k}\right) + N\Gamma\left(m-1-j,\frac{\sigma}{k}\right) - \frac{\sigma}{k}E_1 + \frac{\sigma}{k}E_2,$$

$$Z_{p13,a}^{(2)} \overset{d}{=} \frac{\sigma}{2(l_1+1)}E_1 + \frac{\sigma}{n}E_2 + \mu,$$

$$Z_{p13,b}^{(1)} \overset{d}{=} \Gamma\left(k-m-1+j,\frac{\sigma}{k}\right) + N\Gamma\left(m-1-j,\frac{\sigma}{k}\right) - \frac{\sigma}{k}E_1 + \frac{\sigma}{k}E_2 + \frac{m(T-\mu)}{k},$$

$$Z_{p13,b}^{(2)} \overset{d}{=} \frac{\sigma}{2(l_1+1)}E_1 + \frac{\sigma}{n}E_2 + \frac{T+\mu}{2},$$

$$Z_{p13,c}^{(1)} \overset{d}{=} \Gamma\left(k-m-1+j,\frac{\sigma}{k}\right) + N\Gamma\left(m-1-j,\frac{\sigma}{k}\right) - \frac{\sigma}{k}E_1 + \frac{\sigma}{k}E_2 + \frac{(n-k+l_2+1)(T-\mu)}{k},$$

$$Z_{p13,c}^{(2)} \overset{d}{=} \frac{\sigma}{2(l_1+1)}E_1 + \frac{\sigma}{2(k-m-l_2-1)}E_2 + \mu,$$

$$Z_{p13,d}^{(1)} \overset{d}{=} \Gamma\left(k-m-1+j,\frac{\sigma}{k}\right) + N\Gamma\left(m-1-j,\frac{\sigma}{k}\right) - \frac{\sigma}{k}E_1 + \frac{\sigma}{k}E_2 + \frac{m(T-\mu)}{k},$$

$$Z_{p13,d}^{(2)} \overset{d}{=} \frac{\sigma}{2(l_1+1)}E_1 + \frac{\sigma}{2(k-m-l_2-1)}E_2 + \frac{T+\mu}{2},$$

$$Z_{p13,e}^{(1)} \overset{d}{=} \Gamma\left(k-m-1+j,\frac{\sigma}{k}\right) + N\Gamma\left(m-1-j,\frac{\sigma}{k}\right) - \frac{\sigma}{k}E_1 + \frac{(m-l_1-1)\sigma}{(m+l_1+1)k}E_2,$$

$$Z_{p13,e}^{(2)} \overset{d}{=} \frac{\sigma}{2(l_1+1)}E_1 + \frac{\sigma}{m+l_1+1}E_2 + \mu,$$

$$Z_{p13,f}^{(1)} \overset{d}{=} \Gamma\left(k-m-1+j,\frac{\sigma}{k}\right) + N\Gamma\left(m-1-j,\frac{\sigma}{k}\right) - \frac{\sigma}{k}E_1 + \frac{(m-l_1-1)\sigma}{(m+l_1+1)k}E_2 + \frac{(m-l_1-1)(T-\mu)}{k},$$

$$Z_{p13,f}^{(2)} \overset{d}{=} \frac{\sigma}{2(l_1+1)}E_1 + \frac{\sigma}{m+l_1+1}E_2 + T,$$

$$Z_{p13,g}^{(1)} \overset{d}{=} \Gamma\left(k-m-1+j,\frac{\sigma}{k}\right) + N\Gamma\left(m-1-j,\frac{\sigma}{k}\right) - \frac{\sigma}{k}E_1 + \frac{(k-m-l_1-l_2-2)\sigma}{(k-m-l_2+l_1)k}E_2$$
$$+ \frac{(n-k+l_2+1)(T-\mu)}{k},$$

$$Z_{p13,g}^{(2)} \overset{d}{=} \frac{\sigma}{2(l_1+1)}E_1 + \frac{\sigma}{k-m-l_2+l_1}E_2 + \mu,$$

$$Z_{p13,h}^{(1)} \overset{d}{=} \Gamma\left(k-m-1+j,\frac{\sigma}{k}\right) + N\Gamma\left(m-1-j,\frac{\sigma}{k}\right) - \frac{\sigma}{k}E_1 + \frac{(k-m-l_1-l_2-2)\sigma}{(k-m-l_2+l_1)k}E_2$$
$$+ \frac{(m-l_1-1)(T-\mu)}{k},$$

$$Z_{p13,h}^{(2)} \overset{d}{=} \frac{\sigma}{2(l_1+1)}E_1 + \frac{\sigma}{k-m-l_2+l_1}E_2 + T,$$

$$Z_{p13,a}^{*} \overset{d}{=} \Gamma\left(k-1-m+j,\frac{\sigma}{k}\right) + N\Gamma\left(m-j,\frac{\sigma}{k}\right) + E\left(\frac{(m-l_1)\sigma}{(m+l_1)k}\right),$$

$$Z_{p13,b}^{*} \overset{d}{=} \Gamma\left(k-1-m+j,\frac{\sigma}{k}\right) + N\Gamma\left(m-j,\frac{\sigma}{k}\right) + E\left(\frac{(m-l_1)\sigma}{(m+l_1)k}\right) + \frac{(T-\mu)(m-l_1)}{k},$$

$$Z_{p13,c}^{*} \overset{d}{=} \Gamma\left(k-1-m+j,\frac{\sigma}{k}\right) + N\Gamma\left(m-j,\frac{\sigma}{k}\right) + E\left(\frac{(l_2-l_1+1)\sigma}{(l_1+l_2+1)k}\right) + \frac{(T-\mu)(m-l_2-1)}{k},$$

$$Z_{p13,d}^{*} \overset{d}{=} \Gamma\left(k-1-m+j,\frac{\sigma}{k}\right) + N\Gamma\left(m-j,\frac{\sigma}{k}\right) + E\left(\frac{(l_2-l_1+1)\sigma}{(l_1+l_2+1)k}\right) + \frac{(T-\mu)(m-l_1)}{k},$$

$$Z_{p14,a}^{(1)} \overset{d}{=} \Gamma\left(k-2,\frac{\sigma}{k}\right) + \frac{\sigma}{k}E_1 + \frac{\sigma}{k}E_2,$$

$$Z_{p14,a}^{(2)} \overset{d}{=} -\frac{\sigma}{n}E_1 + \frac{\sigma}{n}E_2 + \mu,$$

$$Z_{p14,b}^{(1)} \stackrel{d}{=} \Gamma\left(k-2, \frac{\sigma}{k}\right) + \frac{\sigma}{k}E_1 + \frac{\sigma}{k}E_2 + \frac{m(T-\mu)}{k},$$

$$Z_{p14,b}^{(2)} \stackrel{d}{=} -\frac{\sigma}{n}E_1 + \frac{\sigma}{n}E_2 + \frac{T+\mu}{2},$$

$$Z_{p14,c}^{(1)} \stackrel{d}{=} \Gamma\left(k-2, \frac{\sigma}{k}\right) + \frac{\sigma}{k}E_1 + \frac{\sigma}{k}E_2 + \frac{(T-\mu)(n-k+l+1)}{k},$$

$$Z_{p14,c}^{(2)} \stackrel{d}{=} -\frac{\sigma}{n}E_1 + \frac{\sigma}{2(k-m-l-1)}E_2 + \mu,$$

$$Z_{p14,d}^{(1)} \stackrel{d}{=} \Gamma\left(k-2, \frac{\sigma}{k}\right) + \frac{\sigma}{k}E_1 + \frac{\sigma}{k}E_2 + \frac{(T-\mu)m}{k},$$

$$Z_{p14,d}^{(2)} \stackrel{d}{=} -\frac{\sigma}{n}E_1 + \frac{\sigma}{2(k-m-l-1)}E_2 + \frac{T+\mu}{2},$$

$$Z_{p15,a}^{(1)} \stackrel{d}{=} \Gamma\left(m-1+k-j, \frac{\sigma}{k}\right) + N\Gamma\left(j-m-1, \frac{\sigma}{k}\right) + \frac{\sigma}{k}E_1 + \frac{(m-l_1-1)\sigma}{(m+l_1+1)k}E_2,$$

$$Z_{p15,a}^{(2)} \stackrel{d}{=} -\frac{\sigma}{n}E_1 - \frac{\sigma}{m+l_1+1}E_2 + \mu,$$

$$Z_{p15,b}^{(1)} \stackrel{d}{=} \Gamma\left(m-1+k-j, \frac{\sigma}{k}\right) + N\Gamma\left(j-m-1, \frac{\sigma}{k}\right) + \frac{\sigma}{k}E_1 + \frac{(m-l_1-1)\sigma}{(m+l_1+1)k}E_2 + \frac{(T-\mu)(n-k+l_2+1)}{k},$$

$$Z_{p15,b}^{(2)} \stackrel{d}{=} -\frac{\sigma}{n}E_1 - \frac{\sigma}{m+l_1+1}E_2 + \mu,$$

$$Z_{p16,a}^{(1)} \stackrel{d}{=} \Gamma\left(m-1, \frac{\sigma}{k}\right) + N\Gamma\left(d-m-2, \frac{\sigma}{k}\right) + NE\left(\frac{(m-l_1-1)\sigma}{(k-m-1-l_1+l_2)k}\right) + \frac{\sigma}{k}E_1 + \frac{(m-l_1-1)\sigma}{(m+l_1+1)k}E_2,$$

$$Z_{p16,a}^{(2)} \stackrel{d}{=} -\frac{\sigma}{n}E_1 - \frac{\sigma}{m+l_1+1}E_2 + \mu,$$

$$Z_{p16,b}^{(1)} \stackrel{d}{=} \Gamma\left(m-1, \frac{\sigma}{k}\right) + N\Gamma\left(d-m-2, \frac{\sigma}{k}\right) + NE\left(\frac{(m-l_1-1)\sigma}{(k-m-1-l_1+l_2)k}\right) + \frac{\sigma}{k}E,$$

$$Z_{p16,b}^{(2)} \stackrel{d}{=} -\frac{\sigma}{n}E + NE\left(\frac{\sigma}{k+l_2}\right) + \mu,$$

$$Z_{q1}^{(1)} \stackrel{d}{=} \Gamma\left(d, \frac{\sigma}{d}\right),$$

$$Z_{q1}^{(2)} \stackrel{d}{=} \log\left(\frac{n}{2d}\right)Z_{q1}^{(1)} + T,$$

$$Z_{q2,a}^{(1)} \stackrel{d}{=} \Gamma\left(k-1, \frac{\sigma}{k}\right),$$

$$Z_{q2,a}^{(2)} \stackrel{d}{=} \log\left(\frac{n}{2k}\right)Z_{q2,a}^{(1)} + NE\left(\frac{\sigma}{k+l}\right) + T,$$

$$Z_{q2,b}^{(1)} \stackrel{d}{=} \Gamma\left(k-1, \frac{\sigma}{k}\right),$$

$$Z_{q2,b}^{(2)} \stackrel{d}{=} NE\left(\frac{\sigma}{k+l}\right) + T,$$

$$Z_{q3,e}^{(1)} \stackrel{d}{=} \Gamma\left(m-1, \frac{\sigma}{m}\right) + \frac{\sigma}{m}E,$$

$$Z_{q3,e}^{(2)} \stackrel{d}{=} -\frac{\sigma}{n}E + T,$$

$$Z_{q4,e}^{(1)} \stackrel{d}{=} \Gamma\left(m-1, \frac{\sigma}{k}\right) + \frac{\sigma}{k}E,$$

$$Z_{q4,e}^{(2)} \stackrel{d}{=} -\frac{\sigma}{n}E + NE\left(\frac{\sigma}{m+l+1}\right) + T,$$

$$Z_{q5,e}^{(1)} \stackrel{d}{=} \Gamma\left(m-1, \frac{\sigma}{d}\right) + N\Gamma\left(d-m-1, \frac{\sigma}{d}\right) + \frac{\sigma}{d}E_1 + \frac{(m-l-1)\sigma}{(m+l+1)d}E_2,$$

$$Z_{q5,e}^{(2)} \stackrel{d}{=} -\frac{\sigma}{n}E_1 - \frac{\sigma}{m+l+1}E_2 + T,$$

$$Z_{q6,e}^{(1)} \overset{d}{=} \Gamma\left(m-1, \frac{\sigma}{k}\right) + N\Gamma\left(k-m-2, \frac{\sigma}{k}\right) + \frac{\sigma}{k}E_1 + \frac{(m-l_1-1)\sigma}{(m+l_1+1)k}E_2,$$

$$Z_{q6,e}^{(2)} \overset{d}{=} -\frac{\sigma}{n}E_1 - \frac{\sigma}{m+l_1+1}E_2 + NE\left(\frac{\sigma}{k+l_2}\right) + T,$$

$$Z_{p4,a,o}^{(1)} \overset{d}{=} \Gamma\left(j+d-m-1, \frac{\sigma}{d}\right) + N\Gamma\left(m-j, \frac{\sigma}{d}\right) + \frac{(l_2-l_1)\sigma}{(l_2+l_1+1)d}E + \frac{(T-\mu)(m-l_2)}{d},$$

$$Z_{p4,a,o}^{(2)} \overset{d}{=} \frac{\sigma}{l_2+l_1+1}E + \mu,$$

$$Z_{p4,b,o}^{(1)} \overset{d}{=} \Gamma\left(j+d-m-1, \frac{\sigma}{d}\right) + N\Gamma\left(m-j, \frac{\sigma}{d}\right) + \frac{(l_2-l_1)\sigma}{(l_2+l_1+1)d}E + \frac{(T-\mu)(m-l_1)}{d},$$

$$Z_{p4,b,o}^{(2)} \overset{d}{=} \frac{\sigma}{l_2+l_1+1}E + T,$$

$$Z_{p5,o}^{(1)} \overset{d}{=} \Gamma\left(d-j+m, \frac{\sigma}{d}\right) + N\Gamma\left(j-m-1, \frac{\sigma}{d}\right) + \frac{(m-l_1)\sigma}{(m+l_1+1)d}E + \frac{(T-\mu)(n-d+l_2)}{d},$$

$$Z_{p5,o}^{(2)} \overset{d}{=} -\frac{\sigma}{m+l_1+1}E + \mu,$$

$$Z_{p6,a,o}^{(1)} \overset{d}{=} \Gamma\left(j+k-m-1, \frac{\sigma}{k}\right) + N\Gamma\left(m-j, \frac{\sigma}{k}\right) + \frac{(m-l_1)\sigma}{(m+l_1+1)k}E,$$

$$Z_{p6,a,o}^{(2)} \overset{d}{=} \frac{\sigma}{m+l_1+1}E + \mu,$$

$$Z_{p6,b,o}^{(1)} \overset{d}{=} \Gamma\left(j+k-m-1, \frac{\sigma}{k}\right) + N\Gamma\left(m-j, \frac{\sigma}{k}\right) + \frac{(m-l_1)\sigma}{(m+l_1+1)k}E + \frac{(m-l_1)(T-\mu)}{k},$$

$$Z_{p6,b,o}^{(2)} \overset{d}{=} \frac{\sigma}{m+l_1+1}E + T,$$

$$Z_{p6,c,o}^{(1)} \overset{d}{=} \Gamma\left(j+k-m-1, \frac{\sigma}{k}\right) + N\Gamma\left(m-j, \frac{\sigma}{k}\right) + + \frac{(k-m-l_1-l_2-2)\sigma}{(k-m+l_1-l_2-1)k}E + \frac{(n-k+l_2+1)(T-\mu)}{k},$$

$$Z_{p6,c,o}^{(2)} \overset{d}{=} \frac{\sigma}{k-m-l_2+l_1-1}E + \mu,$$

$$Z_{p6,d,o}^{(1)} \overset{d}{=} \Gamma\left(j+k-m-1, \frac{\sigma}{k}\right) + N\Gamma\left(m-j, \frac{\sigma}{k}\right) + \frac{(k-m-l_1-l_2-2)\sigma}{(k-m+l_1-l_2-1)k}E + \frac{(m-l_1)(T-\mu)}{k},$$

$$Z_{p6,d,o}^{(2)} \overset{d}{=} \frac{\sigma}{k-m+l_1-l_2-1}E + T,$$

$$Z_{p7,a,o}^{(1)} \overset{d}{=} \Gamma\left(m+k-j, \frac{\sigma}{k}\right) + N\Gamma\left(j-m-1, \frac{\sigma}{k}\right) + \frac{(m-l_1)\sigma}{(m+l_1+1)k}E,$$

$$Z_{p7,a,o}^{(2)} \overset{d}{=} -\frac{\sigma}{m+l_1+1}E + \mu,$$

$$Z_{p7,b,o}^{(1)} \overset{d}{=} \Gamma\left(m+k-j, \frac{\sigma}{k}\right) + N\Gamma\left(j-m-1, \frac{\sigma}{k}\right) + \frac{(m-l_1)\sigma}{(m+l_1+1)k}E + \frac{(n-k+1+l_2)(T-\mu)}{k},$$

$$Z_{p7,b,o}^{(2)} \overset{d}{=} -\frac{\sigma}{m+l_1+1}E + \mu,$$

$$Z_{p8,a,o}^{(1)} \overset{d}{=} \Gamma\left(m, \frac{\sigma}{k}\right) + N\Gamma\left(k-m-2, \frac{\sigma}{k}\right) + NE\left(\frac{n-k+l_1+1}{(l_1+l_2+1)k}\right) + \frac{n+1+l_1-k}{(k-l_1-1)k}E,$$

$$Z_{p8,a,o}^{(2)} \overset{d}{=} -\frac{\sigma}{k-l_1-1}E + \mu,$$

$$Z_{p8,b,o}^{(1)} \overset{d}{=} \Gamma\left(m, \frac{\sigma}{k}\right) + N\Gamma\left(k-m-2, \frac{\sigma}{k}\right) + NE\left(\frac{n-k+l_1+1}{(l_1+l_2+1)k}\right),$$

$$Z_{p8,b,o}^{(2)} \overset{d}{=} NE\left(\frac{\sigma}{k+l_2}\right) + \mu,$$

$$Z_{q3,o}^{(1)} \overset{d}{=} \Gamma\left(m, \frac{\sigma}{m+1}\right) + \frac{\sigma}{m+1}E,$$

$$Z_{q3,o}^{(2)} \overset{d}{=} -\frac{\sigma}{m+1}E + T,$$

$$Z_{q5,o}^{(1)} \overset{d}{=} \Gamma\left(m, \frac{\sigma}{d}\right) + N\Gamma\left(d - m - 1, \frac{\sigma}{d}\right) + \frac{(m-l)\sigma}{(m+l+1)d}E,$$

$$Z_{q5,o}^{(2)} \overset{d}{=} -\frac{\sigma}{m+l+1}E + T,$$

$$Z_{q6,o}^{(1)} \overset{d}{=} \Gamma\left(m, \frac{\sigma}{k}\right) + N\Gamma\left(k - m - 2, \frac{\sigma}{k}\right) + \frac{(m-l_1)\sigma}{(m+l_1+1)k}E,$$

$$Z_{q6,o}^{(2)} \overset{d}{=} -\frac{\sigma}{m+l_1+1}E + NE\left(\frac{\sigma}{k+l_2}\right) + T,$$

$$Z_{p16}^{*} \overset{d}{=} \Gamma\left(m, \frac{\sigma}{k}\right) + N\Gamma\left(k - 2 - m, \frac{\sigma}{m}\right) + E\left(\frac{(m-l_1-1)\sigma}{k(m+l_1+1)}\right).$$

List of notation used in Lemma 7.3

$$P_{a1} = \frac{(-1)^l n!}{2^n j!(k-1-j-l)!l!(n-k)!(n-k+l+1)},$$

$$P_{a2} = \frac{(-1)^l n!}{2^{k+l}(k-1)!l!(n-k-l)!(k+l)},$$

$$P_{b1,e} = \frac{(-1)^l n!}{2^n j!(k-1-j-l)!l!(n-k)!(k+l+1)},$$

$$P_{b2} = \frac{(-1)^l n!}{2^{k+l}(k-1)!(n-k-l)!l!(k+l)},$$

$$P_{b1,o} = \frac{(-1)^l n!}{2^n j!(k-1-j-l)!l!(n-k)!(k+l)},$$

$$P_{c1} = \frac{(-1)^l n!}{2^n j!(m-j-1-l)!l!(n-k+1)!(n-k+l+2)},$$

$$P_{c2} = \frac{n!}{2^n m!(n-k+1)!},$$

$$P_{c3} = \frac{(-1)^l n!}{2^{k+l} m!(n-k-l)!l!(l+m+1)},$$

$$P_{d1,e} = \frac{(-1)^{l_1+l_2} n!}{2^n j!(m-1-j-l_1)!l_1!(k-2-m-l_2)!l_2!(n-k)!m(m+l_1+1)(n-k+l_2+1)},$$

$$P_{d1,o} = \frac{(-1)^{l_1+l_2} n!}{2^n j!(m-j-l_1)!l_1!(k-2-m-l_2)!l_2!(n-k)!(m+l_1+1)(n-k+l_2+1)},$$

$$P_{d2} = \frac{(-1)^l n!}{2^n m!(k-2-m-l)!l!(n-k)!(n-r+l+1)m},$$

$$P_{d3} = \frac{(-1)^{l_1+l_2} n!}{2^n m!(j-1-m-l_1)!l_1!(k-1-j-l_2)!l_2!(n-k)!(l_1+m+1)(n-k+1+l_2)},$$

$$P_{d4} = \frac{(-1)^{l_1+l_2} n!}{2^{k+l_1} m!(k-2-m-l_2)!l_2!(n-k-l_1)!l_1!(k+l_1)(m+l_2+1)},$$

$$Z_{a1}^{(1)} \overset{d}{=} \Gamma\left(j, \frac{\sigma}{k}\right) + N\Gamma\left(k-1-j, \frac{\sigma}{k}\right) + \frac{k-1-l}{k(n-r+l+1)}E,$$

$$Z_{a1}^{(2)} \overset{d}{=} \ln\left(\frac{n}{2k}\right)Z_{a1}^{(1)} + \frac{\sigma}{(n-r+l+1)}E + \mu,$$

$$Z_{a2}^{(1)} \overset{d}{=} \Gamma\left(k-1, \frac{\sigma}{k}\right),$$

$$Z_{a2}^{(2)} \overset{d}{=} \ln\left(\frac{n}{2k}\right) Z_{a2}^{(1)} + NE\left(\frac{1}{l+k}\right) + \mu,$$

$$Z_{b1,e}^{(1)} \overset{d}{=} \Gamma\left(j, \frac{\sigma}{k}\right) + N\Gamma\left(k-1-j, \frac{\sigma}{k}\right) + \frac{(k-l-1)\sigma}{k(k+l+1)} E,$$

$$Z_{b1,e}^{(2)} \overset{d}{=} \frac{\sigma}{(k+l+1)} E + \mu,$$

$$Z_{b1,o}^{(1)} \overset{d}{=} \Gamma\left(j, \frac{\sigma}{k}\right) + N\Gamma\left(k-1-j, \frac{\sigma}{k}\right) + \frac{(k-1-l)\sigma}{k(k+l)} E,$$

$$Z_{b1,o}^{(2)} \overset{d}{=} \frac{\sigma}{k+l} E + \mu,$$

$$Z_{b2}^{(1)} \overset{d}{=} \Gamma\left(k-1, \frac{\sigma}{k}\right),$$

$$Z_{b2}^{(2)} \overset{d}{=} NE\left(\frac{\sigma}{k+l}\right) + \mu,$$

$$Z_{c1}^{(1)} \overset{d}{=} \Gamma\left(j, \frac{\sigma}{k}\right) + N\Gamma\left(m-j-1, \frac{\sigma}{k}\right) + \frac{\sigma}{k} E_1 + \frac{(n-k-l)\sigma}{k(n-k+l+2)} E_2,$$

$$Z_{c1}^{(2)} \overset{d}{=} \frac{\sigma}{n} E_1 + \frac{1}{n-k+l+2} E_2 + \mu,$$

$$Z_{c2}^{(1)} \overset{d}{=} \Gamma\left(m-1, \frac{\sigma}{k}\right) + \frac{\sigma}{k} E_1 + \frac{\sigma}{k} E_2,$$

$$Z_{c2}^{(2)} \overset{d}{=} -\frac{\sigma}{n} E_1 + \frac{\sigma}{n} E_2 + \mu,$$

$$Z_{c3}^{(1)} \overset{d}{=} \Gamma\left(m-1, \frac{\sigma}{k}\right) + \frac{\sigma}{k} E,$$

$$Z_{c3}^{(2)} \overset{d}{=} -\frac{\sigma}{n} E + NE\left(\frac{\sigma}{m+l+1}\right) + \mu,$$

$$Z_{d1,e}^{(1)} \overset{d}{=} \Gamma\left(k+j-m-1, \frac{\sigma}{k}\right) + N\Gamma\left(m-1-j, \frac{\sigma}{k}\right) + \frac{\sigma}{k} E_1 + \frac{(m-l_1-1)\sigma}{k(m+l_1+1)} E_2,$$

$$Z_{d1,e}^{(2)} \overset{d}{=} \frac{\sigma}{n} E_1 + \frac{\sigma}{m+l_1+1} E_2 + \mu,$$

$$Z_{d1,o}^{(1)} \overset{d}{=} \Gamma\left(k+j-m-1, \frac{\sigma}{k}\right) + N\Gamma\left(m-j, \frac{\sigma}{k}\right) + \frac{(m-l_1)\sigma}{k(m+l_1+1)} E_2,$$

$$Z_{d1,o}^{(2)} \overset{d}{=} \frac{\sigma}{m+l_1+1} E + \mu,$$

$$Z_{d2}^{(1)} \overset{d}{=} \Gamma\left(k-2, \frac{\sigma}{k}\right) + \frac{\sigma}{k} E_1 + \frac{\sigma}{k} E_2,$$

$$Z_{d2}^{(2)} \overset{d}{=} -\frac{\sigma}{n} E_1 + \frac{\sigma}{n} E_2 + \mu,$$

$$Z_{d3,e}^{(1)} \overset{d}{=} \Gamma\left(k+m-j-1, \frac{\sigma}{k}\right) + N\Gamma\left(j-m-1, \frac{\sigma}{k}\right) + \frac{\sigma}{k} E_1 + \frac{(m-l_1-1)\sigma}{k(m+l_1+1)} E_2,$$

$$Z_{d3,e}^{(2)} \overset{d}{=} -\frac{\sigma}{n} E_1 - \frac{\sigma}{m+l_1+1} E_2 + \mu,$$

$$Z_{d3,o}^{(1)} \overset{d}{=} \Gamma\left(k+m-j, \frac{\sigma}{k}\right) + N\Gamma\left(j-m-1, \frac{\sigma}{k}\right) + \frac{(m-l_1)\sigma}{k(m+l_1+1)} E,$$

$$Z_{d3,o}^{(2)} \stackrel{d}{=} -\frac{\sigma}{m + l_1 + 1}E + \mu,$$

$$Z_{d4,e}^{(1)} \stackrel{d}{=} \Gamma\left(m - 1, \frac{\sigma}{k}\right) + N\Gamma\left(k - 2 - m, \frac{\sigma}{k}\right) + \frac{\sigma}{k}E_1 + \frac{(m - l_2 - 1)\sigma}{k(m + l_2 + 1)}E_2,$$

$$Z_{d4,e}^{(2)} \stackrel{d}{=} -\frac{\sigma}{n}E_1 - \frac{\sigma}{m + l_2 + 1}E_2 + NE\left(\frac{\sigma}{k + l_1}\right) + \mu,$$

$$Z_{d4,o}^{(1)} \stackrel{d}{=} \Gamma\left(m, \frac{\sigma}{k}\right) + N\Gamma\left(k - 2 - m, \frac{\sigma}{k}\right) + \frac{(m - l_2)\sigma}{k(m + l_2 + 1)}E,$$

$$Z_{d4,o}^{(2)} \stackrel{d}{=} -\frac{\sigma}{m + l_2 + 1}E + NE\left(\frac{\sigma}{k + l_1}\right) + \mu.$$

Proof of Lemma 3.4 Here, we will only give the proof for the case when $k > m + 1$, and all other cases can all be proved similarly. We have

$$E\left(e^{t\hat{\sigma}+s\hat{\mu}}, D = k|D > 0\right) = \sum_{j=0}^{m-1} E\left(e^{t\hat{\sigma}+s\hat{\mu}}, D = k, J = j|D > 0\right)$$

$$+ E\left(e^{t\hat{\sigma}+s\hat{\mu}}, D = k, J = m|D > 0\right)$$

$$+ \sum_{j=m+1}^{k-1} E\left(e^{t\hat{\sigma}+s\hat{\mu}}, D = k, J = j|D > 0\right)$$

$$+ E\left(e^{t\hat{\sigma}+s\hat{\mu}}, D = k, J = k|D > 0\right),$$

where J is the number of X's less than μ as defined earlier in Sect. 3. As the derivation of these four expectations are quite similar, we only derive the last one and omit others for the sake of brevity. In this case, we only need to focus on three order statistics $X_{m:n}$, $X_{m+1:n}$ and $X_{k:n}$, and then consider that there exists $m - 1$ i.i.d. observations less than $X_{m:n}$, and $k - m - 2$ i.i.d. observations are between $X_{m+1:n}$ and $X_{k:n}$. We then have

$$E\left(e^{t\hat{\sigma}+s\hat{\mu}}, D = k, J = k|D > 0\right) = \frac{n!}{(m-1)!(k-m-2)!(n-k)!(1-p_0)} \int_{-\infty}^{\mu} \int_{-\infty}^{X_{m+1:n}} \int_{X_{m+1:n}}^{\mu}$$

$$\times \left[\int_{-\infty}^{X_{m:n}} e^{-tx} \frac{1}{2\sigma} e^{\frac{x-\mu}{\sigma}}\right]^{m-1} \left[\int_{X_{m+1:n}}^{X_{k:n}} e^{tx} \frac{1}{2\sigma} e^{\frac{x-\mu}{\sigma}}\right]^{k-m-2}$$

$$\times e^{t(n-k+1)x_{k:n}} \frac{1}{2\sigma} e^{\frac{x_{k:n}-\mu}{\sigma}} \left(1 - \frac{1}{2} e^{\frac{x_{k:n}-\mu}{\sigma}}\right)^{n-k} dx_{k:n}$$

$$\times e^{\left(\frac{s}{2}-t\right)x_{m:n}} \frac{1}{2\sigma} e^{\frac{x_{m:n}-\mu}{\sigma}} dx_{m:n}$$

$$\times e^{\left(\frac{s}{2}+t\right)x_{m+1:n}} \frac{1}{2\sigma} e^{\frac{x_{m+1:n}-\mu}{\sigma}} dx_{m+1:n}$$

$$= \sum_{l_1=0}^{k-m-2} \sum_{l_2=0}^{n-k} \left[p_{16,a} M_{Z_{p16,a}^{(1)}, Z_{p16,a}^{(2)}}(t, s) + p_{16,b} M_{Z_{p16,b}^{(1)}, Z_{p16,b}^{(2)}}(t, s)\right].$$

The last equation is obtained by using the binomial expansion.

In this case, if we are only interested in the marginal distribution of $\hat{\sigma}$, we can then focus only on $X_{m+1:n}$ and $X_{d:n}$ and consider that there exists m i.i.d. observations less then $X_{m+1:n}$ and $k - m - 2$ i.i.d. observations between $X_{m+1:n}$ and $X_{k:n}$. We then have

$$E\left(e^{t\hat{\sigma}}, D = k, J = k | D > 0\right) = \frac{n!}{m!(k - m - 2)!(n - k)!(1 - p_0)} \int_{-\infty}^{\mu} \int_{-\infty}^{X_{k:n}}$$

$$\times \left[\int_{-\infty}^{X_{m+1:n}} e^{-tx} \frac{1}{2\sigma} e^{\frac{x-\mu}{\sigma}}\right]^m \left[\int_{X_{m+1:n}}^{X_{k:n}} e^{tx} \frac{1}{2\sigma} e^{\frac{x-\mu}{\sigma}}\right]^{k-m-2}$$

$$\times e^{tx_{m+1:n}} \frac{1}{2\sigma} e^{\frac{x_{m+1:n}-\mu}{\sigma}} dx_{m+1:n}$$

$$\times e^{t(n-k+1)x_{k:n}} \frac{1}{2\sigma} e^{\frac{x_{k:n}-\mu}{\sigma}} \left(1 - \frac{1}{2} e^{\frac{x_{k:n}-\mu}{\sigma}}\right)^{n-k} dx_{k:n}$$

$$= \sum_{l_1=0}^{k-m-2} \sum_{l_2=0}^{n-k} p_{16}^* M_{Z_{p16}^*}(t),$$

as required.   ∎

References

Arnold, B.C., Balakrishnan, N., Nagaraja, N.H.: A First Course in Order Statistics. Wiley, New York (1992)

Bain, L.J., Engelhardt, M.: Interval estimation for the two-parameter double exponential distribution. Technometrics **15**, 875–887 (1973)

Balakrishnan, N., Chandramouleeswaran, M.P.: Reliability estimation and tolerance limits for Laplace distribution based on censored samples. Mircoelectronics Reliab. **36**, 375–378 (1996)

Balakrishnan, N., Cohen, A.C.: Order Statistics and Inference: Estimation Methods. Academic Press, Boston (1991)

Balakrishnan, N., Cutler, C.D.: Maximum likelihood estimation of Laplace parameters based on Type-II censored samples. In: Nagaraja, H.N., Sen, P.K., Morrison, D.F. (eds.) Statistical Theory and Applications: Papers in Honor of Herbert A. David, pp. 145–151. Springer, New York (1995)

Balakrishnan, N., Kundu, D.: Hybrid censoring: models, inferential results and applications. Comput. Stat. Data Anal. **57**, 166–209 (2013) (with discussions)

Balakrishnan, N., Zhu, X.: Exact likelihood-based point and interval estimation for Laplace distribution based on Type-II right censored samples. J. Stat. Comput. Simul. **86**, 29–54 (2016)

Childs, A., Balakrishnan, N.: Conditional inference procedures for the Laplace distribution based on Type-II right censored samples. Stat. Probab. Lett. **31**, 31–39 (1996)

Childs, A., Balakrishnan, N.: Maximum likelihood estimation of Laplace parameters based on general Type-II censored samples. Stat. Pap. **38**, 343–349 (1997)

Childs, A., Balakrishnan, N.: Conditional inference procedures for the Laplace distribution when the observed samples are progressively censored. Metrika **52**, 253–265 (2000)

Childs, A., Chandrasekar, B., Balakrishnan, N., Kundu, D.: Exact likelihood inference based on Type-I and Type-II hybrid censored samples from the exponential distribution. Ann. Inst. Stat. Math. **55**, 319–330 (2003)

Devore, J.: Probability and Statistics for Engineering and the Sciences. Cengage Learning (2015)

Epstein, B.: Truncated life tests in the exponential case. Ann. Math. Stat. **25**, 555–564 (1954)

Iliopoulos, G., Balakrishnan, N.: Exact likelihood inference for Laplace distribution based on Type-II censored samples. J. Stat. Plan. Inference **141**, 1224–1239 (2011)

Iliopoulos, G., MirMostafaee, S.M.T.K.: Exact prediction intervals for order statistics from the Laplace distribution based on the maximum-likelihood estimators. Statistics **48**, 575–592 (2014)

Johnson, N.L., Kotz, S., Balakrishnan, N.: Continuous Univariate Distributions, vol. 1, 2nd edn. Wiley, New York (1994)

Johnson, N.L., Kotz, S., Balakrishnan, N.: Continuous Univariate Distributions, vol. 2, 2nd edn. Wiley, New York (1995)

Kappenman, R.F.: Conditional confidence intervals for double exponential distribution parameters. Technometrics **17**, 233–235 (1975)

Kappenman, R.F.: Tolerance intervals for the double exponential distribution. J. Am. Stat. Assoc. **72**, 908–909 (1977)

Kotz, S., Kozubowski, T.J., Podgorski, K.: The Laplace Distribution and Generalizations: A Revisit with Applications to Communications, Economics, Engineering, and Finance. Wiley, New York (2001)

Puig, P., Stephens, M.A.: Tests of fit for the Laplace distribution, with applications. Technometrics **42**, 417–424 (2000)

Zhu, X., Balakrishnan, N.: Exact inference for Laplace quantile, reliability and cumulative hazard functions based on Type-II censored data. IEEE Trans. Reliab. **65**, 164–178 (2016)

Zhu, X., Balakrishnan, N.: Exact likelihood-based point and interval estimation for lifetime characteristics of laplace distribution based on a time-constrained life-testing experiment. In: Adhikari, A., Adhikari, M.R., Chaubey, Y.P. (eds.) Mathematical and Statistical Applications in Life Sciences and Engineering, Accepted. Springer, New York (2017) (to appear)

Statistical Inference for Two-Compartment Model Parameters with Bootstrap Method and Genetic Algorithm

Özlem Türkşen and Müjgan Tez

Abstract Two-compartment model has common usage in modeling stage of dynamical systems. It is possible to consider the two-compartment model as a regression model which is intrinsically nonlinear in parameters. Evaluation of the nonlinear model parameters in statistical perspective will help to improve the compartmental system. In this study, statistical inference of two-compartment model parameters is achieved in respect to point estimation and interval estimation. The point estimates of compartment model parameters are obtained according to the nonlinear least squares (NLS) approach. Genetic algorithm (GA), a well-known population-based evolutionary algorithm, is preferred as an optimization tool. The main contribution of the study is obtaining bias-corrected point estimates and bias-corrected accelerated confidence interval (CI) estimates of compartment parameters. In order to obtain the CIs, sampling distribution of parameter estimates is defined with the application of fixed-X nonlinear bootstrap method which preserves the fixed nature of predictor variable. Two bootstrap methods are used for CI calculations: (i) Percentile and (ii) bias-corrected accelerated (BCa). A simulated data set and a real data set from the pharmacokinetic (PK) literature are chosen for application purpose. It is seen from the results that bias-reduced point estimates and sampling distribution of parameter estimates can be obtained by preserving the time-dependent nature of the dynamical system by using fixed-X bootstrapping. Besides, BCa method gives more realistic interval estimates than percentile method.

Keywords Compartment model · Nonlinear regression · Bootstrap method Genetic algorithm (GA)

Ö. Türkşen (✉)
Faculty of Science, Statistics Department, Ankara University, Ankara, Turkey
e-mail: turksen@ankara.edu.tr

M. Tez
Faculty of Arts and Sciences, Statistics Department, Marmara University,
İstanbul, Turkey
e-mail: mtez@marmara.edu.tr

© Springer International Publishing AG 2018   241
M. Tez and D. von Rosen (eds.), *Trends and Perspectives
in Linear Statistical Inference*, Contributions to Statistics,
https://doi.org/10.1007/978-3-319-73241-1_14

1 Introduction

Compartment models are generally used for modeling of dynamical systems in applied science, e.g., biomedicine, engineering, information science, and pharmacokinetics (PK). Among all, one of the mostly used fields of science is PK which is defined as the study of the time course of drug absorption, distribution, metabolism, and excretion. The compartment models are analytically proper to explain how the concentration of a drug in blood plasma declines over time. In general, a compartment model is described in the analytic form of a system of ordinary differential equations which leads to be poly-exponential form. From this point of view, a compartment model can be considered as a nonlinear regression model which is intrinsically nonlinear in parameters similar to studies of Wagner (1975), Lai (1985), Seber and Wild (1989), Bonate (2011). The most common used approach for parameter estimation of nonlinear regression model is nonlinear least squares (NLS) approach which is based on the minimization of sum of squares error (SSE). However, it is hard to obtain the estimates of compartment model parameters analytically by using NLS approach since the normal equations become nonlinear and typically intractable. In this case, compartment model parameter estimation should be achieved numerically which simply means that the SSE is calculated for many combinations of parameter values, and the combination that yields the least sum of squares is selected as the solution.

In order to minimize the SSE, population-based derivative-free optimization algorithm should be preferred as an optimization tool, e.g., genetic algorithm (GA), particle swarm optimization (PSO), ant colony, and bee colony. These algorithms are also called nature-inspired optimization algorithms in the literature (Yang 2014). The most popular of these algorithms is genetic algorithm (GA), developed by Holland (1975), essentially forms the foundations of modern evolutionary computing. The GA is a population-based heuristic algorithm which uses stochastic search process to make exploration.

It is possible to obtain that the point estimates of compartment model parameters by minimizing the SSE through GA in the event of nonlinear model assumptions are satisfied, e.g., zero mean for errors, a finite constant variance, and uncorrelated errors. However, in order to obtain interval estimation of parameters, it is necessary to know sampling distributions of parameter estimates. Bootstrapping, first introduced by Efron (1979), is mainly recommended for estimating sampling distribution. Its working principle is related to the Monte Carlo simulation-based statistical inference. There are two types of bootstrapping: (i) parametric and (ii) nonparametric. For nonlinear models, parametric bootstrapping is not recommended since the distribution of errors is generally difficult to characterize. In contrast, the nonparametric bootstrap method makes no assumptions concerning the distribution of sample or model for the data. In nonparametric bootstrap, we assume that data has empirical distribution and sampling distribution is estimated by selecting resamples with replacement from the original sample. It should be noted that these samples are assumed to be independent and identically (iid) distributed. The most useful reference about theory and applications of bootstrap is Efron and Tibshirani (1994).

Wu (1986) and Midi (2000) explained how to use bootstrap methods in regression models. In general, two methods exist for bootstrap resampling of regression models. These are called random-X bootstrapping (paired bootstrapping) and fixed-X bootstrapping (residual bootstrapping). In random-X bootstrapping, the predictor variables are random and not under the control of experimenter. Bootstrap samples are selected directly using the paired data. However, in fixed-X bootstrapping, the predictor variables are fixed and under the control of experimenter. Residuals are resampled with replacement to obtain bootstrap samples which preserve the fixed nature of predictor variables. In PK studies, generally, time is considered as predictor variable and treated as fixed in compartmental model. Bootstrapping on residuals provides bias-reduced point estimation and more realistic confidence interval (CI) estimation for compartment model parameters by preserving the fixed nature of predictor variable in PK field.

There are some different approaches for calculating the CI by using bootstrapping. These approaches can be classified into three groups: nonpivotal, pivotal, and test inversion. The simplest and the most common form is the nonpivotal method. These methods are (i) percentile method, (ii) bias-corrected (BC) method, and (iii) bias-corrected accelerated (BCa) method. The percentile method is the most common method used to estimate the CI. It involves sorting the bootstrap estimates from the smallest to largest and then finding the $B(\alpha/2)$th and $B(1-\alpha/2)$th observations. These values are declared the lower and upper bounds of CIs. Here, α denotes the first type error probability. This method is criticized when the bootstrap distribution is asymmetrical. The BC method was developed by Efron (1987) to overcome some of the shortcomings of the percentile method. The BC method offers better coverage if the bootstrap distribution is asymmetrical. While the BC method corrects for bias, it does not correct for skewness. The skewness and bias are both considered for CI estimation by the application of BCa method which is proposed by Efron (1992).

In the literature, there are some studies for application of bootstrapping to PK problems. Kundu and Mitra (1998) considered the estimation procedure of linear compartment model parameters and they also applied two bootstrap CIs, percentile and bootstrap-t methods, for interval calculations. Hunt et al. (1998) used bootstrapping methods to obtain a measure of reliability of parameter predictions for a PK model. Contreras and Walter (2000) presented three parameter estimation methods for compartment models. They also bootstrapped the weighted residuals and used BCa approach to obtain CI of parameters. Honda et al. (2008) used bootstrap method for validity of PK model. Ogden and Jiang (2010) presented an application of bootstrap algorithm to compartment model for parameter estimation. Some brief examples about bootstrap application in PK field can be found in Bonate (2011). Thompson (2012) presented bootstrap interval estimation including BCa for two-compartment models with parameters estimated by iteratively reweighted nonlinear least squares. Burns et al. (2014) applied nonparametric bootstrap analysis to estimate the CI of PK parameters without symmetric assumption. Thai et al. (2014)

evaluated the performance of three bootstrap methods (case bootstrap, nonparametric residual bootstrap, and parametric bootstrap) by a simulation study in nonlinear mixed effects models with heteroscedastic error.

One of the main aims of the study is presenting the applicability of fixed-X bootstrapping to obtain bias-corrected point estimates. The other aim is obtaining unbiased CIs with the application of BCa through fixed-X bootstrapping to present how important to consider skewness and bias for CI estimation. The rest of the chapter is organized as follows: Some basic information about two-compartment model is given and parameter estimation with GA is explained in Sect. 2. In Sect. 3, bootstrap method is explained and bootstrap CI types are given in detail. A simulated data set and a real data set from the PK literature are used for application purpose in Sect. 4. In Sect. 5, conclusion is given with future work.

2 Analysis of Two-Compartment Model

Compartment models are commonly used modeling approaches which provide good insight into the underlying behavior of most drugs in PK studies. The two-compartment PK model with elimination from the central compartment is the most common model used to define what the body does to the drugs. A block diagram of two-compartment PK model is presented in Fig. 1.

In Fig. 1, the block diagram represents the two-compartment PK model with first-order transport between the central and peripheral compartments. The first-order drug elimination happens from the central compartment. Here, k_{12} is the first-order transfer rate constant from the central compartment to the peripheral compartment and has units of time^{-1}, k_{21} is the first-order transfer rate constant from the peripheral compartment to the central compartment and has units of time^{-1}, and k_{el} is the first-order elimination rate constant from the central compartment and has units of time^{-1}. After the drug injection to the central compartment, based on the model given in Fig. 1, the rate of change of the amount of the drug in the central compartment (Y) at any time is equal to the rate of drug transfer

Fig. 1 Block diagram of two-compartment PK model

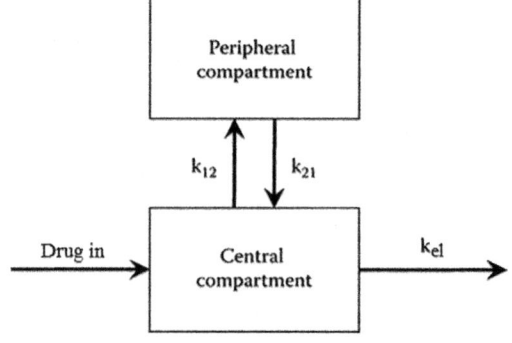

from the peripheral compartment to the central compartment minus the rate of drug transfer from the central compartment to the peripheral compartment, minus the rate of drug elimination. This rate is described by the following differential equation:

$$\frac{dY}{dt} = k_{21}X - k_{12}Y - k_{el}Y,$$

(1)

where X is the amount of the drug in the peripheral compartment. The integrated equation of Eq. (1) is defined as a function of time

$$Y = \frac{\delta(\lambda_1 - k_{21})}{(\lambda_1 - \lambda_2)} e^{-\lambda_1 t} + \frac{\delta(k_{21} - \lambda_2)}{(\lambda_1 - \lambda_2)} e^{-\lambda_2 t}$$

(2)

Equation (2) contains two exponents. One exponent describes the distribution process with the hybrid rate constant, λ_1, and the another describes the elimination process with the hybrid rate constant, λ_2. As a result of integration process, the following two relationships were obtained:

$$\lambda_1 + \lambda_2 = k_{12} + k_{21} + k_{el}$$

(3)

$$\lambda_1 \lambda_2 = k_{21} k_{el},$$

(4)

where

$$\lambda_1 = \frac{1}{2} \left[(k_{12} + k_{21} + k_{el}) + \sqrt{(k_{12} + k_{21} + k_{el})^2 - 4k_{21}k_{el}} \right]$$

(5)

$$\lambda_2 = \frac{1}{2} \left[(k_{12} + k_{21} + k_{el}) - \sqrt{(k_{12} + k_{21} + k_{el})^2 - 4k_{21}k_{el}} \right]$$

(6)

Equation (2) can be simplified to

$$Y = Ae^{-\lambda_1 t} + Be^{-\lambda_2 t},$$

(7)

where

$$A = \frac{\delta(\lambda_1 - k_{21})}{(\lambda_1 - \lambda_2)}$$

(8)

and

$$B = \frac{\delta(k_{21} - \lambda_2)}{(\lambda_1 - \lambda_2)}$$

(9)

in which δ is a constant related to ratio of the dose of the drug and volume of the central compartment. The two-exponential model, given in Eq. (7) with the simplest form, is consistent with the theory of two-compartment model in which the terms of exponential equations are directly translated to PK parameters.

2.1 Modeling Stage for Two-Compartment System

The two-exponential compartment model can be represented as a nonlinear regression model as follows:

$$Y_i = f(t_i, \boldsymbol{\theta}) + \varepsilon_i, i = 1, 2, \ldots, n. \tag{10}$$

Here f is a nonlinear function, defined in Eq. (7); t_i is the time; Y_i is the drug concentration in central compartment; $\boldsymbol{\theta}$ is a vector of unknown compartment model parameters denoted as $\boldsymbol{\theta} = [k_{12} \quad k_{21} \quad k_{el}]'$, and ε_i are random errors, assumed $\varepsilon_i \sim iid(0, \sigma_\varepsilon^2), i = 1, 2, \ldots, n$.

It is well known that the parameter estimates are reliable if the assumptions on error terms are satisfied. In this study, it is assumed that the assumptions of error terms are satisfied to apply the ordinary NLS approach. The estimation of compartment model parameter vector $\boldsymbol{\theta}$ is obtained by minimizing the SSE given below:

$$\varphi(\boldsymbol{\theta}) = \sum_{i=1}^{n} [Y_i - f(t_i, \boldsymbol{\theta})]^2 = \sum_{i=1}^{n} \left[Y_i - \left(Ae^{-\lambda_1 t_i} + Be^{-\lambda_2 t_i} \right) \right]^2 \tag{11}$$

which is considered as objective function. In order to optimize the objective function, given in Eq. (11), derivative-free optimization algorithms should be preferred because of the nonlinearity.

2.2 Optimization Stage for Two-Compartment System

In optimization stage, GA, a well-known population-based metaheuristic algorithm, is preferred to use. It has also been used for estimation of compartment model parameters in the study of Türkşen and Tez (2016). The GA represents an intelligent exploration of random search used to solve many nonlinear problems. It starts with an initial population of artificial chromosomes with the size, n_{pop}. In each generation, the fitness value or the objective function value of every individual in the population is calculated and current population is composed. The best fitness values in the population are selected by using selection operators, e.g., roulette wheel selection and stochastic uniform, for reproduction. The selected individuals are modified by using genetic operators, e.g., crossover and mutation. The crossover and the mutation operators are used with the crossover probability (Pr_{cr}) and mutation probability (Pr_{mut}), respectively. Then, the current population is replaced with the new population. The algorithm runs until the stopping condition, which can be considered as maximum number of generations (*maxgen*), is satisfied. It should be noted that the solution of the optimization problem with GA highly depends on how the tuneable parameters, e.g., n_{pop}, Pr_{cr}, Pr_{mut}, and *maxgen*, are chosen.

3 Application of Bootstrap Method to Two-Compartment Model

3.1 The Nonparametric Bootstrap for Nonlinear Regression Model

Briefly, the nonparametric bootstrap allows a researcher to make statistical inference without making assumptions about the form the population and without deriving the sampling distribution explicitly. Therefore, the nonparametric bootstrap method is more proper to apply for the two-compartment model which is considered as a nonlinear regression model given in Eq. (10). It is clear from Eq. (10) that predictor variable is time which is under the experimenter's control. In this case, application of fixed-X bootstrapping is more proper to obtain sampling distribution of compartment model parameters' estimates. Basically, the algorithmic steps of the fixed-X bootstrapping can be given as follows:

Step 1: Fit the nonlinear model

$$Y_i = f(t_i, \boldsymbol{\theta}) + \varepsilon_i = Ae^{-\lambda_1 t_i} + Be^{-\lambda_2 t_i} + \varepsilon_i, i = 1, 2, \ldots, n \qquad (12)$$

to the original sample of observations to obtain NLS estimate of $\boldsymbol{\theta}$ denoted as $\hat{\boldsymbol{\theta}}$.

Step 2: Calculate the observed residuals

$$e_i = Y_i - f(t_i, \hat{\boldsymbol{\theta}}), i = 1, 2, \ldots, n \qquad (13)$$

Step 3: Resample the residuals $\{e_i, i = 1, 2, \ldots, n\}$ with replacement to obtain a bootstrap sample $\{e_i^*, i = 1, 2, \ldots, n\}$. Then, by holding the predictor variables as fixed, generate the bootstrap dependent variable as

$$Y_i^* = f(t_i, \hat{\boldsymbol{\theta}}) + e_i^*, i = 1, 2, \ldots, n \qquad (14)$$

It should be noted here that correcting for the mean is necessary in non-linear case since the mean of the residuals is not guaranteed to be equal to zero. Therefore, e_i's are resampled from the centered residuals

$\{e_1 - \bar{e}, e_2 - \bar{e}, \ldots, e_n - \bar{e}\}$ instead of resampling from the observed residuals $\{e_1, e_2, \ldots, e_n\}$.

Step 4: Calculate the estimate of model parameters based on the bootstrap sample $(X_i, Y_i^*), i = 1, 2, \ldots, n$ which is denoted as $\hat{\boldsymbol{\theta}}^*$.

Step 5: Repeat the Steps 3 and 4 B times to obtain the bootstrap replications $\hat{\boldsymbol{\theta}}_1^*, \hat{\boldsymbol{\theta}}_2^*, \ldots, \hat{\boldsymbol{\theta}}_B^*$ and bootstrap distribution of parameter estimates. Then, the bias, variance, and mean squared error (MSE) of the parameter estimates can be calculated by

(i)
$$\widehat{Bias}(\hat{\boldsymbol{\theta}}) = \overline{\hat{\boldsymbol{\theta}}^*} - \hat{\boldsymbol{\theta}}, \overline{\hat{\boldsymbol{\theta}}^*} = \frac{1}{B} \sum_{b=1}^{B} \hat{\boldsymbol{\theta}}_b^* \tag{15}$$

(ii)
$$\widehat{Var}(\hat{\boldsymbol{\theta}}) = \frac{1}{B} \sum_{b=1}^{B} \left(\hat{\boldsymbol{\theta}}_b^* - \overline{\hat{\boldsymbol{\theta}}^*}\right)^2 \tag{16}$$

(iii)
$$\widehat{MSE}(\hat{\boldsymbol{\theta}}) = \widehat{Var}(\hat{\boldsymbol{\theta}}) + \widehat{Bias}^2(\hat{\boldsymbol{\theta}}) = \frac{1}{B} \sum_{b=1}^{B} \left(\hat{\boldsymbol{\theta}}_b^* - \hat{\boldsymbol{\theta}}\right)^2 \tag{17}$$

where $\overline{\hat{\boldsymbol{\theta}}^*}$ denotes the bootstrap estimate of model parameters. The square root of Eq. (16) is the estimate for the standard error of the parameter estimates. It is clear that the bootstrap is a useful tool to obtain the estimates of some statistics, e.g., bias, variance, and MSE. However, the bootstrap should not be used to compute point estimates themselves. The computing point estimates may reflect biased estimation from the samples since the sampling distribution of the bootstrap statistics is frequently not symmetric. The bias-corrected point estimates of parameters, $\boldsymbol{\theta}_{adj}$, can be obtained as

$$\boldsymbol{\theta}_{adj} = \hat{\boldsymbol{\theta}} - \widehat{Bias}(\hat{\boldsymbol{\theta}}) = 2\hat{\boldsymbol{\theta}} - \overline{\hat{\boldsymbol{\theta}}^*}. \tag{18}$$

In bootstrapping, a key question that often arises is how large B should be chosen to be valid (Bonate 2011). All the researchers agree on that the larger B leads to more accuracy, e.g., B should be at least 1000 for interval estimation.

3.2 Bootstrap Confidence Interval

The BCa method is proposed by Efron (1992) as a modification of percentile method to adjust the percentiles for bias and skewness. This is achieved by using two coefficients called bias correction and acceleration. The bias correction coefficient adjusts for the skewness in the bootstrap sampling distribution.

The acceleration coefficient adjusts for nonconstant variances within the resampled data sets (Haukoos and Lewis 2005). The advantages of BCa method can be found in Beyaztas et al. (2014) in detail. The algorithmic steps of the BCa method can be given as follows:

Step 1: Let us assume that the bootstrap replicated estimates $\hat{\theta}_1^*, \hat{\theta}_2^*, \ldots, \hat{\theta}_B^*$ are calculated.

Step 2: Count the number of member of B bootstrap estimates that are less than $\hat{\theta}$, calculated from the original data. Call this number r and set $b = \phi^{-1}(r/B)$ where $\phi^{-1}(.)$ is the inversion of the standard normal distribution.

Step 3: Calculate the acceleration constant, denoted as a, based on jackknifing the original data set, $a = \dfrac{\sum_{i=1}^n \left(\hat{\theta}_{(-i)} - \hat{\theta}\right)^3}{6\left[\sum_{i=1}^n \left(\hat{\theta}_{(-i)} - \hat{\theta}\right)^2\right]^{3/2}}$ where $\hat{\theta} = \dfrac{\sum_{i=1}^n \hat{\theta}_{(-i)}}{n}$. $\hat{\theta}_{(-i)}$ represents the value of $\hat{\theta}$ produced when the ith observation is deleted from the sample, called jackknife values of $\hat{\theta}$.

Step 4: Compute the $a_1 = \phi\left(b - \dfrac{Z_{\alpha/2} + b}{1 - a(Z_{\alpha/2} + b)}\right)$ and $a_2 = \phi\left(b + \dfrac{Z_{\alpha/2} + b}{1 - a(Z_{\alpha/2} + b)}\right)$, where ϕ is the standard normal cumulative distribution function.

Step 5: Locate the endpoints of the CIas $lower^* = [\![B \times a_1]\!]$ and $upper^* = [\![B \times a_2]\!]$. Here, the square brackets indicate rounding the nearest integer.

4 Application

In this section, a simulated data set (Wagner 1975) and a real data set (Ağabeyoğlu 1999) are used to present the bootstrap application for point estimation and interval estimation of compartment model parameters. The estimates of the parameters are obtained by minimizing the objective function φ, given in Eq. (11), via the GA. The tuneable parameters of GA are given in Table 1.

Table 1 Tuneable parameter values of GA

Parameters	Values
Population size (n_{pop})	50
Maximum number of generation (*maxgen*)	100
Probability of crossover (Pr_{cr})	0.90
Probability of mutation (Pr_{mut})	0.01
Selection operator	Roulette wheel selection
Crossover operator	Single point crossover

Matlab 7.9 is used for all calculations. In order to evaluate the prediction performance of predicted models, root-mean-square error (*RMSE*) and mean absolute percentage error (*MAPE*) metrics are used. These are defined as

$$RMSE = \sqrt{\frac{1}{n} \sum_{i=1}^{n} \left(Y_i - \hat{Y}_i\right)^2} \tag{19}$$

$$MAPE = \frac{1}{n} \sum_{i=1}^{n} \left| \frac{Y_i - \hat{Y}_i}{Y_i} \right|. \tag{20}$$

In order to obtain the sampling distribution of parameter estimates, fixed-*X* bootstrapping is applied. The CI of compartment model parameters is obtained by using BCa method. The bootstrap sample size, *B*, is chosen equal to 1000 for simulated and real data sets.

4.1 Simulated Data Set

Wagner (1975) studied on a simulated intravenous data set in order to see the effect of bias on the estimates of k_{12}, k_{21}, and k_{el} parameters. In this study, we use the same data set as Wagner (1975) for the application of fixed-*X* bootstrapping to obtain point estimates and interval estimates of parameters. For generating the intravenous data set, the parameter values $k_{12} = 1.162$, $k_{21} = 0.515$, $k_{el} = 0.038$ and $\delta = 100$ were used at times (in hour) 0, 0.5, 1, 1.5, 2, 2.5, 3, 3.5, 4, 5, 6, 7, 9, 11, 15, 18, 24. The appropriate statistical model can be written as

$$Y_i = 70.2419e^{-1.7035t} + 29.758e^{-0.0115t} + \varepsilon_i, i = 1, 2, \ldots, 17 \tag{21}$$

where the error term, ε_i, is assumed independently distributed with zero mean and 0.05 standard deviation, denoted as $\varepsilon_i \sim (0, 0.05)$. The generated drug concentration values are obtained by using Eq. (21) and these values are assumed to be independent. The drug concentration–time plot can be seen in Fig. 2.

The model parameters are estimated by applying the GA with the tuneable parameters given in Table 1. The lower and upper bounds of the compartment parameters are given in Table 2.

The parameter estimates obtained by GA and NONLIN, denoted as $\hat{\theta}$ and $\hat{\theta}_{NONLIN}$, respectively, are given in Table 3.

In Table 3, the bias values of the parameter estimates are presented in parenthesis. It can be easily seen from Table 3 that the parameter estimates, obtained by using GA, are quite unbiased than the estimates obtained by NONLIN method. It is possible to say that the GA can be a preferable optimization tool. The main advantage of using GA is that it does not need to define the initial values of parameters as NONLIN in the beginning of the optimization process.

Fig. 2 The drug concentration–time plot

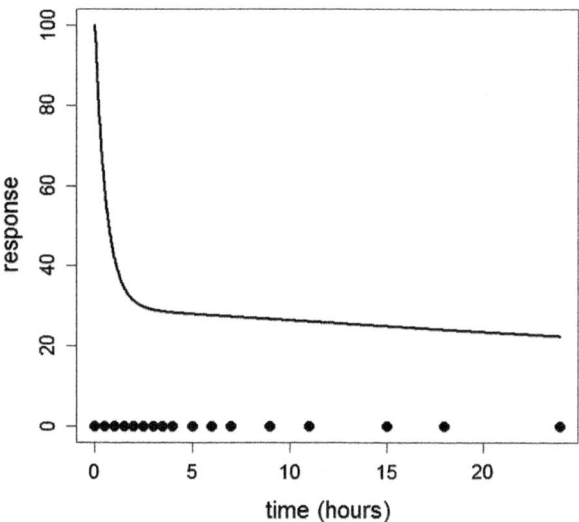

Table 2 Bounds of the compartment parameters

Parameters	k_{12}	k_{21}	k_{el}
Lower bounds	0	0	0
Upper bounds	2	1	1

Table 3 Point estimates and bias values of parameters

Parameters	Point estimates	
θ	θ	θ_{NONLIN}
k_{12}	1.1893 (0.0273)	1.853 (0.691)
k_{21}	0.5372 (0.0222)	0.797 (0.282)
k_{el}	0.0409 (0.0029)	0.068 (0.03)

In order to obtain bias-reduced point estimates of parameters and to get CIs of parameter estimates, sampling distribution of parameter estimates is needed. It is achieved by applying resampling to the sample with fixed-X bootstrapping. The histograms for sampling distribution of parameter estimates obtained by fixed-X bootstrapping are presented in Fig. 3a–c.

It can be seen from Fig. 3a–c that the sampling distributions are not normally distributed. The point estimates, $\hat{\theta}$, and bias-corrected point estimates, θ_{adj}, are presented in Table 4 with the estimates of bias, standard error, and *MSE*. The presented results are averaged over 100 runs. Table 4 also includes the performance metrics of the predicted model, denoted as *RMSE* and *MAPE*, for $\hat{\theta}$ and θ_{adj}.

It can be easily seen from Table 4 that model prediction performance is better with the θ_{adj} than $\hat{\theta}$. Furthermore, bias, standard error, and *MSE* estimates of θ_{adj} are smaller than the $\hat{\theta}$'s.

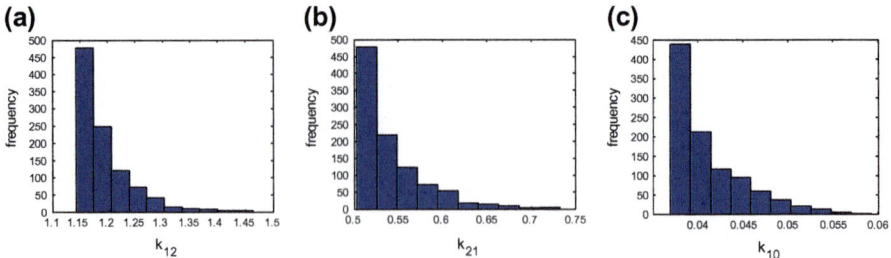

Fig. 3 Histograms of the sampling distributions of two-compartment model parameters estimates: **a** $\hat{k}_{12}$, **b** $\hat{k}_{21}$, and **c** $\hat{k}_{el}$

Table 4 Point estimations and the estimates of bias, standard error, and MSE

Point estimates			Estimated statistics for θ					
θ	θ	θ_{adj}	Bias		SE		MSE	
			θ	θ_{adj}	θ	θ_{adj}	θ	θ_{adj}
k_{12}	1.1893	1.1567	0.0273	0.0053	0.0356	0.0330	0.0020	0.0011
k_{21}	0.5372	0.5120	0.0222	0.0030	0.0285	0.0273	0.0013	0.0008
k_{el}	0.0409	0.0382	0.0029	0.0002	0.0035	0.0036	~0	~0
RMSE	0.2268	0.0593						
MAPE	0.0056	0.0016						

Table 5 Percent coverage and average width of CIs

Parameters	Percent coverage		Average width	
θ	Percentile	BCa	Percentile	BCa
k_{12}	0.14	0.35	0.1795	0.1344
k_{21}	0.30	0.55	0.1347	0.0362
k_{el}	0.60	0.95	0.0145	0.0063

For interval estimation of the $\hat{\theta}$, percentile and BCa methods are applied. The percent coverage and average width of intervals are presented in Table 5. It is seen from Table 5 that the BCa method is more preferable than the percentile method for interval estimation of compartment model parameters due to having higher percent of coverage and smaller average width.

4.2   Real Data Set

The data set, also studied by Türkşen and Tez (2016), is about the amount of the drug concentration in plasma given by Ağabeyoğlu (1999). The time–concentration values are given in Table 6. It should be noted here that the C_p values are assumed to be independent.

Table 6 Drug concentration–time values (Ağabeyoğlu 1999)

No	1	2	3	4	5	6	7	8	9	10	11	12	13
t (hr)	0	0.25	0.50	0.75	1	2	3	4	6	8	12	16	24
C_p (mg/ml)	16.4	14.2	12.53	11.17	10.09	7.56	6.44	5.85	5.16	4.65	3.18	3.12	2.09

According to the drug concentration–time plot in Fig. 4, the concentration values are not linearly dependent on time.

The lower and upper bounds for the two-compartment model parameters are given in Table 7.

By applying the fixed-X bootstrapping to the data set given in Table 6, the bias-corrected point estimates, θ_{adj}, with the estimation of bias, standard error, MSE values of parameter estimates, $\hat{\theta}$, and sampling distribution of the estimates are obtained. The obtained point estimates and estimated statistics for $\hat{\theta}$ are presented in Table 8. Table 8 also includes the performance metrics, $RMSE$, and $MAPE$, of the predicted model for $\hat{\theta}$ and θ_{adj}. It can be easily seen from Table 8 that $MAPE$ values are equal and $RMSE$ values are similar for point estimates. Even though one can prefer to use one of the point estimates according to the model prediction performance metrics, θ_{adj} is preferred for point estimates of compartment model parameters since the unbiased point estimates are the most preferable in PK studies.

The histogram of the sampling distribution for the estimates of parameters is shown in Fig. 5. Figure 5a–c represents the histograms of the parameter estimates,

Fig. 4 The drug concentration–time plot

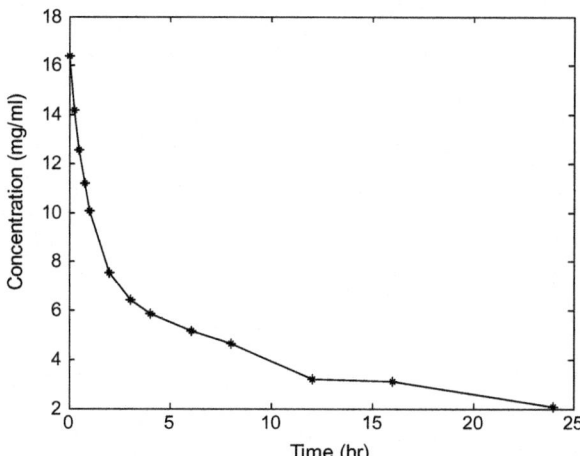

Time (hr)

Table 7 Bounds of the compartment parameters

Parameters	k_{12}	k_{21}	k_{el}
Lower bounds	0	0	0
Upper bounds	1	1	1

Table 8 Point estimations and estimates of bias, standard error, and MSE

θ	Point estimates		Estimated statistics for θ		
	θ	$\hat{\theta}_{adj}$	Bias	SE	MSE
k_{12}	0.5056	0.5010	0.0046	0.0186	0.000368
k_{21}	0.4773	0.4654	0.0119	0.0351	0.00137
k_{el}	0.12	0.1191	0.0009	0.0048	0.0000242
RMSE	0.174	0.175			
MAPE	0.028	0.028			

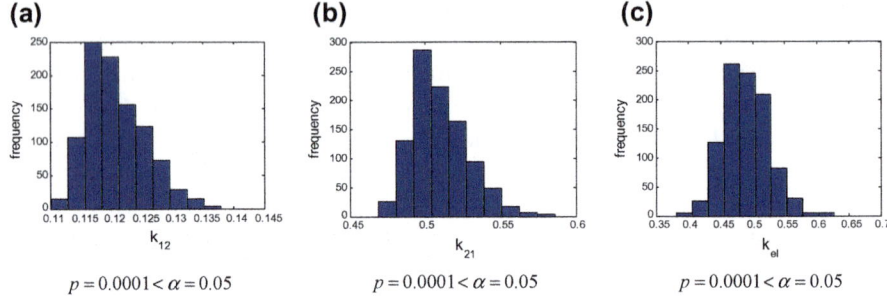

$p = 0.0001 < \alpha = 0.05$ $p = 0.0001 < \alpha = 0.05$ $p = 0.0001 < \alpha = 0.05$

Fig. 5 Histograms of the sampling distributions of two-compartment model parameters estimates: **a** $\hat{k}_{12}$, **b** $\hat{k}_{21}$, and **c** $\hat{k}_{el}$

$\hat{k}_{12}$, $\hat{k}_{21}$, and $\hat{k}_{el}$, respectively. The normality test results of sampling distributions are presented in the bottom of Fig. 5a–c. It can be easily said from the test results that the distribution of parameter estimates is not normal. Here, p and α represent the calculated probability and nominal significance level, respectively.

The bootstrap CI, obtained by the application of BCa, for parameter estimates is given in Table 9. From Table 9, it is clear that the CI of BCa gives more realistic results than percentile method since the bias and skewness are considered during the calculations.

Table 9 CIs and widths for parameter estimates with percentile and BCa methods

θ	Confidence interval		Width	
	Percentile	BCa	Percentile	BCa
k_{12}	[0.1134 0.1315]	[0.1134 0.1311]	0.0181	0.0177
k_{21}	[0.4803 0.5528]	[0.4774 0.5481]	0.0725	0.0707
k_{el}	[0.4202 0.5683]	[0.4075 0.5403]	0.1481	0.1328

5 Conclusion

In this study, we have considered the two-compartment model as a nonlinear regression model. The point estimates of parameters are obtained according to the NLS approach via GA, assuming that the model assumptions are satisfied. In order to obtain bias-corrected point estimates and sampling distribution of parameter estimates, fixed-X bootstrapping is applied by preserving the fixed nature of the predictor variable in nonlinear regression model. Percentile and BCa methods are used for CI calculations. We have performed numerical study on simulated and real data sets. The results of the numerical studies can be summarized as below:

(i)   GA is a convenient optimization tool for nonlinear problems;
(ii) If the data is time-dependent and small-sized, fixed-X bootstrapping should be preferred to obtain sampling distribution of estimates;
(iii) BCa is a proper method which gives more realistic CI by adjusting the skewness in the sampling distribution and nonconstant variances within the resampled data sets.

For future work, it will be more suitable to consider the correlation structure of response measurements during the modeling stage of nonlinear problems. Besides, in compartmental modeling, nonlinear mixed effects models (NLMEM) should be applied for modeling of more than one response measurements to get more realistic point estimation and interval estimation of parameters by using the bootstrap methods.

Acknowledgements This study is supported by The Scientific and Technological Research Council of Turkey (TÜBİTAK-2218). The authors thank the two reviewers for their very helpful comments, useful insights, and suggestions.

References

Ağabeyoğlu, İ.: Basic Pharmacokinetic. Chapter 11, Ankara, Turkey, pp. 183–226 (1999)

Beyaztaş, U., Alın, A., Martin, M.A.: Robust BCa-Jab method as a diagnostic tool for linear regression models. J. Appl. Stat. **41**(7), 1593–1610 (2014)

Bonate, P.L.: Pharmacokinetic-Pharmacodynamic Modeling and Simulation. Springer, New York (2011)

Burns, R.N., Chaturvedula, A., Turner, D.C., Zhang, H., van der Berg, C.M.: Population pharmacokinetic pharmacodynamic modeling of caffeine using visual analogue scales. Pharmacol. Pharm. **5**, 444–454 (2014)

Contreras, M., Walter, G.G.: Numerical Methods for Point and Interval Parameter Estimation in Compartmental Models Used in Small Sample Pharmacokinetic and Epidemiological Studies. Report, numbered BU-1542–M, pp. 1–25 (2000)

Efron, B.: Bootstrap methods: another look at the jackknife. Ann. Stat. **7**(1), 1–26 (1979)

Efron, B.: Better bootstrap confidence intervals. J. Am. Stat. Assoc. **82**(397), 171–185 (1987)

Efron, B.: More accurate confidence intervals in exponential families. Biometrika **79**, 231–245 (1992)

Efron, B., Tibshirani, R.J.: An Introduction to Bootstrap. Chapman and Hall, CRC Press (1994)
Haukoos, J.S., Lewis, R.J.: Advanced statistics: bootstrapping confidence intervals for statistics
    with "Difficult" distributions. Acad. Emerg. Med. 12(4), 360–365 (2005)
Holland, J.: Adaptation in natural and artificial systems. MIT Press, Cambridge, MA, USA (1975)
Honda, N., Nakade, S., Kasai, H., Hashimoto, Y., Ohno, T., Kitagawa, J., Yamauchi, A.,
    Hasegava, C., Kikawa, S., Kunisawa, T., Tanigawara, Y., Miyata, Y.: Population pharma-
    cokinetics of landiolol hydrochloride in healthy subjects. Drug Metab. Pharmacokinet. 23(6),
    447–455 (2008)
Hunt, C.A., Givens, G.H., Guzy, S.: Bootstrapping for pharmacokinetic models: visualization of
    predictive and parameter uncertainty. Pharm. Res. 15(5), 690–697 (1998)
Kundu, D., Mitra, A.: Estimating the parameters of the linear compartment model. J. Stat. Plan.
    Infer. 70, 317–334 (1998)
Lai, T.L.: Regression analysis of compartmental models. J. Res. Natl. Bur. Stan. 90(6), 525–530
    (1985)
Midi, H.: Bootstrap methods in a class of non-linear regression models. Pertanika J. Sci. Technol.
    8(2), 175–189 (2000)
Ogden, R.T., Jiang, H.: Nonparametric evaluation of heterogeneity of brain regions in
    neuroreceptor mapping applications. Stat. Its Interface 3, 59–67 (2010)
Seber, G.A.F., Wild, J.: Nonlinear Regression. Wiley (1989)
Thai, H.T., Mentre, F., Holford, N., Follet, C.V., Comets, E.: Evaluation of bootstrap methods for
    estimating uncertainty of parameters in nonlinear mixed-effects models: a simulation study in
    population pharmacokinetics. J. Pharmacokinet. Pharmacodyn., Springer Verlag 41(1), 15–33
    (2014). (HAL archives-ouvertes)
Thompson, Z.J.: Statistical Estimation of Physiologically-based Pharmacokinetic Models:
    Identifiability, Variation, and Uncertainty with an Illustration of Chronic Exposure to Dioxin
    and Dioxin-like-compounds. Graduate Theses and Dissertations. University of South Florida
    (2012)
Türkşen, Ö., Tez, M.: An application of nelder-mead heuristic-based hybrid algorithms: estimation
    of compartment model parameters. Int. J. Artif. Intell. 14(1), 525–530 (2016)
Wagner, J.G.: Application of the Loo-Riegelman absorption method. J. Pharmacokinet. Biopharm.
    3(1), 51–67 (1975)
Wu, C.F.J.: Jackknife, bootstrap and other resampling methods in regression analysis. Ann. Stat.
    14(4), 1261–1295 (1986)
Yang, X.-S.: Nature-Inspired Optimization Algorithms. Elsevier, USA (2014)

Author Index

© Springer International Publishing AG 2018
M. Tez and D. von Rosen (eds.), *Trends and Perspectives
in Linear Statistical Inference*, Contributions to Statistics,
https://doi.org/10.1007/978-3-319-73241-1

Printed by Printforce, the Netherlands